# The Atom
# - electrons, protons and neutrons

**by**
**Trevor G. Underwood**

By the same author:

*The Surface Temperature of the Earth.* (November 2019);

*Urbain Le Verrier on the Movement of Mercury
- annotated translations.* (November 2021);

*Quantum Electrodynamics – annotated sources.
Volumes I and II.* (April 2023);

*Special Relativity.* (June 2023);

*General Relativity.* (November 2023);

*Gravity.* (March 2024);

*Electricity & Magnetism.* (May 2024);

*Quantum Entanglement.* (June 2024);

*The Standard Model.* (September 2024);

*New Physics.* (October 2024);

*The Cosmological Redshift of Light.* (November 2024);

*Cosmic Microwave Background Radiation.* (January 2025)

*Fundamental Physics.* (May 2025).

all distributed by Lulu.com.

Published by Trevor G. Underwood
18 SE 10th Ave.
Fort Lauderdale, FL 33301

ISBN: 979-8-218-93124-7 (hardcover)
Library of Congress Control Number: 2026903008

Printed and distributed by Lulu Press, Inc.

700 Park Offices Dr.
Ste. 250
Durham, NC, 27713
http://www.lulu.com/shop

# CONTENTS.

# PREFACE

This volume fills a gap left by the eleven volumes that I published between April 2023 and January 2025. Previous volumes addressed subatomic particles. This volume extends my analysis of fundamental physics to the atomic theory of matter; in particular to the discovery of electrons, protons and neutrons. **Part I** provides an introduction to the atomic theory of matter, including to the structure of an atom, chemical elements, the Periodic Table, and molecules. **Part II** describes the discovery of the electron, including the main papers by J. J. Thomson in 1897, 1906 and his Nobel Lecture in 1906; and papers by Neils Bohr in 1913. **Part III** describes the discovery of the proton by Ernest Rutherford, including his papers between 1899 and 1914, and his Nobel Lecture in 1908. **Part IV** describes the discovery of the neutron by James Chadwick, including papers by Rutherford between 1919 and 1920; Bohr's Nobel Lecture in 1922; papers by Werner Heisenberg between 1922 and 1925; papers by Chadwick between 1932 and 1935, and his Nobel Lecture in 1935.

# PART I     The atomic theory of matter.

https://www.scirp.org/pdf/samplechapter_2018081010593995.pdf

**Atoms.**

Democritus, who was a Greek philosopher, introduced the concept of the *atom* around 450 BC, describing it as solid and varying in size, mass, shape, position, and arrangement. According to him, the continued subdivisions of matter would ultimately yield atoms which would not be further divisible. The word 'atom' has been derived from the Greek word 'a-tomio' which means 'uncutable' or 'non-divisible'. These earlier ideas were mere speculations and there was no way to test them experimentally. These ideas remained dormant for a very long time and were revived again by scientists in the nineteenth century.

The *atomic theory of matter* was first proposed on a firm scientific basis in 1808 by John Dalton, a British school teacher. His theory, called *Dalton's atomic theory*, regarded the atom as the ultimate particle of matter. See below: John Dalton, *A New System of Chemical Philosophy*, Part I (Manchester: S. Russell, 1808), 141–144, 211–216. Dalton's atomic theory was able to explain the law of conservation of mass, law of constant composition and law of multiple proportion very successfully. However, it failed to explain the results of many experiments, for example, it was known that substances like glass or ebonite when rubbed with silk or fur generate electricity. Many different kinds of sub-atomic particles were discovered in the twentieth century.

*In 1911, Rutherford provided the modern atomic structure*. Rutherford, E. (May 1911). LXXIX. *The scattering of α and β particles by matter and the structure of the atom. The London, Edinburgh, and Dublin Philosophical Magazine and Journal of Science*. 21 (125): 669–688. doi:10.1080/14786440508637080. ISSN 1941-5982; see below.

Two years later, *in 1913, Bohr took the next step by adapting Rutherford's theory of atomic structure to Planck's quantum hypothesis* creating his **model of the atom**.

See below, and in Underwood, T. G., *Quantum Electrodynamics,* Vol. I, p 108: Bohr, N. (1913). On the Constitution of Atoms and Molecules, Part I. *Phil. Mag.*, 26, 151, 1-24; https://doi.org/10.1080/14786441308634955. He supposed the atom to consist of a nucleus with a positive charge $Ze$ and $Z$ electrons with charge $-e$ each, moving according to the laws of classical mechanics. He introduced the idea that an electron could drop from a higher-energy orbit to a lower one, in the process emitting a quantum of discrete energy. This became a basis for what is now known as the *old quantum theory*. From a set of assumptions concerning the stationary state of an atom and the frequency of the radiation emitted or absorbed when the atom passes from one such state to another, he showed that it is possible to obtain a simple interpretation of the main laws governing the line spectra of the elements, and especially to deduce the Balmer formula for the hydrogen spectrum. Addresses mechanism of the binding of electrons by a positive nucleus in relation to Planck's theory. See below and in Underwood, T. G., *Quantum Electrodynamics,* Vol. I, p 117: Bohr, N. (1913). On the Constitution of Atoms and Molecules, Part II. Systems Containing Only a Single Nucleus. *Phil. Mag.*, 26, 153, 476-502; https://doi.org/10.1080/14786441308634993; application of Bohr's model of the atom to systems containing a single nucleus.

An ***atom*** is made up of a central ***nucleus*** surrounded by a cloud of ***electrons***, forming the basic building block of matter. The ***nucleus***, which is located at the center, contains ***protons*** (positively charged) and ***neutrons*** (neutral particles). The nucleus is dense and carries a positive charge. The ***electron*** cloud surrounding the nucleus, contains electrons (negatively charged) that revolve around the nucleus in defined energy levels or orbits. The arrangement of these subatomic particles determines the atom's properties and behavior.

***Protons*** have a positive electric charge and ***neutrons*** have no charge, so the ***nucleus*** is positively charged. The ***electrons*** are negatively charged, and this opposing charge is what binds them to the ***nucleus*** by ***electromagnetic force***. If the numbers of protons and electrons are equal, as they normally are, then the atom is electrically neutral as a whole.

A charged atom is called an ***ion***. If an atom has more electrons than protons, then it has an overall negative charge and is called a negative ion (or ***anion***). Conversely, if it has more protons than electrons, it has a positive charge and is called a positive ion (or ***cation***).

The ***protons*** and ***neutrons*** in the nucleus are attracted to each other by the ***nuclear force***. This force is usually stronger than the electromagnetic force that repels the positively charged protons from one another. Under certain circumstances, the repelling electromagnetic force becomes stronger than the nuclear force. In this case, the nucleus splits and leaves behind different elements. This is a form of nuclear decay.

***Atoms*** are extremely small, typically around 100 picometers (pm) across ($1\times10^{-12}$ m, or one trillionth of a meter; (He: 62pm; Cs: 520pm). Mass range: $1.67 \times 10^{-27}$ to $4.52 \times 10^{-25}$ kg). A human hair is about a million carbon atoms wide. Atoms are smaller than the shortest wavelength of visible light, which means humans cannot see atoms with conventional microscopes. They are so small that accurately predicting their behavior using classical physics is not possible due to quantum effects. More than 99.94% of an atom's mass is in the ***nucleus***.

*Atoms are the smallest units of matter that retain the properties of their chemical element*, Atoms can exist independently or combine to form *molecules*, which in turn create the matter we observe.

## Structure of an atom

The *nucleus* is the dense central core of an atom, containing two types of subatomic particles: *protons* and *neutrons*.

*Protons*, are positively charged particles. The number of protons in the nucleus, known as the *atomic number*, uniquely defines each element. For instance, hydrogen has one proton, while carbon, which has an atomic number of 6, meaning carbon has 6 protons in its nucleus. The positive charge of protons provides the overall positive charge of the nucleus.

*Neutrons*, are neutral and have no electric charge. The number of neutrons in the nucleus can differ even among atoms of the same *chemical element*, leading to the formation of *isotopes*.

The *chemical elements* are distinguished from each other by the number of *proton*s that are in their atoms. Atoms with the same number of protons but a different number of neutrons are called *isotopes* of the same *element*.

*Isotopes* are different forms of a chemical element that have the same number of protons but varying numbers of neutrons. This results in atoms of the same element having different atomic masses. For instance, Carbon-12 and Carbon-14 are isotopes of carbon.
Isotopes of an element exhibit almost identical chemical behavior but can have different physical properties. Hydrogen has three isotopes: Protium, Deuterium and Tritium.

The large majority of an atom's *mass* comes from the protons and neutrons that make it up. The total number of these particles (called "*nucleons*") in a given atom is called the *mass number*. For example, a carbon atom which has mass number of 12 contains 6 protons and 6 neutrons. The mass number provides information about the *overall mass* of the atom, as it accounts for the fused mass of its protons and neutrons, which are the heaviest subatomic particles.

As even the most massive atoms are far too light to work with directly, chemists instead use the unit of *moles*. One mole of atoms of any element always has the same number of atoms (about $6.022 \times 10^{23}$). This number was chosen so that *if an element has an atomic mass of 1 u, a mole of atoms of that element has a mass close to one gram.*

The *atomic number* and *mass number* are essential because they uniquely identify an element and identify the specific *isotope* of an *element*, which affects the element's atomic mass and nuclear stability.

Surrounding the nucleus is the *electron cloud*, which is composed of negatively charged particles, which called *electrons*. *Electrons* are negatively charged particles that move around the nucleus in regions called *orbitals*, which are organized into *energy levels* or *shells*. These are much smaller and lighter than protons and neutrons.

Atoms lack a well-defined outer boundary, so their dimensions are usually described in terms of an *atomic radius*. This is a measure of the distance out to which the electron cloud extends from the nucleus. This assumes the atom to exhibit a spherical shape, which is only obeyed for atoms in vacuum or free space. Atomic radii may be derived from the distances between two nuclei when the two atoms are joined in a chemical bond.

Elementary particles possess an intrinsic *quantum mechanical property* known as *spin*. This is analogous to the angular momentum of an object that is spinning around its center of mass, although strictly speaking these particles are believed to be point-like and cannot be said to be rotating. *Spin* is measured in units of the reduced Planck constant ($\hbar$), with electrons, protons and neutrons all having spin $1/2$ $\hbar$, or "spin-$1/2$". In an *atom*, electrons in motion around the nucleus possess *orbital angular momentum* in addition to their *spin*, while the nucleus itself possesses angular momentum due to its *nuclear spin*.

*The nucleus of an atom will have no spin when it has even numbers of both neutrons and protons*, but for other cases of odd numbers, the nucleus may have a spin. Normally nuclei with spin are aligned in random directions because of thermal equilibrium, but for certain elements (such as xenon-129) it is possible to polarize a significant proportion of the nuclear spin states so that they are aligned in the same direction—a condition called hyperpolarization.

The idea of a quantized spinning of the electron was put forward for the first time by Compton in August 1921 [*The magnetic electron.*], who pointed out the possible bearing of this idea on the origin of the natural unit of magnetism.

Without being aware of Compton's suggestion Samuel Goudsmit and George Uhlenbeck published a joint paper in November 1925 [*Ersetzung der Hypothese vom unmechanischen Zwang durch eine Forderung bezuglich des inneren Verhaltens jedes einzelnen Elektrons.* (Replacement of the hypothesis of unmechanical coercion by a requirement regarding the internal behavior of each individual electron.). See below.], which noted doublets in the alkali spectra which did not conform to current models of the atom. They proposed applying the model of the spinning electron to interpret a number of features of the quantum theory of the *Anomalous Zeeman effect*. They applied the classical formula for a spherical rotating electron with finite radius and surface charge.

[The *Anomalous Zeeman Effect* refers to the splitting of spectral lines in the presence of a magnetic field, particularly when electron spin is considered, leading to more complex splitting patterns than those observed in the normal Zeeman effect.

The *Normal Zeeman Effect* occurs when there is no electron spin involved, resulting in a straightforward splitting of spectral lines into three components (triplet) due to the interaction of the magnetic field with the orbital angular momentum of the electrons.

The *Anomalous Zeeman Effect* involves the total spin of the electrons, leading to additional splitting patterns. The presence of electron spin complicates the energy level calculations, resulting in a different number of sub-levels and a more intricate spectrum.]

*The presence of an external electric field can cause a comparable splitting and shifting of spectral lines* by modifying the electron energy levels, a phenomenon called the *Stark effect*.

The *magnetic field produced by an atom*—its *magnetic moment* or *magnetic dipole moment* —is determined by these various forms of angular momentum, just as a rotating charged object classically produces a magnetic field, but *the most dominant contribution comes from electron spin*. Due to the nature of electrons to obey the Pauli exclusion principle, in which no two electrons may be found in the same quantum state, *bound electrons pair up with each other, with one member of each pair in a spin up state and the other in the opposite, spin down state*. Thus, these spins cancel each other out, reducing the total magnetic dipole moment to zero in some atoms with even number of electrons.

The *magnetic moment* is a vector quantity which expresses the magnetic force effect of a magnet. The *magnetic field* of a *magnetic dipole* is proportional to its *magnetic dipole moment*. The dipole component of an object's magnetic field is symmetric about the direction of its magnetic dipole moment, and decreases as the inverse cube of the distance from the object. The *magnetic dipole moment* of an object determines the magnitude of torque the object experiences in a given magnetic field. When the same magnetic field is applied, objects with larger magnetic moments experience larger torques. The strength (and direction) of this torque depends not only on the magnitude of the *magnetic moment* but also on its orientation relative to the direction of the magnetic field. Its direction points from the south pole to the north pole of the magnet (i.e., inside the magnet).

*The magnetic properties of materials are mainly due to the magnetic moments of their atoms' orbiting electrons*. In *magnetic materials*, the cause of the *magnetic moment* are the *spin states* of the *electrons*. The magnetic moments of the nuclei of atoms are typically thousands of times smaller than the electrons' magnetic moments, so they are negligible in the context of the magnetization of materials.

***Ordinarily, the enormous number of electrons in a material are arranged such that their magnetic moments cancel out***. This is due, to some extent, to electrons combining into pairs with opposite intrinsic ('spin') magnetic moments as a result of the ***Pauli Exclusion Principle***, and combining into filled subshells with zero net magnetic moments. Moreover, even when the electron configuration is such that there are unpaired electrons and/or non-filled subshells, it is often the case that the various electrons in the solid will contribute magnetic moments that point in different, random directions so that the material will not be magnetic. Sometimes—either spontaneously, or owing to an applied external magnetic field—each of the electron magnetic moments will be, on average, lined up. A suitable material can then produce a strong net magnetic field.

The magnetic behavior of a material depends on its **structure**, particularly its **electron configuration**, and also on the **temperature**. At high temperatures, random thermal motion makes it more difficult for the electrons to maintain alignment.

The **electronic structure** of carbon given above has two unpaired electrons in the p orbitals while nitrogen has three such electrons. Oxygen and fluorine have two and one unpaired electrons respectively, while neon has no unpaired electrons. Elements such as carbon, with **unpaired electrons**, tend to be attracted by a magnet and are said to be **paramagnetic**. Elements such as neon with **all the electrons paired** tend to be repelled by a magnet and are said to be **diamagnetic**.

In **ferromagnetic elements** such as iron, cobalt and nickel, an odd number of electrons leads to an unpaired electron and a net overall magnetic moment. The **orbitals** of neighboring atoms overlap and a lower energy state is achieved when the spins of unpaired electrons are aligned with each other, a spontaneous process known as an exchange interaction. When the **magnetic moments** of ferromagnetic atoms are lined up, the material can produce a measurable macroscopic field. **Paramagnetic materials** have atoms with magnetic moments that line up in random directions when no magnetic field is present, but the magnetic moments of the individual atoms line up in the presence of a field.

### Potential energy of an electron in an atom

The ***potential energy of an electron in an atom*** is negative relative to when the distance from the nucleus goes to infinity; its dependence on the electron's position reaches the minimum inside the nucleus, roughly in inverse proportion to the distance. ***In the quantum-mechanical model, a bound electron can occupy only a set of states centered*** *on the nucleus, and each state corresponds to a specific energy leve*l. An energy level can be measured by the amount of energy needed to unbind the electron from the atom, and is usually given in units of electronvolts (eV). The lowest energy state of a bound electron is called the **ground state**, i.e., stationary state, while an electron transition to a higher level results in an **excited state**. For an electron to transition between two different states, e.g. ground state to first excited state, it must absorb or emit a photon at an energy matching the difference in the potential energy of those levels, according to the Niels Bohr model, what can be precisely calculated by the Schrödinger equation. Electrons jump between orbitals in a particle-like fashion. For example, if a single photon strikes the electrons, only a single electron changes states in response to the photon.

The energy of an emitted photon is proportional to its frequency, so **these specific energy levels appear as distinct bands in the electromagnetic spectrum**. Each element has a characteristic spectrum that can depend on the nuclear charge, subshells filled by electrons, the electromagnetic interactions between the electrons and other factors.

When a continuous spectrum of energy is passed through a gas or plasma, some of the photons are absorbed by atoms, causing electrons to change their energy level. Those excited electrons that remain bound to their atom spontaneously emit this energy as a photon, traveling in a random direction, and so drop back to lower energy levels. Thus, **the atoms behave like a filter that forms a series of dark absorption bands in the energy output**. An observer viewing the atoms from a view that does not include the continuous spectrum in the background, instead sees a series of emission lines from the photons emitted by the atoms. Spectroscopic measurements of the strength and width of atomic spectral lines allow the composition and physical properties of a substance to be determined.

Close examination of the spectral lines reveals that some display a fine structure splitting. This occurs because of **spin–orbit coupling**, which is an interaction between the spin and motion of the outermost electron. When an atom is in an external magnetic field, spectral lines become split into three or more components; a phenomenon called the Zeeman effect. This is caused by the interaction of the magnetic field with the magnetic moment of the atom and its electrons. Some atoms can have multiple electron configurations with the same energy level, which thus appear as a single spectral line. The interaction of the magnetic field with the atom shifts these electron configurations to slightly different energy levels, resulting in multiple spectral lines.

## Chemical Elements

The various **chemical elements** are formally identified by their unique atomic numbers, their accepted names, and their chemical symbols. The concept of an "element" as an indivisible substance has developed through three major historical phases: Classical definitions (such as those of the ancient Greeks), chemical definitions, and atomic definitions.

The *atomic number of an element* is equal to the number of protons in each atom, and defines the element. For example, all carbon atoms contain 6 protons in their atomic nucleus; so, the atomic number of carbon is 6. The number of protons in the nucleus determines its electric charge, which in turn determines the number of bound electrons of an atom in its non-ionized state. The electrons occupy atomic orbitals that determine the atom's chemical properties.

A *chemical element* is a *species of atom defined by its number of protons*. The number of protons is called the *atomic number* of that element. For example, oxygen has an atomic number of 8: each oxygen atom has 8 protons in its nucleus. *Atoms of the same element can have different numbers of neutrons in their nuclei, known as isotopes of the element*. Atoms of one element can be transformed into atoms of a different element in nuclear reactions, which change an atom's atomic number. Almost all baryonic matter in the universe is composed of elements (among rare exceptions are neutron stars).

The standard *atomic weight* (commonly called "atomic weight") *of an element* is the *average of the atomic masses of all the chemical element's isotopes as found in a particular environment, weighted by isotopic abundance, relative to the atomic mass unit*.

Chemists and nuclear scientists have different definitions of a *pure element*. In chemistry, a pure element means a substance whose atoms all (or in practice almost all) have the same atomic number, or number of protons. Nuclear scientists, however, define a pure element as one that consists of *only one isotope*.

The term "chemical element" is also widely used to mean a pure chemical substance consisting of a single element. For example, oxygen gas consists only of atoms of oxygen.

*Isotopes* are atoms of the same *element* (that is, with the same number of protons in their nucleus), but *having different numbers of neutrons*. Thus, for example, there are three main isotopes of carbon. All carbon atoms have 6 protons, but they can have either 6, 7, or 8 neutrons. Since the mass numbers of these are 12, 13 and 14 respectively, said three isotopes are known as carbon-12, carbon-13, and carbon-14 ($^{12}C$, $^{13}C$, and $^{14}C$). Natural carbon is a mixture of $^{12}C$ (about 98.9%), $^{13}C$ (about 1.1%) and about 1 atom per trillion of $^{14}C$. The number of neutrons in a nucleus usually has very little effect on an element's chemical properties. An exception is hydrogen, for which the kinetic isotope effect is significant. Thus, all carbon isotopes have nearly identical chemical properties because they all have six electrons, even though they may have 6 to 8 neutrons. That is why *atomic number*, rather than mass number or atomic weight, is considered the identifying characteristic of an element.

All elements have *radioactive* isotopes (radioisotopes), but many of these radioisotopes are not found in nature due to a low half-life. Radioisotopes typically decay into other elements via *alpha decay*, *beta decay*, or *gamma decay*.

*Alpha decay* occurs when the nucleus emits an *alpha particle*, which is a helium nucleus consisting of two protons and two neutrons. The result of the emission is a new element with a lower atomic number. *Beta decay* (and electron capture) is regulated by the weak force, and results from a transformation of a neutron into a proton, or a proton into a neutron. The neutron to proton transition is accompanied by the emission of an *electron* and an *antineutrino*, while proton to neutron transition (except in electron capture) causes the emission of a *positron* and a *neutrino*. The electron or positron emissions are called *beta particles*. Beta decay either increases or decreases the atomic number of the nucleus by one. Electron capture is more common than positron emission, because it requires less energy. In this type of decay, an electron is absorbed by the nucleus, rather than a positron emitted from the nucleus. A neutrino is still emitted in this process, and a proton changes to a neutron. **Gamma decay** results from a change in the energy level of the nucleus to a lower state, resulting in the emission of *electromagnetic radiation*. The excited state of a nucleus which results in gamma emission usually occurs following the emission of an alpha or a beta particle. Thus, gamma decay usually follows alpha or beta decay.

The *mass number of an element*, A, is the *number of nucleons (protons and neutrons) in the atomic nucleus*. Different isotopes of a given element are distinguished by their mass number, which is written as a superscript on the left-hand side of the chemical symbol (e.g., $^{238}U$). The *mass number* is always an integer and has units of "nucleons". Thus, magnesium-24 (24 is the mass number) is an atom with 24 nucleons (12 protons and 12 neutrons).

Whereas the mass number simply counts the total number of neutrons and protons and is thus an integer, the *atomic mass* of a particular isotope (or "nuclide") of the element is the *mass of a single atom of that isotope*, and is typically expressed in *daltons* (symbol: Da), aka universal atomic mass units (symbol: u), which is defined as 1/12 of the mass of a free neutral carbon-12 atom in the ground state.

The *relative atomic mass* is a dimensionless number equal to the atomic mass divided by the atomic mass constant, which equals 1 Da. In general, the *mass number* of a given nuclide differs in value slightly from its *relative atomic mass*. This *mass deficit* is primarily due to the **nuclear binding energy**. However, the relative atomic mass of each isotope is quite close to its mass number (always within 1%). The only isotope whose atomic mass is exactly a natural number is $^{12}C$, which has a mass of 12 Da; because the **dalton** is defined as 1/12 of the mass of a free neutral carbon-12 atom in the ground state.

*A commonly used basic distinction among the elements is their state of matter (phase), whether solid, liquid, or gas, at standard temperature and pressure (STP)*. Most elements are solids at STP, while several are gases. Only bromine and mercury are liquid at 0 degrees Celsius (32 degrees Fahrenheit) and 1 atmosphere pressure; caesium and gallium are solid at that temperature, but melt at 28.4 °C (83.1 °F) and 29.8 °C (85.6 °F), respectively. *Melting and boiling points, typically expressed in degrees Celsius at a pressure of one atmosphere, are commonly used in characterizing the various elements*.

Atoms of the same element may bond to each other chemically in more than one way, allowing the pure element to exist in multiple chemical structures (spatial arrangements of atoms) which differ in their properties. The ability of an element to polymorph in one of many structural forms is known as 'allotropy'. Non-metallic elements known for polymorphism include carbon, sulfur, phosphorus, oxygen, and nitrogen.

*The density at selected standard temperature and pressure (STP) is often used in characterizing the elements*. The **mass density of an element** depends on the mass of the atomic nucleus and the separation between the atoms created by the bound electrons. Density is given in kilograms per cubic meter (kg/m3), but may also be expressed in grams per cubic centimeter (g/cm$^3$).

Several terms are commonly used to characterize the general physical and chemical properties of the **chemical elements**. A first distinction is between **metals**, which readily conduct electricity, **nonmetals**, which do not, and a small group, the **metalloids**, having intermediate properties and often behaving as semiconductors.

A more refined classification is provided by the **periodic table**. The properties of the elements can be summarized in this form, which powerfully and elegantly organizes the elements by physical and chemical properties. *Each row forms a period of elements that have the same number of electron shells*. There are *18 numbered groups, each forming its own column of elements* whose chemical properties are dominated by the orbital location of the outermost electron. A **block** is a set of elements sharing atomic orbitals that their valence electrons or vacancies occupy.

By November 2016, the International Union of Pure and Applied Chemistry (IUPAC) recognized *a total of 118 elements*. The first 94 occur naturally on Earth, and the remaining 24 are synthetic elements produced in nuclear reactions.

## Periodic Table

Much of the modern understanding of *chemical elements* developed from the work of Dmitri Mendeleev, a Russian chemist who published the first recognizable *periodic table* in 1869. This table organizes the elements by increasing atomic number into rows ("periods") in which the columns ("groups") share recurring ("periodic") physical and chemical properties. The periodic table summarizes various properties of the elements, allowing chemists to derive relationships between them and to make predictions about elements not yet discovered, and potential new compounds.

The *periodic table* groups *elements* with similar chemical and physical properties together. Initially, the Russian chemist, Dmitri Mendeleev grouped elements in order of increasing atomic weights but this arrangement was later changed to one which grouped the elements in order of increasing atomic numbers. It was observed that *if elements were arranged in order of increasing atomic numbers, elements with similar chemical properties recurred periodically and at regular intervals*. This so-called *periodic law* was used in the construction of the *periodic table*.

In that table, there are eighteen (18) vertical columns which are called *groups* and about seven (7) rows, called *periods*. *All the members of a group have the same outer shell electron configuration*, apart from the "n" value which is different. For example, the lithium group elements all have the $ns^1$ outer shell configuration. *Members of a group* also *have similar chemical properties*. *Group 1a elements* are called **alkali elements** (hydrogen is not included in this group). *Group 2a elements* are called *alkaline earth elements*, and these have the $ns^2$ outer shell structure. *Group 7a elements* are called *halogens* with an outer shell configuration of $ns^2np^5$. *Group 8a elements* are called *rare gases or noble gases*.

Another way of naming the different groups of elements depends on the *outer shell orbital* which the different groups fill. Thus, **groups 1a and 2a elements**, in which **s** *orbitals* are filled, are called *s-block elements*. *Groups 3a-8a elements,* in which *p orbitals* are filled, are called *p-block elements*. Groups 1a to 7a which have incompletely filled s and p orbitals of the highest n-value are also referred to as representative (or main group) elements while the noble gases have a completely filled p orbital of highest n-value.

*Groups 1b-8b,* in which the *d orbitals* are filled, are called *d-block elements*. Groups 1b, 3b-8b have incompletely filled d orbitals and these are referred to as transition elements. Group 2b, with a completely filled d orbital is not normally included in the group of transition elements. The elements Ce-Lu, in which the *4f orbitals* are filled, are called *rare earth elements* or *lanthanides*. The elements Th-Lr, in which the *5f orbitals* are filled, are called *actinides* (another group of rare earth elements). The lanthanides and actinides together are called the inner-transition elements or *f-block elements*.

In any *atom* the *K shell* is the lowest in energy followed by the L and then the M and so on. Within a *shell*, the *s orbital* is the lowest in energy followed by the p orbitals and then the d orbitals and so on. The lowest energy shell and the lowest energy orbital in that shell fills first.

The lightest elements are hydrogen and helium, which current theory suggests were both created by Big Bang nucleosynthesis in the first 20 minutes of the universe in a ratio of around 3:1 by mass (or 12:1 by number of atoms), along with tiny traces of the next two elements, lithium and beryllium; almost all other elements found in nature were made by various natural methods of nucleosynthesis. On Earth, small amounts of new atoms are naturally produced in nucleogenic reactions, or in cosmogenic processes, such as cosmic ray spallation. New atoms are also naturally produced on Earth as radiogenic daughter isotopes of ongoing radioactive decay processes such as alpha decay, beta decay, spontaneous fission, cluster decay, and other rarer modes of decay.

*There are now 118 known elements.* "Known" here means observed well enough, even from just a few decay products, to have been differentiated from other elements. Most recently, the synthesis of element 118 (since named oganesson) was reported in October 2006, and the synthesis of element 117 (tennessine) was reported in April 2010. Of these 118 elements, the first 94 elements have been detected directly on Earth as primordial *nuclides* (species of atom characterized by the constitution of its nucleus containing a certain number of neutrons and protons) present from the formation of the Solar System, or as naturally occurring fission or transmutation products of uranium and thorium. Six of these occur in extreme trace amounts: technetium, atomic number 43; promethium, number 61; astatine, number 85; francium, number 87; neptunium, number 93; and plutonium, number 94. These 94 elements have been detected in the universe at large, in the spectra of stars, as well as neutron star mergers and supernovae, where short-lived radioactive elements are newly being made.

*Elements with atomic numbers 83 through 94 are unstable enough that radioactive decay of all isotopes can be detected.* Some of these elements, notably bismuth (atomic number 83), thorium (atomic number 90), and uranium (atomic number 92), have one or more isotopes with half-lives long enough to survive from before the Solar System formed. The remaining longest-lived isotopes have half-lives too short for them to have been present at the beginning of the Solar System, and are therefore "*transient elements*". Of these 11 transient elements, five (polonium, radon, radium, actinium, and protactinium) are relatively common decay products of thorium and uranium. The remaining six transient elements (technetium, promethium, astatine, francium, neptunium, and plutonium) occur only rarely, as products of rare decay modes or nuclear reaction processes involving uranium or other heavy elements.

*The remaining 24 heaviest elements (those beyond plutonium, element 94) are radioactive, with half-lives so short that they are not found on Earth* and must be synthesized. Five have been discovered in the spectrum of Przybylski's star, from element 95 (americium) to 99 (einsteinium). These are thought to be neutron capture products of uranium and thorium. The lightest radioactive isotope is tritium, which undergoes Beta decay with a half-life of 12.3 years. At $2 \times 10^{19}$ years, over $10^9$ times the estimated age of the universe, bismuth-209 has the longest known alpha decay half-life of any nuclide, and is almost always considered on par with the 80 stable elements.

*Electrons occupy specific energy levels or shells around the nucleus*. The arrangement of electrons in these levels is called the *electron configuration*. *The electron configuration of an element is mainly responsible for its chemical properties*. The electron configuration of an element describes the distribution of electrons in an atom's orbitals, following the Aufbau principle, Pauli exclusion principle, and Hund's rule.

*Quantum mechanics* causes the bound electrons to be organized into a set of layered shells. Each shell can only contain a fixed number of electrons occupying paired orbitals.

*The electron configurations of these shells mediate the interaction with neighboring atoms and determine the chemical properties of atoms*. The shell configuration *determines the structure of the periodic table*. Electrons that are confined to an atom are only allowed to assume certain discrete energy levels. This restriction, known as quantization, is a fundamental facet of the quantum mechanics theory that predicts the wave-like behavior of particles and energy at the smallest scale. For atoms, these energy levels are represented by *electron subshells*, where the wave form of an electron is held in a type of standing wave with a specific wavelength. Each whole number of wavelengths yields one or more atomic orbitals, which describe each the electron's charge distribution at that energy level. Every orbital can hold a pair of electrons. The arrangement of electrons in an atom's orbitals is called the *electron configuration*.

These orbitals and their sub-shells are grouped together in **shells, with each shell having a principal quantum number that indicates the energy level**. Each shell can only have a fixed number of electrons, which is given by the formula $2n^2$, where n is 1, 2, 3, 4, and so on. Hence, the count of electrons each shell can contain are 2, 8, 18, 32, and so forth. A shell is represented by a row on the periodic table.

*Electrons* occupy different *energy levels*, starting with the lowest energy level closest to the nucleus.

> The first energy level (n=1) can hold up to 2 electrons.
> The second energy level (n=2) can hold up to 8 electrons.
> The third energy level (n=3) can hold up to 18 electrons, and so on.

Within each energy level, electrons are distributed among different types of *orbitals* (s, p, d, f), each with a specific shape and orientation.

> **s-*orbital***: Spherical shape, can hold up to 2 electrons.
> **p-*orbital***: Dumbbell shape, can hold up to 6 electrons (3 orbitals, each holding 2 electrons).
> **d-*orbital***: More complex shapes, can hold up to 10 electrons (5 orbitals).
> **f-*orbital***: Even more complex shapes, can hold up to 14 electrons (7 orbitals).

Electrons fill orbitals starting with the lowest energy level, **with each orbital holding a maximum of two electrons with opposite spins**. Electrons fill degenerate orbitals singly before pairing. The notation involves the energy level and orbital type (e.g., $^1s$, $^2p$) with superscripts indicating the number of electrons (e.g., $^1s^2$). For example, carbon (atomic number 6) has the configuration $^1s^{22}s^{22}p^2$. A condensed notation uses the nearest noble gas as shorthand, such as sodium (Na) with $[Ne]^3s^1$.

When classical physics was applied to atoms and other smaller particles, the results obtained were contrary to those expected. In 1900, Max Planck solved this mystery by stating that the energy of an object is not continuous; and that the object will only acquire enough energy to take it from one energy level to another.

> See Underwood, T. G., *Quantum Electrodynamics*, Vol. I, pp. 80-89: Planck, M. (1901). *Ueber das Gesetz der Energieverteilung im Normalspektrum*. (On the Law of Distribution of Energy in the Normal Spectrum.) *Ann. Physik*, 309, 3, 553-63; https://doi.org/10.1002/ andp.19013090310; translated in Ando, K. *On the Law of Distribution of Energy in the Normal Spectrum*. https://web.archive.org/web/ 20150222133544/ http://theochem.kuchem.kyoto-u.ac.jp/Ando/planck1901.pdf; in November 1900 Planck revised his first approach, relying on Boltzmann's statistical interpretation of the second law of thermodynamics as a way of gaining a more

fundamental understanding of the principles behind his radiation law. The central assumption behind his new derivation, presented to the DPG on December 14, 1900, was the supposition, now known as the Planck postulate, that electromagnetic energy could be emitted only in quantized form, in other words, the energy could only be a multiple of an elementary unit E = hv, where h is Planck's constant, also known as Planck's action quantum (introduced already in 1899), and v is the frequency of the radiation. From this Planck arrived at the famous formula for the density of black body radiation at temperature T.

If we give energy to an object in state 1, the object will only move to state 2 if it has acquired enough energy equal to $\Delta E = E2 - E1$ (where E2 is the energy of state 2 and E1 that of state 1). The energy in this case is in form of a packet or bundle and so, we speak of the energy being quantized. Thus, the object can only have a certain definite energy value. Planck called this packet of energy ($=\Delta E$) a quantum of energy. If the energy of the object in energy level 1 is less than $\Delta E$, the object will not move. This started a new branch of physics called quantum physics whose laws apply to tiny particles like atoms, electrons, etc.

The energy of an electron is governed by an integer called the ***Principal Quantum Number*** (n). The principal quantum numbers are just arbitrary numbers which are introduced in order to make predictions match experiment. They are given numbers 1, 2, 3, 4, etc., increasing as an electron moves further away from the nucleus. In simple terms, the ***principal quantum numbers*** are the shell numbers; thus, the K shell is principal quantum number 1, L shell is principal quantum number 2 and so on.

The ***modern quantum theory*** says that electrons move around the nucleus in a definite region of space. The region of space in which the electron spends at least 95% of its time is called an ***orbital***. The orbitals are arranged in shells around the nucleus. For each ***principal quantum number*** (n), there are $n^2$ ***orbitals***.

[2] The maximum number of electrons allowed for each n-value is $2n^2$.

These are shown in Table 1.2. below. Note that numbering starts from the innermost shell since the K shell is shell number 1 (that is ***principal quantum number*** 1).

**Table 1.2. Shell symbols, orbitals and the maximum number of electrons in them for each principal quantum number.**

| Shell number (n) | Symbol | Number of orbitals | Maximum number of electrons[2] |
|---|---|---|---|
| 1 | K | 1 | 1 |
| 2 | L | 4 | 8 |
| 3 | M | 9 | 18 |
| 4 | N | 16 | 32 |
| 5 | O | 25 | 50 |
| 6 | P | 36 | 72 |

Apart from the **K shell** which has one orbital (the s orbital), both theory and experiments have shown that the group of orbitals in each of the L, M, N, etc., shells do not have the same energies especially in a many-electron atom. In a hydrogen atom, which has only one electron, the orbitals in each group have the same energy. Such orbitals having the same energy are said to be **degenerate**.

In a many-electron atom, the **L shell** has two energy levels; the lower one has one orbital which is called an s orbital, the higher one has three identical orbitals which are **degenerate** and these are called p orbitals. The s orbital is spherical whereas the p orbitals are dumb-bell shaped situated at right angles to each other and in three dimensions along the x-, y- and z-axes; hence designated $np_x$, $np_y$ and $np_z$ respectively, (where n is the **principal quantum number**).

The **M shell** has three energy levels; the lowest one consists of one orbital (also an s orbital), the second one has three orbitals which are identical and **degenerate** (also called p orbitals); and the third level consists of five **degenerate** orbitals called d orbitals. The N shell has four energy levels, s, p and d just like the M shell, but in addition, the highest one has seven **degenerate** orbitals called f orbitals.

Therefore, within a given shell, the energies increase in the order s < p < d < f[3], etc.

[3] The symbols s, p, d and f arise from electronic transitions in spectroscopy; s for Sharp, p for Principal, d for Diffuse and f for Fundamental

It was stated earlier that the energy of the electron is governed by the **principal quantum number** n; this number also governs the **size** of the **orbitals**. An s orbital of principal quantum number one (denoted 1s) is smaller than an s orbital of principal quantum number two (denoted 2s). The orbitals can be specified by the "n" values in which they appear;

23

thus, we would talk of a 1s orbital to mean an s orbital of principal quantum number one. Similarly, a 2s or 3s orbital would mean an s orbital in principal quantum numbers two and three respectively. Analogously, 2p orbitals are those of principal quantum number two ($2p_x$, $2p_y$ or $2p_z$). In the same way, we can also specify an electron in a particular orbital. Thus, a 1s electron or a 3d electron means an electron in the s orbital of principal quantum number one or an electron in the d orbital of principal quantum number three, and so on.

**Pauli Exclusion Principle**

Electrons are added to each atom, one at a time across a period. The lowest energy orbitals fill first and once those are full, the next lower energy ones fill and so on. Structures which show the n-values and their orbitals that are occupied, together with the number of electrons in them are called *electronic structures*. The question one can ask at this point is: "*how many electrons can a single orbital accommodate?*". The answer to this question was arrived at by Wolfgang Pauli, who introduced a restriction on the number of electrons which can go into a single orbital. The restriction called the *Pauli Exclusion Principle* states that "any orbital will not hold more than two electrons". This means that any orbital can hold 0, 1 or 2 electrons but not more than 2. Therefore, the maximum number of electrons allowed for each n value will be equal to $2n^2$ (see Table 1.2).

With this and the *Pauli Exclusion Principle* in mind, we can now write the electronic structures for some of the atoms. We begin with the simplest atom, hydrogen. This atom has got only one electron, and this will therefore, go into the s orbital of the K shell. The electronic structure we can write for hydrogen is therefore, $1s^1$, where 1 indicates the *principal quantum number* (1 for the K shell), s indicates the **orbital filling**, and [1] indicates the *number of electrons*.

The lowest energy level of an atom is called the *ground state* of that atom. For example, we have seen above that in hydrogen the electron normally resides in the 1s orbital. The electronic state $1s^1$ is the lowest energy state for hydrogen and it is therefore the ground state. However, the electron in hydrogen can also move to higher levels, say to a 2s orbital. Such states of higher energy are called *excited states* of the atom.

Note that the s orbital of the K shell of hydrogen is half-full so that in the next atom, helium, an added electron will go into the same orbital and the electronic structure for helium is therefore $1s^2$. At this point, the s orbital of the K shell is full in accord with *Pauli Exclusion Principle*. In the next atom, lithium (Li), the L shell starts filling and the electron will go into the s orbital of this shell (n = 2). The electronic structure of lithium is therefore, $1s^2 2s^1$ and that of beryllium (Be) is $1s^2 2s^2$. The s orbital of the L shell is now filled up, and in the next atom boron (B), the p orbitals of the L shell start filling so that the electronic structure

24

of boron is $1s^2 2s^2 2p^1$. Since there are three p orbitals, six electrons are required to fill them completely and the atoms B, C, C, O, F and Ne all fill up the 2p orbitals with Ne having the electronic structure $1s^2 2s^2 2p^6$.

The trend is the same in the ***third row*** of the ***periodic table***. The s orbital fills first for this row (note: this is the s orbital of the M shell) in Na and Mg, followed by the p orbitals. The d orbitals are not filled. In the ***fourth row***, the s orbital of the N shell fills first (in K and Ca). In the next atom scandium (Sc), the d orbitals of row three become lower in energy than the p orbitals of row four, and so they start filling. The electronic structure we can write for scandium is therefore, $1s^2 2s^2 2p^6 3s^2 3p^6 4s^2 3d^1$. The d orbitals require ten electrons (for the five of them) and the ten atoms Sc to Zn fill these orbitals. After the 3d orbitals are filled, the p orbitals of the fourth row start filling. Again, note that the d and f orbitals of the fourth row are not filled at this point. The d orbitals for row four are filled in the ***fifth row*** after the s orbital of that row is filled. In the ***sixth row***, the s orbital of that row fills first and then the next electron in La (atom 57) enters the 5d orbital after which, the f orbitals of row four (elements Ce-Lu) start filling. These require fourteen electrons (for the seven of them) and so we see a group of fourteen atoms normally written at the bottom of the table. After the 4f orbitals are filled, the 5d orbitals continue filling in the remaining nine atoms (Hf-Hg; note that one 5d electron went into La) followed by the 6p orbitals. The filling up of electrons into orbitals becomes complicated as we go down the table so that we will not continue any further. From the foregoing discussion, we note that the orbitals are filled in the order:

$$1s < 2s < 2p < 3s < 3p < 4s3d < 4p < 5s < 4d < 5p < 6s < 4f < 5d < 6p\cdots$$

The ***electronic structures*** are also written in the same order. There are three points to be noted in these structures. The first is that ***the sum of the superscripts in the electronic structure gives the atomic number of the atom***. For example, in the electronic structure of oxygen, $1s^2 2s^2 2p^4$, the sum of the superscripts is 8 (2 + 2 + 4), so the atomic number of oxygen is 8. This means that once we know the atomic number of an atom, we can easily write its electronic structure and vice versa. The second is that ***the sum of the electrons in the highest principal quantum number***, especially for the main group atoms, gives the ***group number*** of that atom. For example, in the electronic structure of oxygen, the highest principal quantum number is 2, and there are 6 electrons there (2 + 4 = 6). So, oxygen is a group 6A element. Finally, ***the highest principal quantum number in the electronic structure gives the period in which that atom appears***. In the example of oxygen, the highest principal quantum number in its electronic structure is 2, so oxygen is in ***period 2***.

## Hund's Rule

The *Pauli Exclusion Principle* stated earlier tells us exactly how many electrons can go into a particular orbital; an orbital will not have more than two electrons. With this principle, we can straight away write the *electronic structure* of for example, carbon as, $1s^2 2s^2 2p^2$. We saw that there are three *degenerate* p orbitals designated $np_x$, $np_y$, $np_z$. The question now arises: "how do the two 2p electrons of carbon enter the three p orbitals?". There are two possibilities; either both can go into the $2p_x$ orbital (i.e., $1s^2 2s^2 2p_x^2$, with the $2p_y$ and $2p_z$ orbitals empty) or one can go into the $2p_x$ and the other one into the $2p_y$ (i.e., $1s^2 2s^2 2p_x^1 2p_y^1$ with the $2p_z$ orbital empty). Friedrich Hund came up with a rule that is used to decide which electronic structure will be the most stable. The rule, called *Hund's rule states that "electrons will occupy degenerate orbitals singly if those orbitals are empty"*.

This means that *as long as there is a set of orbitals with the same energy, the electrons will go in separate orbitals until each such orbital has one electron*. After that, additional electrons will start pairing up in accord with the *Pauli Principle*. With this, we can write the most stable electronic structure of carbon as: $1s^2 2s^2 2p_x^1 2p_y^1$ (note: the choice of which orbital is occupied i.e., whether the $2p_x$, $2p_y$ or $2p_z$, is purely arbitrary). The *electronic structure* of nitrogen, oxygen, fluorine and neon are:

N: $1s^2 2s^2 2p_x^1 2p_y^1 2p_z^1$,
O: $1s^2 2s^2 2p_x^2 2p_y^1 2p_z^1$,
F: $1s^2 2s^2 2p_x^2 2p_y^2 2p_z^1$,
Ne: $1s^2 2s^2 2p_x^2 2p_y^2 2p_z^2$.

Note that nitrogen has one electron in each of the p orbitals in accord with Hund's rule and that pairing starts in oxygen. The same trend applies to the d orbitals and also, the f orbitals.

## Nuclear fusion and nuclear fission

During the *nuclear fusion* of lower mass atoms such as hydrogen, the *net change in mass deficit* is released as energy, as determined by the mass–energy equivalence relationship. *This process of fusing hydrogen atoms into helium is what drives the energy output of the Sun*. Over time, the result is an increasing concentration of helium at the stellar core. During the evolution of stars much more massive than the Sun, increasingly massive nuclei are then formed through a type of fusion called the alpha process, until iron-52 is reached. *The binding energy of a nucleus reaches its peak value for isotopes of iron and nickel*. Hence, beyond that point, further fusion results in a lower binding energy, so energy is absorbed rather than released. As a result, an inert iron core forms that does not contribute to the star's energy output.

In the ***nuclear fission process***, the resulting particles have a higher net binding energy. This change in the net mass deficit again results in a release of energy. Hence, highly radioactive elements such as uranium-235 can be useful sources of energy production.

## Molecules

Concepts similar to molecules have been discussed since ancient times, but modern investigation into the nature of molecules and their bonds began in the 17th century. Refined over time by scientists such as Robert Boyle, Amedeo Avogadro, Jean Perrin, and Linus Pauling, the study of molecules is today known as molecular physics or molecular chemistry.

The definition of the molecule has evolved as knowledge of the structure of molecules has increased. Earlier definitions were less precise, defining molecules as the smallest particles of pure chemical substances that still retain their composition and chemical properties. This definition often breaks down since many substances in ordinary experience, such as rocks, salts, and metals, are composed of large crystalline networks of chemically bonded atoms or ions, but are not made of discrete molecules.

The modern concept of molecules can be traced back towards pre-scientific and Greek philosophers such as Leucippus and Democritus who argued that all the universe is composed of atoms and voids. Circa 450 BC Empedocles imagined fundamental elements (fire (), earth (), air (), and water ()) and "forces" of attraction and repulsion allowing the elements to interact.

A fifth element, the incorruptible quintessence aether, was considered to be the fundamental building block of the heavenly bodies. The viewpoint of Leucippus and Empedocles, along with the aether, was accepted by Aristotle and passed to medieval and renaissance Europe.

In a more concrete manner, however, the concept of aggregates or units of bonded atoms, i.e. "molecules", traces its origins to Robert Boyle's 1661 hypothesis, in his famous treatise *The Sceptical Chymist*, that matter is composed of clusters of particles and that chemical change results from the rearrangement of the clusters. Boyle argued that matter's basic elements consisted of various sorts and sizes of particles, called "corpuscles", which were capable of arranging themselves into groups. In 1789, William Higgins published views on what he called combinations of "ultimate" particles, which foreshadowed the concept of valency bonds. If, for example, according to Higgins, the force between the ultimate particle of oxygen and the ultimate particle of nitrogen were 6, then the strength of the force would be divided accordingly, and similarly for the other combinations of ultimate particles.

Amedeo Avogadro created the word "molecule". His 1811 paper "*Essay on Determining the Relative Masses of the Elementary Molecules of Bodies*", he essentially states, i.e. according to Partington's *A Short History of Chemistry*, that: The smallest particles of gases are not necessarily simple atoms, but are made up of a certain number of these atoms united by attraction to form a single *molecule*.

In coordination with these concepts, in 1833 the French chemist Marc Antoine Auguste Gaudin presented a clear account of Avogadro's hypothesis, regarding atomic weights, by making use of "volume diagrams", which clearly show both semi-correct molecular geometries, such as a linear water molecule, and correct molecular formulas, such as $H_2O$.

In 1917, an unknown American undergraduate chemical engineer named Linus Pauling was learning the Dalton hook-and-eye bonding method, which was the mainstream description of bonds between atoms at the time. Pauling, however, was not satisfied with this method and looked to the newly emerging field of quantum physics for a new method. In 1926, French physicist Jean Perrin received the Nobel Prize in physics for proving, conclusively, *the existence of molecules*. He did this by calculating the Avogadro constant using three different methods, all involving liquid phase systems. First, he used a gamboge soap-like emulsion, second by doing experimental work on Brownian motion, and third by confirming Einstein's theory of particle rotation in the liquid phase.

In 1927, the physicists Fritz London and Walter Heitler applied the new *quantum mechanics* to the deal with the saturable, nondynamic forces of attraction and repulsion, i.e., exchange forces, of the *hydrogen molecule*. Their valence bond treatment of this problem, in their joint paper, was a landmark in that it brought chemistry under quantum mechanics. Their work was an influence on Pauling, who had just received his doctorate and visited Heitler and London in Zürich on a Guggenheim Fellowship.

Subsequently, in 1931, building on the work of Heitler and London and on theories found in Lewis' famous article, Pauling published his ground-breaking article "*The Nature of the Chemical Bond*" in which *he used quantum mechanics to calculate properties and structures of molecules*, such as angles between bonds and rotation about bonds.

The science of molecules is called *molecular chemistry* or *molecular physics*, depending on whether the focus is on chemistry or physics. *Molecular chemistry* deals with the laws governing the interaction between molecules that results in the formation and breakage of chemical bonds, while *molecular physics* deals with the laws governing their structure and properties. In practice, however, this distinction is vague. In molecular sciences, a molecule consists of a stable system (bound state) composed of two or more atoms. Polyatomic ions may sometimes be usefully thought of as electrically charged molecules. The term unstable molecule is used for very reactive species, i.e., short-lived assemblies (resonances) of electrons and nuclei, such as radicals, molecular ions, Rydberg molecules, transition states, van der Waals complexes, or systems of colliding atoms as in Bose–Einstein condensate.

*The study of molecules is largely based on quantum mechanics which is essential for the understanding of the chemical bond*. The simplest of molecules is the hydrogen molecule-ion, $H_2+$, and the simplest of all the chemical bonds is the one-electron bond. $H_2+$ is composed of two positively charged protons and one negatively charged electron, which means that the Schrödinger equation for the system can be solved more easily due to the lack of electron–electron repulsion.

*A molecule is a group of two or more atoms that are held together by attractive forces known as chemical bonds*; depending on context, the term may or may not include ions that satisfy this criterion. *A molecule consists of one or more types of atoms combined in a fixed ratio*, making it a more complex structure than a single atom. Molecules can be made up of the same type of atoms (like $O_2$) or different types (like $H_2O$). The formation of *molecules* occurs through *chemical bonds* that hold the atoms together, resulting in various compounds and elements. This distinction is fundamental in understanding the composition of matter in chemistry.

*Molecules are generally held together by covalent bonding*. Several non-metallic elements exist only as molecules in the environment either in compounds or as homonuclear molecules, not as free atoms: for example, hydrogen. *A covalent bond is a chemical bond that involves the sharing of electron pairs between atoms*. These electron pairs are termed shared pairs or bonding pairs, and the stable balance of attractive and repulsive forces between atoms, when they share electrons, is termed covalent bonding.

[In contrast, *ionic bonding* is a type of chemical bond that involves the *electrostatic attraction* between oppositely charged ions, and is the primary interaction occurring in *ionic compounds*. The ions are atoms that have lost one or more electrons (termed cations) and atoms that have gained one or more electrons (termed anions). This transfer of electrons is termed *electrovalence* in contrast to *covalence*. In the simplest case, the cation is a metal atom and the anion is a nonmetal atom, but these ions can be of a more complicated nature, e.g. molecular ions like $NH_4+$ or $SO_{42}-$. At normal temperatures and pressures, *ionic bonding mostly creates solids (or occasionally liquids) without separate identifiable molecules*, but the vaporization/sublimation of such materials does produce separate molecules where electrons are still transferred fully enough for the bonds to be considered *ionic* rather than *covalent*.]

*Chemical compounds* are substances made of atoms of different elements; they can have molecular or non-molecular structure. *Mixtures* are materials containing different chemical substances; that means (in case of molecular substances) that they contain different types of molecules. When different elements undergo chemical reactions, atoms are rearranged into new compounds held together by chemical bonds. Less than twenty elements, including the gold, platinum, iron group metals, can sometimes be found uncombined as relatively pure native element minerals. Nearly all other naturally occurring elements exist in the Earth as *compounds* or *mixtures*. Air is mostly a mixture of molecular nitrogen and oxygen, though it does contain compounds including carbon dioxide and water, as well as atomic argon, a noble gas which is chemically inert and therefore does not undergo chemical reactions.

Molecules as components of matter are common. They also make up most of the oceans and atmosphere. *Most organic substances are molecules*. The substances of life are molecules, e.g. proteins, the amino acids of which they are composed, the nucleic acids (DNA and RNA), sugars, carbohydrates, fats, and vitamins. However, the majority of familiar solid substances on Earth are made partly or completely of crystals or *ionic compounds*, which are not made of molecules. These include all of the minerals that make up the substance of the Earth, sand, clay, pebbles, rocks, boulders, bedrock, the molten interior, and the core of the Earth. All of these contain many chemical bonds, but are not made of identifiable molecules.

No typical molecule can be defined for salts nor for covalent crystals, although these are often composed of repeating unit cells that extend either in a plane, e.g. graphene; or three-dimensionally e.g. diamond, quartz, sodium chloride. The theme of repeated unit-cellular-structure also holds for most metals which are condensed phases with metallic bonding. Thus, solid metals are not made of molecules. In glasses, which are solids that exist in a vitreous disordered state, the atoms are held together by chemical bonds with no presence of any definable molecule, nor any of the regularity of repeating unit-cellular-structure that characterizes salts, covalent crystals, and metals.

Most molecules are far too small to be seen with the naked eye, although molecules of many polymers can reach macroscopic sizes, including biopolymers such as DNA. Molecules commonly used as building blocks for organic synthesis have a dimension of a few angstroms (Å) to several dozen Å, or around one billionth of a meter. Single molecules cannot usually be observed by light (as noted above), but small molecules and even the outlines of individual atoms may be traced in some circumstances by use of an atomic force microscope. Some of the largest molecules are macromolecules or supermolecules. The smallest molecule is the diatomic hydrogen ($H_2$), with a bond length of 0.74 Å.

## Dalton, J. (1766–1844)

John Dalton, a prominent chemist and physicist, was born to a Quaker family in the far north of England. In those days, law and medicine, his first career choices, were closed to Dissenters, so he studied science and mathematics. From the remarkable age of twelve, he taught at local Quaker schools and helped administer one with his brother from age fifteen.

After ten years, Dalton moved to the thriving industrial city of Manchester to become a professor of mathematics and philosophy at New College, a dissenting academy open to non-Anglicans. He was also active in Manchester's Literary and Philosophical Society. He recorded detailed meteorological and other scientific observations throughout his life, performed countless experiments, and repeatedly presented and published his findings.

He achieved many breakthroughs, including the first scientific interpretation of colorblindness (formerly called Daltonism), the definition of partial and independent pressures of gaseous mixtures (Dalton's law), and, arguably his most important work, ***an atomic theory asserting that every molecule contains distinct proportions of atoms***. He also devised one of the first tables of atomic weights.

## Dalton, J. (1808). A New System of Chemical Philosophy, Part I

*S. Russell*, Manchester, 141–144, 211–216.

In the passages below, excerpted from a collection of papers previously published or presented at meetings of learned societies, Dalton explains some of his key findings in physics and chemistry.

### CHAP. II
### ON THE CONSTITUTION OF BODIES

THERE are three distinctions in the kinds of bodies, or three states, which have more especially claimed the attention of philosophical chemists; namely, those which are marked by the terms elastic fluids, liquids, and solids. A very familiar instance is exhibited to us in water, of a perfect liquid, and in ice a complete solid. These observations have tacitly led to the conclusion which seems universally adopted, that all bodies of sensible magnitude, whether liquid or solid, are constituted of a vast number of extremely small particles, or **atoms of matter** bound together by a force of attraction, which is more or less powerful according to circumstances, and which as it endeavors to prevent their separation, is very properly called in that view, attraction of cohesion; but as it collects them from a dispersed state (as from steam into water) it is called, attraction of aggregation, or more simply, affinity. Whatever names it may go by, they still signify one and the same power. It is not my design to call in question this conclusion, which appears completely satisfactory; but to show that we have hitherto made no use of it, and that the consequence of the neglect has been a very obscure view of chemical agency, which is daily growing more so in proportion to the new lights attempted to be thrown upon it. The opinions I more particularly allude to, are those of Berthollet[2] on the **Laws of chemical affinity**; such as that chemical agency is proportional to the mass, and that in all chemical unions there exist insensible gradations in the proportions of the constituent principles.

> [2] Claude Louis Berthollet (1748–1822) was a French chemist known for his work on chemical equilibria.

The inconsistence of these opinions, both with reason and observation, cannot, I think, fail to strike everyone who takes a proper view of the phenomena.

Whether the ultimate particles of a body, such as water, are all alike, that is, of the same figure, weight, &c., is a question of some importance. From what is known, we have no reason to apprehend a diversity in these particulars: if it does exist in water, it must equally exist in the elements constituting water, namely, hydrogen and oxygen. Now it is scarcely possible to conceive how the aggregates of dissimilar particles should be so uniformly the same. If some of the particles of water were heavier than others, if a parcel of the liquid on any occasion were constituted principally of these heavier particles, it must be supposed to affect the specific gravity of the mass, a circumstance not known. Similar observations may be made on other substances. Therefore, we may conclude that the ultimate particles of all homogeneous bodies are perfectly alike in weight, figure, &c. In other words, every particle of water is like every other particle of water; every particle of hydrogen is like every other particle of hydrogen, &c.

Besides the *force of attraction*, which, in one character or another, belongs universally to ponderable bodies, we find another force that is likewise universal, or acts upon all matter which comes under our cognizance, namely, a *force of repulsion*. This is now generally, and I think properly, ascribed to the agency of heat. An atmosphere of this subtle fluid constantly surrounds the atoms of all bodies, and prevents them from being drawn into actual contact. This appears to be satisfactorily proved by the observation, that the bulk of a body may be diminished by abstracting some of its heat: But from what has been stated in the last section, it should seem that enlargement and diminution of bulk depend perhaps more on the arrangement, than on the size of the ultimate particles. Be this as it may, we cannot avoid inferring from the preceding doctrine on heat, and particularly from the section on the natural zero of temperature, that solid bodies such as ice, contain a large potion, perhaps ⅘ of the heat which the same are found to contain in an elastic state, as steam.

*We are now to consider how these two great antagonist powers of attraction and repulsion are adjusted, so as to allow of the three different states of elastic fluids, liquids, and solids.* We shall divide the subject into Sections; namely, first on the constitution of pure elastic fluids; second, on the constitution of mixed elastic fluids; third, on the constitution of liquids, and fourth, on the constitution of solids.   . . .

## CHAP. III
## ON CHEMICAL SYNTHESIS

WHEN any body exists in the elastic state, its ultimate particles are separated from each other to a much greater distance than in any other state; each particle occupies the center of a comparatively large sphere, and supports its dignity by keeping all the rest, which by their gravity, or otherwise, are disposed to encroach upon it, at a respectful distance. When we attempt to conceive the number of particles in an atmosphere, it is somewhat like attempting to conceive the number of stars in the universe. We are confounded with the thought. But if we limit the subject, by taking a given volume of any gas, we seem persuaded that, let the divisions be ever so minute, the number of particles must be finite; just as in a given space of the universe, the number of stars and planets cannot be infinite.

Chemical analysis and synthesis[4] go no farther than to the separation of particles one from another, and to their reunion.

[4] The combination of two chemical entities to form a new entity.

No new creation or destruction of matter is within the reach of chemical agency. We might as well attempt to introduce a new planet into the solar system, or to annihilate one already in existence, as to create or destroy a particle of hydrogen. All the changes we can produce, consist in separating particles that are in a state of cohesion or combination, and joining those that were previously at a distance.

In all chemical investigations it has justly been considered an important object to ascertain the relative weights of the simples which constitute a compound. But unfortunately, the inquiry has terminated here; whereas from the relative weights in the mass, the relative weights of the ultimate particles or atoms of the bodies might have been inferred, from which their number and weight in various other compounds would appear, in order to assist and to guide future investigations, and to correct their results. ***Now it is one great object of this work to show the importance and advantage of ascertaining the relative weights of the ultimate particles, both of simple and compound bodies, the number of simple elementary particles which constitute one compound particle, and the number of less compound particles which enter into the formation of one more compound particle.***

If there are two bodies, A and B, which are disposed to combine, the following is the order in which the combinations may take place, beginning with the simplest: namely,

> 1 atom of A + 1 atom of B = 1 atom of C, binary,
> 1 atom of A + 2 atoms of B = 1 atom of D, ternary.
> 2 atoms of A + 1 atom of B = 1 atom of E, ternary.
> 1 atom of A + 3 atoms of B = 1 atom of F, quaternary.
> 3 atoms of A + 1 atom of B = 1 atom of G, quaternary. &c. &c.

The following general rules may be adopted as guides in all our investigations respecting chemical synthesis:

> 1st. When only one combination of two bodies can be obtained, it must be presumed to be a *binary* one, unless some cause appears to the contrary.
> 2d. When two combinations are observed, they must be presumed to be a *binary* and a *ternary*.

3d. When three combinations are obtained, we may expect one to be a *binary*, and the other two ternary.

4th. When four combinations are observed, we should expect one *binary*, two *ternary*, and one *quaternary*, &c.

5th. A *binary* compound should always be specifically heavier than the mere mixture of its two ingredients.

6th. A *ternary* compound should be specifically heavier than the mixture of a *binary* and a simple, which would, if combined, constitute it; &c.

7th. The above rules and observations equally apply, when two bodies, such as C and D, D and E, &c., are combined.

From the application of these rules, to the chemical facts already well ascertained, we deduce the following conclusions;

*1st*. That water is a binary compound of hydrogen and oxygen, and the relative weights of the two elementary atoms are as 1:7, nearly;

*2d.* That ammonia[5] is a binary compound of hydrogen and azote, and the relative weights of the two atoms are as 1:5, nearly;

[5] A chemical compound of nitrogen and hydrogens with the chemical formula $NH_3$.

*3d*. That nitrous gas is a binary compound of azote and oxygen, the atoms of which weigh 5 and 7 respectively; that nitric acid[6]

[6] A highly corrosive acid with the chemical formula $HNO_3$.

is a binary or ternary compound according as it is derived, and consists of one atom of azote and two of oxygen, together weighing 19; that nitrous oxide[7]

[7] A chemical compound of nitrogen and oxygen with the chemical formula $N_2O$.

is a compound similar to nitric acid, and consists, of one atom of oxygen and two of azote, weighing 17; that nitrous acid is a binary compound of nitric acid and nitrous gas, weighing 31; that oxynitric acid is a binary compound of nitric acid and oxygen, weighing 26;

*4th*. That carbonic oxide is a binary compound, consisting of one atom of charcoal, and of oxygen, weighing 19; &c., &c. In all these cases the weights are expressed in atoms of hydrogen, each of which is denoted by unity.

In the sequel, the facts and experiments from which these conclusions are derived, will be detailed; as well as a great variety of others from which are inferred the constitution and weight of the ultimate particles of the principal acids, the alkalies,[8] the earths,[9] the metals, the metallic oxides and sulphurets, the long train of neutral salts, and in short, all the chemical compounds which have hitherto obtained a tolerably good analysis.

[8] An alkali is an ionic salt of an alkali metal.
[9] Alkaline earth metals are a group of metals that are somewhat reactive at standard temperature.

Several of the conclusions will be supported by original experiments.

From the novelty as well as importance of the ideas suggested in this chapter, it is deemed expedient to give plates, exhibiting the mode of combination in some of the simpler cases. A specimen of these accompanies this first part. The *elements* or *atoms* of such bodies as are conceived at present to be simple, are denoted by a small circle, with some distinctive mark; and the combinations consist in the juxtaposition of two or more of these; when three or more particles of elastic fluids are combined together in one, it is to be supposed that the particles of the same kind repel each other, and therefore take their stations accordingly.

## PART II    The electron

## Thomson, J. J. (December 18, 1856 – August 30, 1940).

Thomson was a British physicist whose study of *cathode rays* led to his *discovery of the electron*, a subatomic particle with a negative electric charge. *In 1897, he showed that cathode rays were composed of previously unknown negatively charged particles (now called electrons)*, which he calculated must have bodies much smaller than atoms and a very large charge-to-mass ratio.

> [*Cathode rays* are *streams of electrons* emitted from the negative electrode (cathode) in a vacuum tube. They were first observed in 1869 by Sir William Crookes and are formed when electrons are accelerated through a voltage difference between electrodes. *Cathode rays* travel from the cathode to the positively charged electrode (anode) and are also referred to as electron beams.]

*In 1906, Thomson was awarded the Nobel Prize in Physics "in recognition of the great merits of his theoretical and experimental investigations on the conduction of electricity by gases."*

Thomson is credited with finding the first evidence for isotopes of a stable (non-radioactive) element in 1912, as part of his exploration into the composition of *canal rays* (positive ions).

> [*Canal rays*, also known as *anode rays*, are *streams of positively charged particles or ions* created in a discharge tube when a high voltage is applied across a gas at low pressure. They were first observed in 1886 by Eugen Goldstein during experiments with Crookes tubes. *Canal rays* originate from the anode and are distinct from cathode rays, which are negatively charged.]

His experiments to determine the nature of positively charged particles, with Francis William Aston, were the first use of mass spectrometry and led to the development of the mass spectrograph.

Thomson was born on December 18, 1856 in Cheetham Hill, Manchester. His mother, Emma Swindells, came from a local textile family. His father, Joseph James Thomson, ran an antiquarian bookshop founded by Thomson's great-grandfather. Joseph John had a brother, Frederick Vernon Thomson, who was two years younger than he was. Thomson was a reserved yet devout Anglican.

Thomson's early education was in small private schools where he demonstrated outstanding talent and interest in science. In 1870, he was admitted to Owens College in Manchester (now the University of Manchester) at the unusually young age of 14, and came under the influence of Balfour Stewart, Professor of Physics, who initiated him into physical research. He began experimenting with contact electrification and soon published his first scientific paper. His parents planned to enroll him as an apprentice engineer to Sharp, Stewart & Co, a locomotive manufacturer, but these plans were cut short when his father died in 1873.

In 1876, Thomson moved on to Trinity College, Cambridge. In 1880, he received his B.A. in mathematics (Second Wrangler in the Tripos and 2nd Smith's Prizeman). He applied for and became a Fellow of Trinity College the following year. He obtained an M.A. (Adams Prizeman) in 1883.

Thomson's prize-winning master's work, (1883), *Treatise on the motion of vortex rings*, shows his early interest in atomic structure. In it, Thomson mathematically described the motions of Lord Kelvin's vortex theory of the atom.

On December 22, 1884, Thomson was appointed Cavendish Professor of Physics at the University of Cambridge. This appointment caused considerable surprise; candidates such as Osborne Reynolds and Richard Glazebrook were older and more experienced in laboratory work, whereas Thomson was known for his work as a mathematician—being recognized as an exceptional talent.

Thomson published a number of papers addressing both mathematical and experimental issues of electromagnetism. He examined the electromagnetic theory of light of James Clerk Maxwell, introduced the concept of electromagnetic mass of a charged particle, and demonstrated that a moving charged body would apparently increase in mass.

Much of his work in mathematical modelling of chemical processes can be thought of as early computational chemistry. In further work, published in book form as *Applications of dynamics to physics and chemistry* (1888), Thomson addressed the transformation of energy in mathematical and theoretical terms, suggesting that all energy might be kinetic.

In 1890, Thomson married Rose Elisabeth Paget at the church of St. Mary the Less. Rose, who was the daughter of Sir George Edward Paget, a physician and then Regius Professor of Physic at Cambridge, was interested in physics. Beginning in 1882, women could attend demonstrations and lectures at the University of Cambridge. Rose attended demonstrations and lectures, among them Thomson's, leading to their relationship. They had two children: George Paget Thomson, won the 1937 Nobel Prize in Physics for proving the wave-like properties of electrons; and Joan Paget Thomson (later Charnock), who became an author—writing children's books, non-fiction, and biographies.

His next book, *Notes on recent researches in electricity and magnetism* (1893), built upon Maxwell's Treatise upon electricity and magnetism, and was sometimes referred to as "the third volume of Maxwell." In it, Thomson emphasized physical methods and experimentation and included extensive figures and diagrams of apparatus, including a number for the passage of electricity through gases. His third book, *Elements of the mathematical theory of electricity and magnetism* (1895) was a readable introduction to a wide variety of subjects, and achieved considerable popularity as a textbook.

Several scientists, such as William Prout and Norman Lockyer, had suggested that atoms were built up from a more fundamental unit, but they envisioned this unit to be the size of the smallest atom, hydrogen. ***Thomson in 1897 was the first to suggest that one of the fundamental units of the atom was more than 1,000 times smaller than an atom, suggesting the subatomic particle now known as the electron***.

Thomson discovered this through his explorations on the properties of ***cathode rays***. Thomson made his suggestion on ***April 30, 1897*** following his discovery that cathode rays (at the time known as Lenard rays) could travel much further through air than expected for an atom-sized particle. ***He estimated the mass of cathode rays by measuring the heat generated when the rays hit a thermal junction and comparing this with the magnetic deflection of the rays.*** His experiments suggested not only that cathode rays were over 1,000 times lighter than the hydrogen atom, but also that their mass was the same in whichever type of atom they came from. ***He concluded that the rays were composed of very light, negatively charged particles which were a universal building block of atoms***. He called the particles "***corpuscles***", but later scientists preferred the name ***electron***, which had been suggested by George Johnstone Stoney in 1891, prior to Thomson's discovery. [Thomson, J. J. (October, 1897). *On the cathode rays. Phil. Mag.*, 44, 293-316; https://web.lemoyne.edu/~giunta/thomson1897.html.]

In April 1897, Thomson had only early indications that the ***cathode rays*** could be deflected electrically (previous investigators such as Heinrich Hertz had thought they could not be). A month after Thomson's announcement of the corpuscle, he found that he could reliably deflect the rays by an electric field if he evacuated the discharge tube to a very low pressure. By comparing the deflection of a beam of ***cathode rays*** by electric and magnetic fields he obtained more robust measurements of the ***mass-to-charge ratio*** that confirmed his previous estimates. This became the classic means of measuring the charge-to-mass ratio of the electron. Later in 1899 he measured the charge of the electron to be of $6.8 \times 10^{-10}$ esu.

Thomson believed that the corpuscles emerged from the atoms of the trace gas inside his cathode-ray tubes. He thus concluded that atoms were divisible, and that the corpuscles were their building blocks. ***In 1904, Thomson suggested a model of the atom, hypothesizing that it was a sphere of positive matter within which electrostatic forces determined the positioning of the corpuscles***. To explain the overall neutral charge of the atom, ***he proposed that the corpuscles were distributed in a uniform sea of positive charge***. In this "**plum pudding model**", the electrons were seen as embedded in the positive charge like raisins in a plum pudding (although in Thomson's model they were not stationary, but orbiting rapidly).

The name ***electron*** was adopted for these particles by the scientific community, mainly due to the advocation by George Francis FitzGerald, Joseph Larmor, and Hendrik Lorentz. ***The term was originally coined by George Johnstone Stoney in 1891 as a tentative name for the basic unit of electrical charge*** (which had then yet to be discovered). For some years Thomson resisted using the word "electron" because he didn't like how some physicists talked of a "positive electron" that was supposed to be the elementary unit of positive charge just as the "negative electron" is the elementary unit of negative charge. Thomson preferred to stick with the word "corpuscle" which he strictly defined as negatively charged. He relented by 1914, using the word "***electron***" in his book *The Atomic Theory*.

In 1905, Thomson discovered the natural radioactivity of potassium. In 1906, Thomson demonstrated that hydrogen had only a single electron per atom. Previous theories allowed various numbers of electrons.

Thomson was an influential teacher, and seven of his students went on to win Nobel Prizes: Ernest Rutherford (Chemistry 1908), Lawrence Bragg (Physics 1915), Charles Barkla (Physics 1917), Francis Aston (Chemistry 1922), Charles Thomson Rees Wilson (Physics 1927), Owen Richardson (Physics 1928) and Edward Appleton (Physics 1947).
Thomson was knighted in 1908 and appointed to the Order of Merit in 1912. At Oxford, he gave the 1914 Romanes Lecture titled *The Atomic Theory*. In 1918, he became Master of Trinity College, Cambridge, a position he held until his death on 30 August 1940. His ashes rest in Westminster Abbey, near the graves of Isaac Newton and his former student, Ernest Rutherford.

# Thomson, J. J. (October, 1897). On the cathode rays.

*Phil. Mag.*, 44, 293-316; https://web.lemoyne.edu/~giunta/thomson1897.html.

The experiments* discussed in this paper were undertaken in the hope of gaining some information as to the nature of the Cathode Rays.

> *Some of these experiments have already been described in a paper read before the Cambridge Philosophical Society (*Proceedings,* ix. 1897), and in a Friday Evening Discourse at the Royal Institution (*Electrician*, May 21, 1897).

The most diverse opinions are held as to these rays; according to the almost unanimous opinion of German physicists they are due to some process in the aether to which-- inasmuch as in a uniform magnetic field their course is circular and not rectilinear--no phenomenon hitherto observed is analogous: another view of these rays is that, so far from being wholly aethereal, they are in fact wholly material, and that they mark the paths of ***particles of matter charged with negative electricity***. It would seem at first sight that it ought not to be difficult to discriminate between views so different, yet experience shows that this is not the case, as amongst the physicists who have most deeply studied the subject can be found supporters of either theory.

The electrified-particle theory has for purposes of research a great advantage over the aethereal theory, since it is definite and its consequences can be predicted; with the aethereal theory it is impossible to predict what will happen under any given circumstances, as on this theory we are dealing with hitherto unobserved phenomena in the aether, of whose laws we are ignorant.

The following experiments were made to test some of the consequences of the electrified-particle theory.

## Charge carried by the Cathode Rays

If these rays are negatively electrified particles, then when they enter an enclosure, they ought to carry into it a charge of negative electricity. This has been proved to be the case by Perrin, who placed in front of a plane cathode two coaxial metallic cylinders which were insulated from each other: the outer of these cylinders was connected with the earth, the inner with a gold-leaf electroscope. These cylinders were closed except for two small holes, one in each cylinder, placed so that the cathode rays could pass through them into the inside of the inner cylinder. Perrin found that when the rays passed into the inner cylinder the electroscope received a charge of negative electricity, while no charge went to the electroscope when the rays were deflected by a magnet so as no longer to pass through the hole.

This experiment proves that something charged with negative electricity is shot off from the cathode, travelling at right angles to it, and that this something is deflected by a magnet; it is open, however, to the objection that it does not prove that the cause of the electrification in the electroscope has anything to do with the cathode rays. Now the supporters of the aethereal theory do not deny that electrified particles are shot off from the cathode; they deny, however, that these charged particles have any more to do with the cathode rays than a rifle-ball has with the flash when a rifle is fired. I have therefore repeated Perrin's experiment in a form which is not open to this objection. The arrangement used was as follows: --Two coaxial cylinders (fig. 1) with slits in them are placed in a bulb connected with the discharge-tube; the cathode rays from the cathode A pass into the bulb through a slit in a metal plug fitted into the neck of the tube; this plug is connected with the anode and is put to earth. The cathode rays thus do not fall upon the cylinders unless they are deflected by a magnet. The outer cylinder is connected with the earth, the inner with the electrometer. When the cathode rays (whose path was traced by the phosphorescence on the glass) did not fall on the slit, the electrical charge sent to the electrometer when the induction-coil producing the rays was set in action was small and irregular; when, however, the rays were bent by a magnet so as to fall on the slit there was a large charge of negative electricity sent to the electrometer. I was surprised at the magnitude of the charge; on some occasions, enough negative electricity went through the narrow slit into the inner cylinder in one second to alter the potential of a capacity of 1.5 microfarads by 20 volts. If the rays were so much bent by the magnet that they overshot the slits in the cylinder, the charge passing into the cylinder fell again to a very small fraction of its value when the aim was true. Thus, this experiment shows that however we twist and deflect the cathode rays by magnetic forces, the negative electrification follows the same path as the rays, and that this negative electrification is indissolubly connected with the cathode rays.

When the rays are turned by the magnet so as to pass through the slit into the inner cylinder, the deflection of the electrometer connected with this cylinder increases up to a certain value, and then remains stationary although the rays continue to pour into the cylinder. This is due to the fact that the gas in the bulb becomes a conductor of electricity when the cathode rays pass through it, and thus, though the inner cylinder is perfectly insulated when the rays are not passing, yet as soon as the rays pass through the bulb the air between the inner cylinder and the outer one becomes a conductor, and the electricity escapes from the inner cylinder to the earth. Thus, the charge within the inner cylinder does not go on continually increasing; the cylinder settles down into a state of equilibrium in which the rate at which it gains negative electricity from the rays is equal to the rate at which it loses it by conduction through the air. If the inner cylinder has initially a positive charge it rapidly

43

loses that charge and acquires a negative one; while if the initial charge is a negative one, the cylinder will leak if the initial negative potential is numerically greater than the equilibrium value.

**Deflection of the Cathode Rays by an Electrostatic Field.**

An objection very generally urged against the view that the cathode rays are negatively electrified particles, is that hitherto no deflection of the rays has been observed under a small electrostatic force, and though the rays are deflected when they pass near electrodes connected with sources of large differences of potential, such as induction-coils or electrical machines, the deflection in this case is regarded by the supporters of the aethereal theory as due to the discharge passing between the electrodes, and not primarily to the electrostatic field. Hertz made the rays travel between two parallel plates of metal placed inside the discharge-tube, but found that they were not deflected when the plates were connected with a battery of storage-cells; on repeating this experiment I at first got the same result, but subsequent experiments showed that the absence of deflection is due to the conductivity conferred on the rarefied gas by the cathode rays. On measuring this conductivity, it was found that it diminished very rapidly as the exhaustion increased; it seemed then that on trying Hertz's experiment at very high exhaustions there might be a chance of detecting the deflection of the cathode rays by an electrostatic force.

The apparatus used is represented in fig. 2.

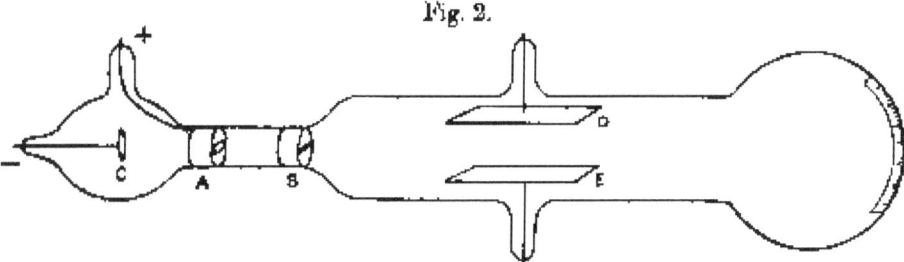

Fig. 2.

The rays from the cathode C pass through a slit in the anode A, which is a metal plug fitting tightly into the tube and connected with the earth; after passing through a second slit in another earth-connected metal plug B, they travel between two parallel aluminum plates about 5 cm. long by 2 broad and at a distance of 1.5 cm. apart; they then fall on the end of the tube and produce a narrow well-defined phosphorescent patch. A scale pasted on the outside of the tube serves to measure the deflection of this patch. At high exhaustions the rays were deflected when the two aluminum plates were connected with the terminals of a battery of small storage cells; the rays were depressed when the upper plate was connected with the negative pole of the battery, the lower with the positive, and raised when the upper plate was connected with the positive, the lower with the negative pole. The deflection was proportional to the difference of potential between the plates, and I could detect the deflection when the potential-difference was as small as two volts. It was only when the vacuum was a good one that the deflection took place, but that the absence of deflection is due to the conductivity of the medium is shown by what takes place when the vacuum has just arrived at the stage at which the deflection begins. At this stage there is a deflection of the rays when the plates are first connected with the terminals of the battery, but if this connection is maintained the patch of the phosphorescence gradually creeps back to its undeflected position. This is just what would happen if the space between the plates were a conductor, though a very bad one, for then the positive and negative ions between the plates would slowly diffuse, until the positive plate became coated with negative ions, the negative plate with positive ones; thus, the electric intensity between the plates would vanish and the cathode rays be free from electrostatic force. Another illustration of this is afforded by what happens when the pressure is low enough to show the deflection and a large difference of potential, say 200 volts, is established between the plates; under these circumstances there is a large deflection of the cathode rays, but the medium under the large electromotive force breaks down every now and then and a bright discharge passes between the plates; when this occurs the phosphorescent patch produced by the cathode rays jumps back to its undeflected position. When the cathode rays are deflected by the electrostatic field, the phosphorescent band breaks up into several bright bands separated by comparatively dark spaces; the phenomena are exactly analogous to those observed by Birkeland when the cathode rays are deflected by a magnet, and called by him the *magnetic spectrum*.

A series of measurements of the deflection of the rays by the electrostatic force under various circumstances will be found later on in the part of the paper which deals with the velocity of the rays and the ratio of the mass of the electrified particles to the charge carried by them. It may, however, be mentioned here that the deflection gets smaller as the pressure diminishes, and when in consequence the potential-difference in the tube in the neighborhood of the cathode increases.

**Conductivity of a Gas through which Cathode Rays are passing.**

The conductivity of the gas was investigated by means of the apparatus shown in fig. 2. The upper plate D was connected with one terminal of a battery of small storage-cells, the other terminal of which was connected with the earth; the other plate E was connected with one of the coatings of a condenser of one microfarad capacity, the other coating of which was to earth; one pair of quadrants of an electrometer was also connected with E, the other pair of quadrants being to earth. When the cathode rays are passing between the plates, the two pairs of quadrants of the electrometer are first connected with each other, and then the connection between them was broken. If the space between the plates were a non-conductor, the potential of the pair of quadrants not connected with the earth would remain zero and the needle of the electrometer would be deflected. There is always a deflection of the electrometer, showing that a current passes between the plates. The magnitude of the current depends very greatly upon the pressure of the gas; so much so, indeed, that it is difficult to obtain consistent readings in consequence of the changes which always occur in the pressure when the discharge passes through the tube.

We shall first take the case when the pressure is only just low enough to allow the phosphorescent patch to appear at the end of the tube; in this case the relation between the current between the plates and the initial difference of potential is represented by the curve shown in fig. 3. In this figure the abscissae represent the initial difference of potential between the plates, each division representing two volts. The quantity of electricity which has passed between the plates in one minute is the quantity required to raise 1 microfarad to the potential- difference shown by the curve. The upper and lower curve relates to the case when the upper plate is connected with the negative and positive pole respectively of the battery.

Fig. 3.

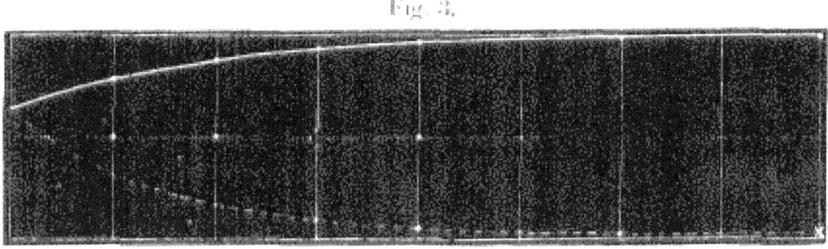

Even when there is no initial difference of potential between the plates the lower plate acquires a negative charge from the impact on it of some of the cathode rays.

We see from the curve that the current between the plates soon reaches a value where it is only slightly affected by an increase in the potential-difference between the plates; this is a feature common to conduction through gases traversed by Röntgen rays, by uranium rays, by ultra-violet light, and, as we now see, by cathode rays. The rate of leak is not greatly different whether the upper plate be initially positively or negatively electrified.

The current between the plates only lasts for a short time; it ceases long before the potential of the lower plate approaches that of the upper. Thus, for example, when the potential of the upper plate was about 400 volts above that of the earth, the potential of the lower plate never rose above 6 volts: similarly, if the upper plate were connected with the negative pole of the battery, the fall in potential of the lower plate was very small in comparison with the potential-difference between the upper plate and the earth.

These results are what we should expect if the gas between the plates and the plug B (fig. 2) were a very much better conductor than the gas between the plates, for the lower plate will be in a steady state when the current coming to it from the upper plate is equal to the current going from it to the plug; now if the conductivity of the gas between the plate and the plug is much greater than that between the plates, a small difference of potential between the lower plate and the plug will be consistent with a large potential-difference between the plates.

So far, we have been considering the case when the pressure is as high as is consistent with the cathode rays reaching the end of the tube; we shall now go to the other extreme and consider the case when the pressure is as low as is consistent with the passage of a discharge through the bulb. In this case, when the plates are not connected with the battery, we get a negative charge communicated to the lower plate, but only very slowly in comparison with the effect in the previous case. When the upper plate is connected with the negative pole of a battery, this current to the lower plate is only slightly increased even when the difference of potential is as much as 400 volts: a small potential-difference of about 20 volts seems slightly to decrease the rate of leak. Potential-differences much exceeding 400 volts cannot be used, as though the dielectric between the plates is able to sustain them for some little time, yet after a time an intensely bright arc flashes across between the plates and liberates so much gas as to spoil the vacuum. The lines in the spectrum of this glare are chiefly mercury lines; its passage leaves very peculiar markings on the aluminum plates.

If the upper plate was charged positively, then the negative charge communicated to the lower plate was diminished, and stopped when the potential-difference between the plates was about 20 volts; but at the lowest pressure, however great (up to 400 volts) the potential-difference, there was no leak of positive electricity to the lower plate at all comparable with the leak of negative electricity to this plate when the two plates were disconnected from the battery. In fact at this very low pressure all the facts are consistent with the view that the effects are due to the negatively electrified particles travelling along the cathode rays, the rest of the gas possessing little conductivity. Some experiments were made with a tube similar to that shown in fig. 2, with the exception that the second plug B was absent, so that a much greater number of cathode rays passed between the plates. When the upper plate was connected with the positive pole of the battery a luminous discharge with well-marked striations passed between the upper plate and the earth-connected plug through which the cathode rays were streaming; this occurred even though the potential-difference between the plate and the plug did not exceed 20 volts. Thus, it seems that if we supply cathode rays from an external source to the cathode a small potential-difference is sufficient to produce the characteristic discharge through a gas.

**Magnetic Deflection of the Cathode Rays in Different Gases.**

The deflection of the cathode rays by the magnetic field was studied with the aid of the apparatus shown in fig. 4. The cathode was placed in a side-tube fastened on to a bell-jar; the opening between this tube and the bell-jar was closed by a metallic plug with a slit in it; this plug was connected with the earth and was used as the anode. The cathode rays passed through the slit in this plug into the bell-jar, passing in front of a vertical plate of glass ruled into small squares. The bell-jar was placed between two large parallel coils arranged as a Helmholtz galvanometer. The course of the rays was determined by taking photographs of the bell-jar when the cathode rays were passing though it; the divisions on the plate enabled the path of the rays to be determined. Under the action of the magnetic field the narrow beam of cathode rays spreads out into a broad fan-shaped luminosity in the gas. The luminosity in this fan is not uniformly distributed, but is condensed along certain lines. The phosphorescence on the glass is also not uniformly distributed; it is much spread out, showing that the beam consists of rays which are not all deflected to the same extent by the magnet. The luminosity on the glass is crossed by bands along which the luminosity is very much greater than in the adjacent parts. These bright and dark bands are called by Birkeland, who first observed them, the magnetic spectrum. The brightest spots on the glass are by no means always the terminations of the brightest streaks of luminosity in the gas; in fact, in some cases a very bright spot on the glass is not connected with the cathode by any appreciable luminosity, though there may be plenty of luminosity in other parts of the gas. One very interesting point brought out by the photographs is that in a given magnetic field, and with a given mean potential- difference between the terminals, the path of the rays is independent of the nature of the gas. Photographs were taken of the discharge in hydrogen, air, carbonic acid, methyl iodide, *i.e.*, in gases whose densities range from 1 to 70, and yet, not only were the paths of the most deflected rays the same in all cases, but even the details, such as the distribution of the bright and dark spaces, were the same; in fact, the photographs could hardly be distinguished from each other. It is to be noted that the pressures were not the same; the pressures in the different gases were adjusted so that the mean potential-differences between the cathode and the anode were the same in all the gases. When the pressure of a gas is lowered, the potential-difference between the terminals increases, and the deflection of the rays produced by a magnet diminishes, or at any rate the deflection of the rays when the phosphorescence is a maximum diminishes. If an air-break is inserted an effect of the same kind is produced.

In the experiments with different gases, the pressures were as high as was consistent with the appearance of the phosphorescence on the glass, so as to ensure having as much as possible of the gas under consideration in the tube.

As the cathode rays carry a charge of negative electricity, are deflected by an electrostatic force as if they were negatively electrified, and are acted on by a magnetic force in just the way in which this force would act on a negatively electrified body moving along the path of these rays, *I can see no escape from the conclusion that they are charges of negative electricity carried by particles of matter*. The question next arises, *What are these particles?* are they atoms, or molecules, or matter in a still finer state of subdivision? To throw some light on this point, I have made a series of measurements of the ratio of the mass of these particles to the charge carried by it. To determine this quantity, I have used two independent methods. The first of these is as follows: --Suppose we consider a bundle of homogeneous cathode rays. Let $m$ be the mass of each of the particles, $e$ the charge carried by it. Let N be the number of particles passing across any section of the beam in a given time; then *Q the quantity of electricity carried by these particles* is given by the equation

$$Ne = Q.$$

We can measure Q if we receive the cathode rays in the inside of a vessel connected with an electrometer. When these rays strike against a solid body, the temperature of the body is raised; the kinetic energy of the moving particles being converted into heat; if we suppose that all this energy is converted into heat, then if we measure the increase in the temperature of a body of known thermal capacity caused by the impact of these rays, we can determine W, the kinetic energy of the particles, and if $v$ is the velocity of the particles,
$$(1/2) Nmv^2 = W.$$
If $\rho$ is the radius of curvature of the path of these rays in a uniform magnetic field H, then

$$mv/e = H\rho = I,$$

where I is written for H$\rho$ for the sake of brevity. From these equations we get

$$(1/2)(m/e)v^2 = W/Q.$$
$$v = 2W/QI,$$
$$m/e = I^2Q/2W.$$

Thus, if we know the values of Q, W, and I, we can deduce the values of $v$ and $m/e$.

To measure these quantities, I have used tubes of three different types. The first I tried is like that represented in fig. 2, except that the plates E and D are absent, and two coaxial cylinders are fastened to the end of the tube. The rays from the cathode C fall on the metal plug B, which is connected with the earth, and serves for the anode; a horizontal slit is cut in this plug. The cathode rays pass through this slit, and then strike against the two coaxial cylinders at the end of the tube; slits are cut in these cylinders, so that the cathode rays pass into the inside of the inner cylinder. The outer cylinder is connected with the earth, the inner cylinder, which is insulated from the outer one, is connected with an electrometer, the deflection of which measures Q, the quantity of electricity brought into the inner cylinder by the rays. A thermo-electric couple is placed behind the slit in the inner cylinder; this couple is made of very thin strips of iron and copper fastened to very fine iron and copper wires. These wires passed through the cylinders, being insulated from them, and through the glass to the outside of the tube, were they were connected with a low-resistance galvanometer, the deflection of which gave data for calculating the rise of temperature of the junction produced by the impact against it of the cathode rays. The strips of iron and copper were large enough to ensure that every cathode ray which entered the inner cylinder struck against the junction. In some of the tubes the strips of iron and copper were placed end to end, so that some of the rays struck against the iron, and others against the copper; in others, the strip of one metal was placed in front of the other; no difference, however, could be detected between the results got with these two arrangements. The strips of iron and copper were weighed, and the thermal capacity of the junction calculated. In one set of junctions this capacity was $5 \times 10^{-3}$, in another $3 \times 10^{-3}$. If we assume that the cathode rays which strike against the junction give their energy up to it, the deflection of the galvanometer gives us W or $(1/2) Nmv^2$.

The value of I, *i.e.*, Hρ, where ρ is the curvature of the path of the rays in a magnetic field of strength H was found as follows:--The tube was fixed between two large circular coils placed parallel to each other, and separated by a distance equal to the radius of either; these coils produce a uniform magnetic field, the strength of which is got by measuring with an ammeter the strength of the current passing through them. The cathode rays are thus in a uniform field, so that their path is circular. Suppose that the rays, when deflected by a magnet, strike against the glass of the tube at E (fig. 5), then, if ρ is the radius of the circular path of the rays,

$$2\rho = CE^2/AC + AC;$$

thus, if we measure CE and AC we have the means of determining the radius of curvature of the path of the rays.

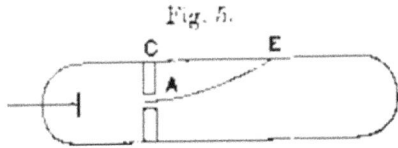

Fig. 5.

The determination of ρ is rendered to some extent uncertain, in consequence of the pencil of rays spreading out under the action of the magnetic field, so that the phosphorescent patch at E is several millimeters long; thus, values of ρ differing appreciably from each other will be got by taking E at different points of this phosphorescent patch. Part of this patch was, however, generally considerably brighter than the rest; when this was the case,

E was taken as the brightest point; when such a point of maximum brightness did not exist, the middle of the patch was taken for E. The uncertainty in the value of $\rho$ thus introduced amounted sometimes to about 20 per cent.; by this I mean that if we took E first at one extremity of the patch and then at the other, we should get values of $\rho$ differing by this amount.

The measurement of Q, the quantity of electricity which enters the inner cylinder, is complicated by the cathode rays making the gas through which they pass a conductor, so that though the insulation of the inner cylinder was perfect when the rays were off, it was not so when they were passing through the space between the cylinders; this caused some of the charge communicated to the inner cylinder to leak away so that the actual charge given to the cylinder by the cathode rays was larger than that indicated by the electrometer. To make the error from this cause as small as possible, the inner cylinder was connected to the largest capacity available, 1.5 microfarad, and the rays were only kept on for a short time, about 1 or 2 seconds, so that the alteration in potential of the inner cylinder was not large, ranging in the various experiments from about .5 to 5 volts. Another reason why it is necessary to limit the duration of the rays to as short a time as possible, is to avoid the correction for the loss of heat from the thermo-electric junction by conduction along the wires; the rise in temperature of the junction was of the order 2°C.; a series of experiments showed that with the same tube and the same gaseous pressure Q and W were proportional to each other when the rays were not kept on too long.

Tubes of this kind gave satisfactory results, the chief drawback being that sometimes in consequence of the charging up of the glass of the tube, a secondary discharge started from the cylinder to the walls of the tube, and the cylinders were surrounded by glow; when this glow appeared, the readings were very irregular; the glow could, however, be got rid of by pumping and letting the tube rest for some time. The results got with this tube are given in the Table under the heading Tube 1.

The second type of tube was like that used for photographing the path of the rays (fig. 4); double cylinders with a thermo-electric junction like those used in the previous tube were placed in the line of fire of the rays, the inside of the bell-jar was lined with copper gauze connected with the earth. This tube gave very satisfactory results; we were never troubled with any glow round the cylinders, and the readings were most concordant; the only drawback was that as some of the connections had to be made with sealing-wax, it was not possible to get the highest exhaustions with this tube, so that the range of pressure for this tube is less than that for tube 1. The results got with this tube are given in the Table under the heading Tube 2.

The third type of tube was similar to the first, except that the openings in the two cylinders were made very much smaller; in this tube the slits in the cylinders were replaced by small

51

holes, about 1.5 mm in diameter. In consequence of the smallness of the openings, the magnitude of the effects was very much reduced; in order to get measurable results, it was necessary to reduce the capacity of the condenser in connection with the inner cylinder to .15 microfarad, and to make the galvanometer exceedingly sensitive, as the rise in temperature of the thermo-electric junction was in these experiments only about .5° C on average. The results obtained in this tube are given in the Table under the heading Tube 3.

The results of a series of measurements with these tubes are given in the following Table:

| Gas. | Value of W/Q. | I. | $m/e$ | $v.$ |
|---|---|---|---|---|
| | | Tube 1. | | |
| Air | $4.6 \times 10^{11}$ | 230 | $.57 \times 10^{-7}$ | $4 \times 10^9$ |
| Air | $1.8 \times 10^{12}$ | 350 | $.34 \times 10^{-7}$ | $1 \times 10^{10}$ |
| Air | $6.1 \times 10^{11}$ | 230 | $.43 \times 10^{-7}$ | $5.4 \times 10^9$ |
| Air | $2.5 \times 10^{12}$ | 400 | $.32 \times 10^{-7}$ | $1.2 \times 10^{10}$ |
| Air | $5.5 \times 10^{11}$ | 230 | $.48 \times 10^{-7}$ | $4.8 \times 10^9$ |
| Air | $1 \times 10^{12}$ | 285 | $.4 \times 10^{-7}$ | $7 \times 10^9$ |
| Air | $1 \times 10^{12}$ | 285 | $.4 \times 10^{-7}$ | $7 \times 10^9$ |
| Hydrogen | $6 \times 10^{12}$ | 205 | $.35 \times 10^{-7}$ | $6 \times 10^9$ |
| Hydrogen | $2.1 \times 10^{12}$ | 460 | $.5 \times 10^{-7}$ | $9.2 \times 10^9$ |
| Carbonic acid | $8.4 \times 10^{11}$ | 260 | $.4 \times 10^{-7}$ | $7.5 \times 10^9$ |
| Carbonic acid | $1.47 \times 10^{12}$ | 340 | $.4 \times 10^{-7}$ | $8.5 \times 10^9$ |
| Carbonic acid | $3.0 \times 10^{12}$ | 480 | $.39 \times 10^{-7}$ | $1.3 \times 10^{10}$ |
| | | Tube 2. | | |
| Air | $2.8 \times 10^{11}$ | 175 | $.53 \times 10^{-7}$ | $3.3 \times 10^9$ |
| Air | $4.4 \times 10^{11}$ | 195 | $.47 \times 10^{-7}$ | $4.1 \times 10^9$ |
| Air | $3.5 \times 10^{11}$ | 181 | $.47 \times 10^{-7}$ | $3.8 \times 10^9$ |
| Hydrogen | $2.8 \times 10^{11}$ | 175 | $.53 \times 10^{-7}$ | $3.3 \times 10^9$ |
| Air | $2.5 \times 10^{11}$ | 160 | $.51 \times 10^{-7}$ | $3.1 \times 10^9$ |
| Carbonic acid | $2 \times 10^{11}$ | 148 | $.54 \times 10^{-7}$ | $2.5 \times 10^9$ |
| Air | $1.8 \times 10^{11}$ | 151 | $.63 \times 10^{-7}$ | $2.3 \times 10^9$ |
| Hydrogen | $2.8 \times 10^{11}$ | 175 | $.53 \times 10^{-7}$ | $3.3 \times 10^9$ |
| Hydrogen | $4.4 \times 10^{11}$ | 201 | $.46 \times 10^{-7}$ | $4.4 \times 10^9$ |
| Air | $2.5 \times 10^{11}$ | 176 | $.61 \times 10^{-7}$ | $2.8 \times 10^9$ |
| Air | $4.2 \times 10^{11}$ | 200 | $.48 \times 10^{-7}$ | $4.1 \times 10^9$ |

Tube 3.

| | | | | |
|---|---|---|---|---|
| Air | $2.5 \times 10^{11}$ | 220 | $.9 \times 10^{-7}$ | $2.4 \times 10^{9}$ |
| Air | $3.5 \times 10^{11}$ | 225 | $.7 \times 10^{-7}$ | $3.2 \times 10^{9}$ |
| Hydrogen | $3 \times 10^{11}$ | 250 | $1.0[\text{sic-CJG}] \times 10^{-7}$ | $2.5 \times 10^{9}$ |

It will be noticed that the value of $m/e$ is considerably greater for Tube 3, where the opening is a small hole, than for Tubes 1 and 2, where the opening is a slit of much greater area. I am of the opinion that the values of $m/e$ got from Tubes 1 and 2 are too small, in consequence of the leakage from the inner cylinder to the outer by the gas being rendered a conductor by the passage of the cathode rays.

It will be seen from these tables that the value of $m/e$ is independent of the nature of the gas. Thus, for the first tube the mean for air is $.40 \times 10^{-7}$, for hydrogen $.42 \times 10^{-7}$, and for carbonic acid gas $.4 \times 10^{-7}$; for the second tube the mean for air is $.52 \times 10^{-7}$, for hydrogen $.50 \times 10^{-7}$, and for carbonic acid gas $.54 \times 10^{-7}$.

Experiments were tried with electrodes made of iron instead of aluminum; this altered the appearance of the discharge and the value of $v$ at the same pressure, the values of $m/e$ were, however, the same in the two tubes; the effect produced by different metals on the discharge will be described later on.

In all the preceding experiments, the cathode rays were first deflected from the cylinder by a magnet, and it was then found that there was no deflection either of the electrometer or the galvanometer, so that the deflections observed were entirely due to the cathode rays; when the glow mentioned previously surrounded the cylinders there was a deflection of the electrometer even when the cathode rays were deflected from the cylinder.

Before proceeding to discuss the results of these measurements I shall describe another method of measuring the quantities $m/e$ and $v$ of an entirely different kind from the preceding; this method is based upon the deflection of the cathode rays in an electrostatic field. If we measure the deflection experienced by the rays when traversing a given length under a uniform electric intensity, and the deflection of the rays when they traverse a given distance under a uniform magnetic field, we can find the values of $m/e$ and $v$ in the following way: --

Let the space passed over by the rays under a uniform electric intensity F be $l$, the time taken for the rays to traverse this space is $l/v$, the velocity in the direction of F is therefore

$$(\mathrm{F}e/m)(l/v),$$

so that θ, the angle through which the rays are deflected when they leave the electric field and enter a region free from electric force, is given by the equation

$$\theta = (Fe/m)(l/v^2).$$

If, instead of the electric intensity, the rays are acted on by a magnetic force H at right angles to the rays, and extending across the distance $l$, the velocity at right angles to the original path of the rays is

$$(Hev/m)(l/v),$$

so that φ, the angle through which the rays are deflected when they leave the magnetic field, is given by the equation

$$\varphi = (He/m)(l/v).$$

From these equations we get

$$v = (\varphi/\theta)(F/H)$$

and

$$m/e = H^2\theta l/F\varphi^2.$$

In the actual experiments H was adjusted so that φ = θ; in this case the equations become

$$v = F/H,$$
$$m/e = H^2 l/F\theta.$$

The apparatus used to measure $v$ and $m/e$ by this means is that represented in fig. 2. The electric field was produced by connecting the two aluminum plates to the terminals of a battery of storage-cells. The phosphorescent patch at the end of the tube was deflected, and the deflection measured by a scale pasted to the end of the tube. As it was necessary to darken the room to see the phosphorescent patch, a needle coated with luminous paint was placed so that by a screw it could be moved up and down the scale; this needle could be seen when the room was darkened, and it was moved until it coincided with the phosphorescent patch. Thus, when light was admitted, the deflection of the phosphorescent patch could be measured.

The magnetic field was produced by placing outside the tube two coils whose diameter was equal to the length of the plates; the coils were placed so that they covered the space occupied by the plates, the distance between the coils was equal to the radius of either. The mean value of the magnetic force over the length $l$ was determined in the following way: a narrow coil C whose length was $l$, connected with a ballistic galvanometer, was placed between the coils; the plane of the windings of C was parallel to the planes of the coils; the cross section of the coil was a rectangle 5 cm. by 1 cm. A given current was sent through the outer coils and the kick $\alpha$ of the galvanometer observed when this current was reversed. The coil C was then placed at the center of two very large coils, so as to be in a field of uniform magnetic force: the current through the large coils was reversed and the kick $\beta$ of the galvanometer again observed; by comparing $\alpha$ and $\beta$ we can get the mean value of the magnetic force over a length $l$; this was found to be

$$60 \times \iota,$$

where $\iota$ is the current flowing through the coils.

A series of experiments was made to see if the electrostatic deflection was proportional to the electric intensity between the plates; this was found to be the case. In the following experiments the current through the coils was adjusted so that the electrostatic deflection was the same as the magnetic: --

| Gas. | $\theta$. | H. | F. | $l$. | $m/e$. | $v$. |
|---|---|---|---|---|---|---|
| Air | 8/110 | 5.5 | $1.5 \times 10^{10}$ | 5 | $1.3 \times 10^{-7}$ | $2.8 \times 10^{9}$ |
| Air | 9.5/110 | 5.4 | $1.5 \times 10^{10}$ | 5 | $1.1 \times 10^{-7}$ | $2.8 \times 10^{9}$ |
| Air | 13/110 | 6.6 | $1.5 \times 10^{10}$ | 5 | $1.2 \times 10^{-7}$ | $2.3 \times 10^{9}$ |
| Hydrogen | 9/110 | 6.3 | $1.5 \times 10^{10}$ | 5 | $1.5 \times 10^{-7}$ | $2.5 \times 10^{9}$ |
| Carbonic Acid | 11/110 | 6.9 | $1.5 \times 10^{10}$ | 5 | $1.5 \times 10^{-7}$ | $2.2 \times 10^{9}$ |
| Air | 6/110 | 5 | $1.8 \times 10^{10}$ | 5 | $1.3 \times 10^{-7}$ | $3.6 \times 10^{9}$ |
| Air | 7/110 | 3.6 | $1. \times 10^{10}$ | 5 | $1.1 \times 10^{-7}$ | $2.8 \times 10^{9}$ |

The cathode in the first five experiments was aluminum, in the last two experiments it was made of platinum; in the last experiment Sir William Crookes's method of getting rid of the mercury vapor by inserting tubes of pounded sulphur, sulphur iodide, and copper filings between the bulb and the pump was adopted. In the calculation of $m/e$ and $v$ no allowance has been made for the magnetic force due to the coil in the region outside the plates; in this region the magnetic force will be in the opposite direction to that between the plates, and will tend to bend the cathode rays in the opposite direction: thus the effective value of H will be smaller than the value used in the equations, so that the values of $m/e$ are larger, and those of $v$ less than they would be if this correction were applied. This method of determining the values of $m/e$ and $v$ is much less laborious and probably more accurate than the former method; it cannot, however, be used over so wide a range of pressures.

From these determinations we see that the value of $m/e$ is independent of the nature of the gas, and that its value $10^{-7}$ is very small compared with the value $10^{-4}$, which is the smallest value of this quantity previously known, and which is the value for the hydrogen ion in electrolysis.

Thus, for the carriers of the electricity in the cathode rays, $m/e$ is very small compared with its value in electrolysis. The smallness of $m/e$ may be due to the smallness of $m$ or the largeness of $e$, or to a combination of these two. That the carriers of the charges in the cathode rays are small compared with ordinary molecules is shown, I think, by Lenard's results as to the rate at which the brightness of the phosphorescence produced by these rays diminishes with the length of path travelled by the ray. If we regard this phosphorescence as due to the impact of the charged particles, the distance through which the rays must travel before the phosphorescence fades to a given fraction (say $1/e$, where $e = 2.71$) of its original intensity, will be some moderate multiple of the mean free path. Now Lenard found that this distance depends solely upon the density of the medium, and not upon its chemical nature or physical state. In air at atmospheric pressure the distance was about half a centimeter, and this must be comparable with the mean free path of the carriers through air at atmospheric pressure. But the mean free path of the molecules of air is a quantity of quite a different order. The carrier, then, must be small compared with ordinary molecules.

The two fundamental points about these carriers seem to me to be (1) that these carriers are the same whatever the gas through which the discharge passes, (2) that the mean free paths depend upon nothing but the density of the medium traversed by these rays.

It might be supposed that the independence of the mass of the carriers of the gas through which the discharge passes was due to the mass concerned being the quasi mass which a charged body possesses in virtue of the electric field set up in its neighborhood; moving the body involves the production of a varying electric field, and, therefore, of a certain amount of energy which is proportional to the square of the velocity. This causes the charged body to behave as if its mass were increased by a quantity, which for a charged sphere is $(1/5)e^2/\mu a$ ('*Recent Researches in Electricity and Magnetism*'), where $e$ is the charge and $a$ the radius of the sphere. If we assume that it is this mass which we are concerned with in the cathode rays, since $m/e$ would vary as $e/a$, it affords no clue to the explanation of either of the properties (1 and 2) of these rays. This is not by any means the only objection to this hypothesis, which I only mention to show that it has not been overlooked.

The explanation which seems to me to account in the most simple and straightforward manner for the facts is founded on a view of the constitution of the chemical elements which has been favorably entertained by many chemists: this view is that the atoms of the different chemical elements are different aggregations of atoms of the same kind. In the form in which this hypothesis was enunciated by Prout, the atoms of the different elements were hydrogen atoms; in this precise form the hypothesis is not tenable, but if we substitute for hydrogen some unknown primordial substance X, there is nothing known which is inconsistent with this hypothesis, which is one that has been recently supported by Sir Norman Lockyer for reasons derived from the study of the stellar spectra.

If, in the very intense electric field in the neighborhood of the cathode, the molecules of the gas are dissociated and are split up, not into the ordinary chemical atoms, but into these primordial atoms, which we shall for brevity call **corpuscles**; and **if these corpuscles are charged with electricity and projected from the cathode by the electric field, they would behave exactly like the cathode rays**. They would evidently give a value of $m/e$ which is independent of the nature of the gas and its pressure, for the carriers are the same whatever the gas may be; again, the mean free paths of these corpuscles would depend solely upon the density of the medium through which they pass. For the molecules of the medium are composed of a number of such corpuscles separated by considerable spaces; now the collision between a single corpuscle and the molecule will not be between the corpuscles and the molecule as a whole, but between this corpuscle and the individual corpuscles which form the molecule; thus the number of collisions the particle makes as it moves through a crowd of these molecules will be proportional, not to the number of the molecules in the crowd, but to the number of the individual corpuscles. The mean free path is inversely proportional to the number of collisions in unit time, and so is inversely proportional to the number of corpuscles in unit volume; now as these corpuscles are all of the same mass, the number of corpuscles in unit volume will be proportional to the mass of unit volume, that is the mean free path will be inversely proportional to the density of the gas. We see, too, that so long as the distance between neighboring corpuscles is large compared with the linear dimensions of a corpuscle the mean free path will be independent of the way they are arranged, provided the number in unit volume remains constant, that is the mean free path will depend only on the density of the medium traversed by the corpuscles, and will be independent of its chemical nature and physical state: this from Lenard's very remarkable measurements of the absorption of the cathode rays by various media, must be a property possessed by the carriers of the charges in the cathode rays.

Thus on this view we have in the cathode rays matter in a new state, a state in which the subdivision of matter is carried very much further than in the ordinary gaseous state: a state in which all matter--that is, matter derived from different sources such as hydrogen, oxygen, &c.--is of one and the same kind; this matter being the substance from which all the chemical elements are built up.

With appliances of ordinary magnitude, the quantity of matter produced by means of the dissociation at the cathode is so small as to almost to preclude the possibility of any direct chemical investigation of its properties. Thus, the coil I used would, I calculate, if kept going uninterruptedly night and day for a year, produce only about one three-millionth part of a gramme of this substance.

The smallness of the value of $m/e$ is, I think, due to the largeness of $e$ as well as the smallness of $m$. There seems to me to be some evidence that the charges carried by the corpuscles in the atom are large compared with those carried by the ions of an electrolyte. In the molecule of HCl, for example, I picture the components of the hydrogen atoms as held together by a great number of tubes of electrostatic force; the components of the chlorine atom are similarly held together, while only one stray tube binds the hydrogen atom to the chlorine atom. The reason for attributing this high charge to the constituents of the atom is derived from the values of the specific inductive capacity of gases: we may imagine that the specific inductive capacity of a gas is due to the setting in the electric field of the electric doublet formed by the two oppositely electrified atoms which form the molecule of the gas. The measurements of the specific inductive capacity show, however, that this is very approximately an additive quantity: that is, that we can assign a certain value to each element, and find the specific inductive capacity of HCl by adding the value for hydrogen to the value for chlorine; the value of $H_2O$ by adding twice the value for hydrogen to the value for oxygen, and so on. Now the electrical moment of the doublet formed by a positive charge on one atom of the molecule and a negative charge on the other atom would not be an additive property; if, however, each atom had a definite electrical moment, and this were large compared with the electrical moment of the two atoms in the molecule, then the electrical moment of any compound, and hence its specific inductive capacity, would be an additive property. For the electrical moment of the atom, however, to be large compared with that of the molecule, the charge on the corpuscles would have to be very large compared with those on the ion.

If we regard the chemical atom as an aggregation of a number of primordial atoms, the problem of finding the configurations of stable equilibrium for a number of equal particles acting on each other according to some law of force--whether that of Boscovich, where the force between them is a repulsion when they are separated by less than a certain critical distance, and an attraction when they are separated by less than a certain critical distance, and an attraction when they are separated by a greater distance, or even the simpler case of a number of mutually repellent particles held together by a central force--is of great interest in connection with the relation between the properties of an element and its atomic weight. Unfortunately, the equations which determine the stability of such a collection of particles increase so rapidly in complexity with the number of particles that a general mathematical investigation is scarcely possible. We can, however, obtain a good deal of insight into the general laws which govern such configurations by the use of models, the simplest of which is the floating magnets of Professor Mayer. In this model the magnets arrange themselves in equilibrium under the mutual repulsions and a central attraction caused by the pole of a large magnet placed above the floating magnets.

A study of the forms taken by these magnets seems to me to be suggestive in relation to the periodic law. Mayer showed that when the number of floating magnets did not exceed 5, they arranged themselves at the corners of a regular polygon--5 at the corners of a pentagon, 4 at the corners of a square, and so on. When the number exceeds 5, however, this law no longer holds: thus 6 magnets do not arrange themselves at the corners of a hexagon, but divide into two systems, consisting of 1 in the middle surrounded by 5 at the corners of a pentagon. For 8 we have two in the inside and 6 outside; this arrangement in two systems, an inner and an outer, lasts up to 18 magnets. After this we have three systems: an inner, a middle, and an outer; for a still larger number of magnets, we have four systems, and so on.

Mayer found the arrangement of magnets was as follows, where, for example, 1.6.10.12 means an arrangement with one magnet in the middle, then a ring of six, then a ring of ten, and a ring of twelve outside: -

| 1. | 2. | 3. | 4. | 5. |
|---|---|---|---|---|
| 1.5 | 2.6 | 3.7 | 4.8 | 5.9 |
| 1.6 | 2.7 | 3.8 | 4.9 | |
| 1.7 | | | | |
| 1.5.9 | 2.7.10 | 3.7.10 | 4.8.12 | 5.9.12 |
| 1.6.9 | 2.8.10 | 3.7.11 | 4.8.13 | 5.9.13 |
| 1.6.10 | 2.7.11 | 3.8.10 | 4.9.12 | |
| 1.6.11 | | 3.8.11 | 4.9.13 | |
| | | 3.8.12 | | |
| | | 3.8.13 | | |
| 1.5.9.12 | 2.7.10.15 | 3.7.12.13 | 4.9.13.14 | |
| 1.5.9.13 | 2.7.12.14 | 3.7.12.14 | 4.9.13.15 | |
| 1.6.9.12 | | 3.7.13.14 | 4.9.14.15 | |
| 1.6.10.12 | | 3.7.13.15 | | |
| 1.6.10.13 | | | | |
| 1.6.11.12 | | | | |
| 1.6.11.13 | | | | |
| 1.6.11.14 | | | | |
| 1.6.11.15 | | | | |
| 1.7.12.14 | | | | |

Now suppose that a certain property is associated with two magnets forming a group by themselves; we should have this property with 2 magnets, again with 8 and 9, again with 19 and 20, and again with 34, 35, and so on. If we regard the system of magnets as a model of an atom, the number of magnets being proportional to the atomic weight, we should have this property occurring in elements of atomic weight 2, (8,9), 19, 20, (34, 35). Again, any property conferred by three magnets forming a system by themselves would occur with atomic weights 3, 10, and 11; 20, 21, 22, 23, and 24; 35, 36, 37 and 39; in fact, we should have something quite analogous to the periodic law, the first series corresponding to the arrangement of the magnets in a single group, the second series to the arrangement in two groups, the third series in three groups, and so on.

## Velocity of the Cathode Rays.

*The velocity of the cathode rays is variable, depending upon the potential-difference between the cathode and anode*, which is a function of the pressure of the gas--the velocity increases as the exhaustion improves; the measurements given above show, however, that at all the pressures at which experiments were made the velocity exceeded $10^9$ cm./sec. This velocity is much greater than the value of $2 \times 10^7$ which I previously obtained (*Phil. Mag.* Oct. 1894) by measuring directly the interval which separated the appearance of luminosity at two places on the walls of the tube situated at different distances from the cathode.

In my earlier experiments the pressure was higher than in the experiments described in this paper, so that the velocity of the cathode rays would on this account be less. The difference between the two results is, however, too great to be wholly explained in this way, and I attribute the difference to the glass requiring to be bombarded by the rays for a finite time before becoming phosphorescent, this time depending upon the intensity of the bombardment. As this time diminishes with the intensity of bombardment, the appearance of phosphorescence at the piece of glass most removed from the cathode would be delayed beyond the time taken for the rays to pass from one place to the other by the difference in time taken by the glass to become luminous; the apparent velocity measured in this way would thus be less than the true velocity. In the former experiments endeavors were made to diminish this effect by making the rays strike the glass at the greater distance from the cathode less obliquely than they struck the glass nearer to the cathode; the obliquity was adjusted until the brightness of the phosphorescence was approximately equal in the two cases. In view, however, of the discrepancy between the results obtained in this way and those obtained by the later method, I think that it was not successful in eliminating the lag caused by the finite time required by the gas to light up.

## Experiments with Electrodes of Different Materials.

In the experiments described in this paper the electrodes were generally made of aluminum. Some experiments, however, were made with iron and platinum electrodes.

Though the value of $m/e$ came out the same whatever the material of the electrode, the appearance of the discharge varied greatly; and as the measurements showed, the potential-difference between the cathode and anode depended greatly upon the metal used for the electrode; the pressure being the same in all cases.

To test this point further I used a tube like that shown in fig. 6, where *a, b, c* are cathodes made of different metals, the anodes being in all cases platinum wires. The cathodes were disks of aluminum, iron, lead, tin, copper, mercury, sodium amalgam, and silver chloride; the potential-difference between the cathode and anode was measured by Lord Kelvin's vertical voltmeter, and also by measuring the length of spark in air which, when placed in parallel with the anode and cathode, seemed to allow the discharge to go as often through the spark-gap as through the tube. With this arrangement the pressures were the same for all the cathodes. The potential-difference between the anode and cathode and the equivalent spark-length depended greatly upon the nature of the cathode. The extent of the variation in potential may be estimated from the following table: --

**Cathode.    Mean Potential-Difference between Cathode and Anode.**

| Cathode | Mean Potential-Difference |
|---|---|
| Aluminum | 1800 volts. |
| Lead | 2100 " |
| Tin | 2400 " |
| Copper | 2600 " |
| Iron | 2900 " |

The potential-difference when the cathode was made of sodium amalgam or silver chloride was less even than that of aluminum.

The order of many of the metals changed about very capriciously, experiments made at intervals of a few minutes frequently giving quite different results. From the abrupt way in which these changes take place I am inclined to think that gas absorbed by the electrode has considerable influence on the passage of the discharge.

I have much pleasure in thanking Mr. Everitt for the assistance he has given me in the preceding investigation.

Cambridge, Aug. 7, 1897.

## Thomson, J. J. (June 1906). LXX. On the Number of Corpuscles in an Atom.

*Phil. Mag.*, 11, 66, 769-81; doi:10.1080/14786440609463496.

[Thomson referred to *electrons* as "*corpuscles*" and resisted the use of "*electron*", but later scientists preferred the name *electron*.]

*I consider in this paper three methods of determining the number of corpuscles in an atom of an elementary substance, all of which lead to the conclusion that this number is of the same order as the atomic weight of the substance.* Two of these methods show in addition that *the ratio of the number of corpuscles in the atom to the atomic weight of the element is the same for all elements.* The data at present available indicate that the number of *corpuscles* in the atom is equal to the *atomic weight*. As, however, the evidence is rather indirect and the data are not very numerous, further investigation is necessary before we can be sure of this equality; the evidence at present available seems~ however, sufficient to establish the conclusion that the number of *corpuscles* is not greatly different from the atomic weight.

It will be seen that the methods are very different and deal with widely separated physical phenomena; and although no one of the methods can, I think, be regarded as quite conclusive by itself, the evidence becomes very strong when we find that such different methods lead to practically identical results.

To enable the argument to be more easily followed, I shall begin with a general description of the methods and the results to which they lead, and postpone the details of the theory of each of the methods to the latter part of the paper.

**The first method founded on the dispersion of light by gases.**

If we regard an atom as consisting of a number of *corpuscles* dispersed through a sphere of uniform positive electrification, *it is evident that the dispersive power of a medium consisting of these atoms will depend upon the mass of the positive electrification as well as upon the mass of the corpuscles,* and will vanish if either of these masses is zero. For consider what takes place when the electric force in the light-wave strikes the atom. Since the wave-length is large compared with an atom, the latter may be regarded as being in a uniform electric field; under this field the *corpuscles* will be displaced in one direction, the positive sphere in the opposite; and if the force persists long enough, this displacement will go on until the force exerted on the *corpuscles* in their displaced position by the positive electricity is equal and opposite to the force exerted on them by the electric force in the light-wave. The displacement of the *corpuscles* relatively to the sphere of positive electricity will polarize the atom, and the collection of polarized atoms will increase the specific inductive capacity, and therefore the refractive index of the medium. If the mass of either the positive electricity or of the *corpuscles* were zero, then, however short the time for which the electric force in the light-wave acted, the *corpuscles* and the positive electricity would adjust themselves in exactly the same way as if the electric force were continuous; so that the specific inductive capacity and the refractive index would be the same for short waves as for long, and there would be no dispersion. If, however, the masses of both the positive electricity and the *corpuscles* are finite, the relative displacement of the *corpuscles* and the positive electricity will depend upon the period of the electric force; and since the specific inductive capacity and refractive index depend upon this displacement, the refractive index of the medium will depend upon the period of the electric force, and there will be dispersion.

In the latter part of the paper the expression for the refractive index of a monatomic gas is investigated; and it is shown that if g is the refractive index of such a gas for light of frequency p, then

$$(\mu^2 - 1)/(\mu^2 + 2) = 4/3 \ \pi NE(Me + mE)/[4/3 \ \pi\rho(Me + mE) - Mmp^2], \qquad (1)$$

where N is the number of atoms in unit volume of the gas, *m* the mass of a *corpuscle*, M the mass of the sphere of positive electrification, e the charge on a *corpuscle*, E the whole charge on the sphere of positive electrification, $\rho$ the density of the electrification in this sphere; e, E, and $\rho$ are expressed in electrostatic measure.

If the term $p^2$ is small, equation (1) may be written

$$(\mu^2 - 1)/(\mu^2 + 2) = NE/\rho \ (1 + Mm/Ee)[p^2/(M + nm)][3E/4\pi\rho]),$$

since E = *ne*, where *n* is the number of *corpuscles* in the atom. If *a* is the radius of the sphere of positive electrification, E = $(4/3)\pi\rho a^3$, i.e. (NE/$\rho$) = N(4/3) $\pi a^3$ = volume of the atoms per cubic centimeter of gas; this is the value of $(\mu^2 - 1)/(\mu^2 + 2)$ when p = 0, i.e. for infinitely long waves. Writing $P_0$ for this quantity, we see, if $\lambda$ is the wave-length of the light,

$$(\mu^2 - 1)/(\mu^2 + 2) = P_0 + P_0^2 \ Mm/E'e' \ [1/N(M + nm)][3\pi/\lambda^2]), \qquad (2)$$

where E' and e' are the values of E and e in electromagnetic measure. ***This is the expression for the refractive index of a monatomic gas.*** I have not been able to find any determinations of the dispersion of such gases. Lord Rayleigh, however, found that the dispersion of helium was of the same order as that of diatomic gases. If the atoms in the molecules of a diatomic gas are not charged, the preceding expression will hold for the refractive index of such a gas; if, however, the atoms carry electrical charges, the theory subsequently given shows that this expression has to be modified. We know that as a matter of fact the atomic refraction of some elements, oxygen for example, depends upon the kind of compound in which the oxygen is found. This variation may be ascribed to the charges carried by the atoms in the molecules. The atomic refraction of hydrogen seems, however, to be constant; and I shall, in the absence of data for monatomic gases, apply the preceding formula to this gas.

From Ketteler's measurements of the refractive index of hydrogen for light of different wave-lengths, we find that for hydrogen at atmospheric pressure

$$(\mu^2 - 1)/(\mu^2 + 2) = 1/3 \; \{2.8014 \times 10^{-4} + 2 \times 10^{-14}/\lambda^2\}$$

Comparing this with the equation (2)

$$(\mu^2 - 1)/(\mu^2 + 2) = P_0 + P_0^2 \; Mm/E'e' \; [1/N(M + nm)][3\pi/\lambda^2]),$$

we find

$$Mm/E'e' \; [1/N(M + nm)] = 6 \times 10^{-8};$$

but $m/e' = 1/.7 \times 10^7$ and $Ne' = 0.8$;

hence

$$[M/(M + nm)] \; [e'/E'] = 1, \text{ approximately};$$

or, since $E' = ne'$

$$[M/(M + nm)] \; [1/n] = 1, \text{ approximately}.$$

This result shows (1) that $n$ cannot differ much from unity, and (2) that M, the mass of the carriers of positive electricity, cannot be small compared with $nm$, the mass of the carriers of negative electricity. Hence, we infer that $n$, the number of *corpuscles* in a hydrogen atom, is not much greater than unity. This result has been deduced from the consideration of the properties of a diatomic molecule ; and if the atoms in the molecule were charged, the expression for $(\mu^2 - 1)/(\mu^2 + 2)$ would have to be modified; but since the dispersion of helium is by Lord Rayleigh's result comparable with that of hydrogen,

64

we see, since the dispersion is proportional to $[M/(M + nm)]$ $[1/n]$, that there cannot be a very large number of *corpuscles* in the helium atom; for if $n$ were large, the dispersion of helium would be far too small.

**2nd Method. Scattering of Rontgen Radiation by Gases.**

It is shown in my *Conduction of Electricity through Gases*, [(2013), Cambridge, UK: Cambridge University Press] that when Rontgen rays pass through a medium in which there are N *corpuscles* per cubic centimeter, the energy in the radiation scattered per cubic centimeter of the medium is $8\pi/3$ $[Ne^4/m^2]E$, where E is the energy of the primary radiation passing through the cubic centimeter, e the charge, and $m$ the mass of the *corpuscle*. Barkla has shown that in the case of gases the energy in the scattered radiation always bears, for the same gas, a constant ratio to the energy in the primary whatever be the nature of the rays, i.e. whether they are hard or soft; and secondly, that the scattered energy is proportional to the mass of the gas. The first of these results is a confirmation of the theory, as the ratio of the energy scattered to that in the primary rays is $8\pi/3$ $[Ne^4/m^2]$ and is independent of the nature of the rays; the second result shows that the number of *corpuscles* per cub. centim, is proportional to the mass of the gas: from this it follows that the number of *corpuscles* in an atom is proportional to the mass of the atom, i. e. to the atomic weight. Barkla measured the ratio of the energy in the scattered radiation to that in the primary in the case of air, and found that it was equal to $2.4 \times 10^{-4}$. Thus, for air

$$8\pi/3 \ [Ne^4/m^2] = 2.4 \times 10^{-4}.$$

Now $e/m = 1.7 \times 10^7$ and $e = 1.1 \times 10^{-20}$; hence

$$Ne = 10.$$

But if n is the number of molecules per c.c.,

$$ne = 0.4$$

hence $N = 25n.$

From this we deduce that there are 25 *corpuscles* in each molecule of air, and this indicates that the number of *corpuscles* in the atom is equal to the *atomic weight*; for the scattering by air is very nearly the same as that by nitrogen, and 25, the number of *corpuscles* in the molecule deduced from Barkla's experiment, is near to 28, the number in each molecule if the number in the atom were equal to the atomic weight.

**3rd Method. Absorption of β Rays.**

We regard the absorption of the *β rays* as due to the effect of the collisions between these rays and the *corpuscles* which they meet with in their path through the absorbing substance....

…

i.e. *n* is of the same order as M/M', the atomic weight of the element.

***Thus, these three very different methods all lead to the result that the number of corpuscles in the atom of an element is of the same order as the atomic weight of the element***; and from the first method we conclude that the mass of the carrier of unit positive charge is large compared with that of the carrier of unit negative charge. If we suppose the whole mass of an atom to be that of its charged parts, e/m for positive unit charge would be of the order $10^4$.

An obvious argument against the number of *corpuscles* in the atom being as small as these results indicate, is that the number of lines showing the Zeeman effect, which must therefore be due to the vibrations of *corpuscles*, in the spectrum, say, of iron is very much greater than the atomic weight of iron. This objection would be conclusive if it could be shown that all these lines are due to the vibrations of *corpuscles* inside the normal atom of iron; but I submit that there is no evidence that this is the case. When an atom of an element is giving out its spectrum either in a flame or in an electric discharge, it is surrounded by a swarm of *corpuscles*; and combinations, not permanent indeed, but lasting sufficiently long for the emission of a large number of vibrations, might be expected to be formed. These systems would give out characteristic spectrum-lines; but these lines would be due, not to the vibrations of *corpuscles* inside the atom, but of *corpuscles* vibrating in the field of force outside the atom. Such lines would not be reversed by cold vapor, though they might be by very hot vapors, by the vapors in flames or in the neighborhood of an electric discharge: the number of lines showing the Zeeman effect reversed by cold vapors is, however, very limited.

**First method: Index of Refraction of a Collection of Atoms.**

We shall now proceed to consider the theory of the method on which the preceding results are based.

If an atom consisting of *corpuscles* dispersed through a sphere of uniform positive electrification is in the path of a wave of light, the electric force in the wave will displace the *corpuscles* in the atom; the motion of these charged *corpuscles* will produce a magnetic field in addition to that in the wave before it struck the *corpuscle*; the existence of this field will alter the velocity of propagation of the wave by an amount which we shall attempt to calculate.

**Index of refraction of a monatomic gas whose atoms contain as much positive as negative electricity** - Consider an element of volume so small that throughout it the electric force in the wave may be regarded as constant. Throughout this volume the atoms will all be affected by the electric force in the same way.

If $\xi_r$, $\eta_r$, $\zeta_r$ are the displacements parallel to the axes of x, y, z of the rth *corpuscle* of an atom, x the displacement of the center of the sphere of positive electrification, e the negative charge on a *corpuscle*, E the charge of positive electrification in the sphere, N the number of atoms per unit volume; then X', Y', Z', the components of the electric force due to the displacement of the *corpuscles*, are given by the equations

$$X' = 4/3\ \pi N(Ex - \textstyle\sum e\xi_r)$$
$$Y' = 4/3\ \pi N(Ey - \textstyle\sum e\eta_r) \qquad\qquad (1)$$
$$Z' = 4/3\ \pi N(Ez - \textstyle\sum e\zeta_r);$$

the summation is for all the *corpuscles* in one atom.

The equations of motion for the *corpuscles* and sphere of positive electrification are

$$M\ d^2x/dt^2 = (X + X')E - 4/3\ \pi\rho e + mE\ \textstyle\sum(x - \xi_r),$$
$$M\ d^2\textstyle\sum \xi_r/dt^2 = -(X + X')E + 4/3\ \pi\rho e\ \textstyle\sum(x - \xi_r),$$

where m is the mass of a *corpuscle* and M that of the sphere of positive electrification.

If all the quantities vary as $e^{4pt}$ we get from these equations

$$x = (X + X')Em/\{4/3\ \pi\rho(Me + mE) - mMp^2\},$$
$$\textstyle\sum \xi_r = -(X + X')EM/\{4/3\ \pi\rho(Me + mE) - mp^2\},$$

...

In consequence of the motion of the charged *corpuscles*, the current is no longer the polarization-current, $Ko/4\pi\ dX/dt$, where Ko is the specific inductive capacity of the aether; but this current *plus* the convection current; thus, u the total current parallel is given by the equation

$$u = Ko/4\pi\ dX/dt + 3/4\pi\ dX'/dt.$$

...

Hence, if $\mu$ is the refractive index,

$$\mu^2 = 1 + 3P/(1 - P),$$

or     $(\mu^2 - 1)/(\mu^2 + 2) = P = 4/3 \; \pi N(mE^2 + mEe)/\{4/3 \; \pi\rho(Me + mE) - mMp^2\}$     (2)

For very long waves, when the term $mMp^2$ may be neglected, we have

$$(\mu^2 - 1)/(\mu^2 + 2) = P = 4/3 \; \pi NE/[4/3 \; \pi\rho];$$

or, since $E = (4/3)\pi\rho a^3$, where $a$ is the radius of the sphere positive electrification,

$$P = (\mu^2 - 1)/(\mu^2 + 2) = 4/3 \; \pi N a^3.$$

This is the value given by Mossotti's theory when the atoms are assumed to be perfectly conducting spheres of radius $a$.

[In 1850, Italian physicist Ottaviano-Fabrizio Mossotti published a groundbreaking paper deriving a relation between the refractive index of gases and their molecular density, framing optical dispersion as the cumulative effect of individual molecular interactions on light propagation. Mossotti modeled gas molecules as discrete entities exerting additive influences on the velocity of light waves passing through the medium, treating them as induced dipoles that scatter and modify the electromagnetic field. This approach built on contemporary understandings of molecular theory, positing that the overall refractive behavior arises from the collective polarizability of these molecules without invoking continuous media assumptions.

The core insight of Mossotti's analysis was that dispersion in dilute gases results from the sum of independent molecular contributions, yielding a precursor relation to the modern Clausius–Mossotti form. Specifically, he arrived at an expression linking the refractive index $n$ to the number density $N$ and an effective polarizability $\alpha'$, approximated for gases, where $n \approx 1$ as
$$(n - 1)/(n + 2) \approx 4\pi/3 \; N\alpha'.$$

This equation highlighted how increased molecular density enhances the refractive index through enhanced polarization, providing a quantitative bridge between microscopic molecular properties and macroscopic optical phenomena. Mossotti's derivation employed electrostatic analogies to describe the field's interaction with molecules, emphasizing the role of induced dipoles in altering light's speed.

Mossotti grounded his theoretical framework in experimental observations of refractive index variations in gases like air, where density changes—induced by pressure or temperature—correlated directly with measurable shifts in optical properties. These empirical links validated the assumption of additive molecular effects in low-density regimes.]

**Case of a diatomlc molecule, the atoms carrying electrical charges equal in magnitude and opposite in sign.**

In this case the refractivity P will contain a term due to the motion of the charged atoms relatively to each other under the electric field due to the light-wave. We can calculate this term as follows …

…

… The consideration of this expression shows that the part of the refractivity arising from the coupling of two atoms together may easily be comparable with the part due to the *corpuscles* within the atom. Thus, to take the case when the waves are so long that we may neglect the term in $p^2$, the contribution to the refractivity due to the coupling is

…

If we compare this with $2N'a^3$, the value due to the *corpuscles* inside the atom, $a$ being the radius of the atom, we see that unless the force between the two atoms varies very rapidly with the distance, a considerable part of the refractivity may be due to the coupling between the atoms.

If $\Delta P_o$ is the part of $(\mu^2 - 1)/(\mu^2 + 2)$ when $\lambda$ is infinite due to the charges on the atoms in the diatomic molecule, we see from equation (3) that the part of $(\mu^2 - 1)/(\mu^2 + 2)$ due to these charges is approximately equal to

$$\Delta P_o + (\Delta P_o)^2 \, M_1 M_2 / E'_1 E'_2 \, 1/(M_1 + M_2) \, 4\pi^2/N\lambda^2, \tag{5}$$

where $E_l'$ is the value of the charge in electromagnetic units.

In the case of a ***molecule consisting of two charged atoms***, the charge on the negative atom will be due to the presence on the atom of extra *corpuscles* which can move freely about.

…

… Thus, unless $DP_0$ is very small compared with $P_o$, ***the dispersion of the gas will depend more on the charges of the atoms in a diatomic gas than on the corpuscles inside the individual atoms***.

69

**Second method: scattering of Rontgen radiation by gases.**

The theory of the *second method* is given in my *'Conduction of Electricity through Gases.'*

**Third method: the absorption by matter of rapidly moving corpuscles.**

We proceed to the consideration of the *third method*, which depends on the absorption by matter of rapidly moving *corpuscles*.

If we suppose that an atom consists of a number of *corpuscles* distributed through positive electrification, we can find an expression for the absorption experienced by the *corpuscles* when they pass through a collection of a large number of such atoms. The rapidly moving *corpuscle* will penetrate the atom, and will be deflected when it comes near an inter-atomic *corpuscle* by the repulsion between the *corpuscles*. This deflection will produce an absorption of the cathode particles. If the *corpuscle* in the atom is held fixed by the forces acting upon it, the colliding *corpuscle* will, after the collision, have the same velocity as before, though the direction of its motion will be deflected. If the interatomic *corpuscle* A is not fixed, the colliding corpuscle B will communicate some energy to it and will itself go off with diminished energy. Without solving the very complicated problem which presents itself when we take into account the forces exerted on A by the other *corpuscles*, we can form some idea of the effects produced by the constraint introduced by such forces by following the effects produced by increasing the mass of A. The general effect of great constraint would be represented by supposing the mass of A to be very large, while absence of constraint would be represented by supposing the mass of A to be equal to that of B.

Let $M_1$, $M_2$ be the masses of the *corpuscles* A and B respectively. We shall suppose that the velocity of the colliding *corpuscle* is so great that in comparison the *corpuscles* in the atom may be regarded as at rest. Let V be the velocity of A before the collision, b the perpendicular let fall from A on V. If $2\theta$ is the angle through which the direction of relative motion is deflected by the collision, we can easily show that

$$\sin^2 \theta = 1/[1 + (b^2 V^4/e^4)\{M_1 M_2/(M_1 + M_2)^2\}];$$

the force between two *corpuscles* separated by a distance r being assumed equal to $e^2/r^2$. Hence, if u, u' are the velocities of B parallel to x before and after the collision,

$$u' - u = -M_1 u/(M_1 + M_2) [2 \sin^2 \theta] + M_1/(M_1 + M_2) \sin 2\theta \cos \phi \sqrt{(V^2 - u^2)}$$

where $\phi$ is the angle between the plane containing b and V and that containing V and x. Averaging, the term containing $\cos \phi$ will disappear, and we have

$$u' - u = - 2M_1u/(M_1 + M_2) [1/\{1 + (b^2V^4/e^4) M_1M_2/(M_1 + M_2)^2\}].$$

If there are N of the interatomic *corpuscles* per unit volume, the number of collisions in which b is between b and b + db made by a *corpuscle* B when it travels over a distance $\Delta x$ is $N\Delta x . 2\pi b . db$. Hence, if U is the sum of the values of u for the B *corpuscles* per unit volume, and $\Delta(U)$ the change in U in the distance $\Delta x$,

$$\Delta(U) = -2UN . \Delta x . M_1/(M_1 + M_2) \int_0^{b1} [1/\{1 + (b^2V^4/e^4) M_1M_2/(M_1 + M_2)^2\}]. \qquad (6)$$

the upper limit being determined by the condition that B comes into collision with the A *corpuscles* one at a time, so that the shortest distance between B and the *corpuscle* with which it comes into collision must be small compared with *a*, the distance between two *corpuscles*. If r is the shortest distance between the A and B *corpuscles*, we can easily show that
$$1 - b^2/r^2 = 2e^2/V^2r (M_1 + M_2)/M_1M_2.$$

Putting r = *a*, we see that b' is of order

$$a \{1 - 2e^2/V^2a (M_1 + M_2)/M_1M_2\}^{1/2}.$$

Integrating the expression on the right-hand side of equation (6), we get

$$dU/dx = 2U N . M_1/(M_1 + M_2) e^4(M_1 + M_2)^2/V^4(M_1M_2)^2$$
$$\log \{1 + (b_1^2V^4/e^4)M_1M_2/(M_1 + M_2)^2\}.$$

Since the logarithmic term only varies slowly, we may put for $b_1$ any quantity of the same order without greatly affecting the result, putting $b' = a \{1 - 2V^2/e^2a (M_1 + M_2)/M_1M_2\}^{1/2}$.

$$d/dx (U) = - U 4\pi Ne^4/V^4 (M_1 + M_2)/M_1M_2^2 \log \{aV^2/e^2) M_1M_2/(M_1 + M_2) -1\}.$$

Thus U, the number of *corpuscles* crossing unit area in unit time, varies as $e^{-\lambda x}$, where

$$\lambda = 4\pi Ne^4/V^4 (M_1 + M_2)/M_1M_2^2 \log \{aV^2/e^2) M_1M_2/(M_1 + M_2) -1\}.$$

*is the coefficient of absorption.* If $M_1 = M_2$, i.e. if the *corpuscles* are quite free to move in the atom,

$$\lambda = 8\pi Ne^4/V^4M_2^2 \log \{1/2 \ M_2 \ aV^2/e^2 - 1\}.$$

If $M_1$ is infinite, i.e. if the *corpuscles* are held fixed by the forces between them, we have

$$\lambda = 4\pi Ne^4/V^4M_2^2 \log \{1/2 \ M_2 \ aV^2/e^2 - 1,$$

the value used in *Method 3*.

We can get an approximate value of $\lambda$ very simply by the following method. Consider a stream of *corpuscles* moving horizontally; the forward motion of any particle will be stopped by a collision in which its direction of motion is "turned through an angle equal to or greater than a right angle; i. e. if $\theta$ is equal to or greater than $\pi/4$ or $\sin^2 \theta > 1/2$, $\sin^2 \theta$ will be greater than $1/2$ if

$$b^2 < e^4/V^4 \ \{(M_1 + M_2)/M_1M_2\}^2$$

The number of collisions made by a *corpuscle* for which b is not greater than this value, as the corpuscle moves over a distance $\Delta x$, is

$$\pi Ne^4/V^4 \ \{(M_1 + M_2)/M_1M_2\}^2 \ \Delta x.$$

Hence, if U is the number of *corpuscles* crossing unit area in unit time, we have, if we neglect the effect of collisions which do not result in a total stoppage of the particle,

$$dU/dx = - U \cdot N\pi e^4/V^4 \ \{(M_1 + M_2)/M_1M_2\}^2,$$

or, if $\lambda$ is the coefficient of absorption,

$$\lambda = N\pi e^4/V^4 \ \{(M_1 + M_2)/M_1M_2\}^2.$$

# Thomson, J. J. (1906). Carriers of Negative Electricity.

Nobel Lecture, December 11, 1906; J.J. Thomson – Nobel Lecture. NobelPrize.org. <https://www.nobelprize.org/prizes/physics/1906/thomson/lecture/>

The Nobel Prize in Physics 1906 was awarded to Joseph John Thomson "in recognition of the great merits of his theoretical and experimental investigations on the conduction of electricity by gases". https://www.nobelprize.org/prizes/physics/1906/summary/>. The idea that electricity is transmitted by a tiny particle related to the atom was first forwarded in the 1830s. In the 1890s, J.J. Thomson managed to estimate its magnitude by performing experiments with charged particles in gases. In 1897 he showed that cathode rays (radiation emitted when a voltage is applied between two metal plates inside a glass tube filled with low-pressure gas) consist of particles— *electrons*— that conduct electricity. Thomson also concluded that *electrons* are part of atoms.

## Introductory

In this lecture I wish to give an account of some investigations which have led to the conclusion that the carriers of negative electricity are bodies, which I have called *corpuscles*, having a mass very much smaller than that of the atom of any known element, and are of the same character from whatever source the negative electricity may be derived.

The first place in which *corpuscles* were detected was a highly exhausted tube through which an electric discharge was passing. When an electric discharge is sent through a highly exhausted tube, the sides of the tube glow with a vivid green phosphorescence. That this is due to something proceeding in straight lines from the cathode - the electrode where the negative electricity enters the tube - can be shown in the following way (the experiment is one made many years ago by Sir William Crookes): A Maltese cross made of thin mica is placed between the cathode and the walls of the tube. When the discharge is past, the green phosphorescence no longer extends all over the end of the tube, as it did when the cross was absent. There is now a well-defined cross in the phosphorescence at the end of the tube; the mica cross has thrown a shadow and the shape of the shadow proves that the phosphorescence is due to something travelling from the cathode in straight lines, which is stopped by a thin plate of mica. The green phosphorescence is caused by *cathode rays* and at one time there was a keen controversy as to the nature of these rays. Two views were prevalent: one, which was chiefly supported by English physicists, was that the rays are *negatively electrified bodies* shot off from the cathode with great velocity; the other view, which was held by the great majority of German physicists, was that the rays are some kind of *ethereal vibration or waves*.

The arguments in favor of the rays being ***negatively charged particles*** are primarily that they are deflected by a magnet in just the same way as moving, negatively electrified particles. We know that such particles, when a magnet is placed near them, are acted upon by a force whose direction is at right angles to the magnetic force, and also at right angles to the direction in which the particles are moving.

Thus, if the particles are moving horizontally from east to west, and the magnetic force is horizontal from north to south, the force acting on the negatively electrified particles will be vertical and downwards.

When the magnet is placed so that the magnetic force is along the direction in which the particle is moving, the latter will not be affected by the magnet.

The next step in the proof that cathode rays are negatively charged particles was to show that when they are caught in a metal vessel they give up to it a charge of negative electricity. This was first done by Perrin. This experiment was made conclusive by placing the catching vessel out of the path of the rays, and bending them into it by means of a magnet, when the vessel became negatively charged.

**Electric deflection of the rays**

If the rays are charged with negative electricity they ought to be deflected by an electrified body as well as by a magnet. In the earlier experiments made on this point no such deflection was observed. The reason of this has been shown to be that when cathode rays pass through a gas, they make it a conductor of electricity, so that if there is any appreciable quantity of gas in the vessel through which the rays are passing, this gas will become a conductor of electricity and the rays will be surrounded by a conductor which will screen them from the effect of electric force, just as the metal covering of an electroscope screens off all external electric effects.

By exhausting the vacuum tube until there was only an exceedingly small quantity of air left in to be made a conductor, I was able to get rid of this effect and to obtain the electric deflection of the cathode rays. This deflection had a direction which indicated a negative charge on the rays.

Thus, cathode rays are deflected by both magnetic and electric forces, just as negatively electrified particles would be.

Hertz showed, however, that cathode particles possess another property which seemed inconsistent with the idea that they are particles of matter, for he found that they were able to penetrate very thin sheets of metal, e.g. pieces of gold leaf, and produce appreciable luminosity on glass behind them. The idea of particles as large as the molecules of a gas passing through a solid plate was a somewhat startling one, and this led me to investigate more closely the nature of the particles which form the cathode rays.

The principle of the method used is as follows: When a particle carrying a charge $e$ is moving with velocity $v$ across the lines of force in a magnetic field, placed so that the lines of magnetic force are at right angles to the motion of the particle, then, if $H$ is the magnetic force, the moving particle will be acted on by a force equal to $Hev$. This force acts in the direction which is at right angles to the magnetic force and to the direction of motion of the particle. If also we have an electric field of force $X$, the cathode ray will be acted upon by a force $Xe$. If the electric and magnetic fields are arranged so that they oppose each other, then, when the force $Hev$ due to the magnetic field is adjusted to balance the force due to the electric field $Xe$, the green patch of phosphorescence due to the cathode rays striking the end of the tube will be undisturbed, and we have

$$Hev = Xe$$

or

$$v = X/H.$$

Thus, if we measure, as we can do without difficulty, the values of $X$ and $H$ when the rays are not deflected, we can determine the value of $v$, the velocity of the particles. In a very highly exhausted tube this may be 1/3 the velocity of light, or about 60,000 miles per second; in tubes not so highly exhausted it may not be more than 5,000 miles per second, but in all cases when the cathode rays are produced in tubes their velocity is much greater than the velocity of any other moving body with which we are acquainted. It is, for example, many thousand times the average velocity with which the molecules of hydrogen are moving at ordinary temperatures, or indeed at any temperature yet realized.

**Determination of $e/m$**

Having found the velocity of the rays, let us now subject them to the action of the electric field alone. Then the particles forming the rays are acted upon by a constant force and the problem is like that of a bullet projected horizontally with a velocity $v$ and falling under gravity. We know that in time $t$, the bullet will fall a depth equal to $1/2gt^2$, where g is the acceleration due to gravity. In our case the acceleration due to the electric field is equal to $Xe/m$, where $m$ is the mass of the particle. The time t = $l/v$, where $l$ is the length of path, and $v$ the velocity of projection.

Thus, the displacement of the patch of phosphorescence where the rays strike the glass is equal to

$$\frac{1}{2} Xe/m \,.\, l^2/v^2.$$

We can easily measure this displacement $d$, and we can thus find $e/m$ from the equation

$$e/m = 2d/X \,.\, v^2/l^2.$$

The results of the determinations of the values of $e/m$ made by this method are very interesting, for ***it is found that, however the cathode rays are produced, we always get the same value of $e/m$ for all the particles in the rays***. We may, for example, by altering the shape of the discharge tube and the pressure of the gas in the tube, produce great changes in the velocity of the particles, but unless the velocity of the particles becomes so great that they are moving nearly as fast as light, when other considerations have to be taken into account, the value of $e/m$ is constant. The value of $e/m$ is not merely independent of the velocity. What is even more remarkable is that it is independent of the kind of electrodes we use and also of the kind of gas in the tube. ***The particles which form the cathode rays must come either from the gas in the tube or from the electrodes; we may, however, use any kind of substance we please for the electrodes and fill the tube with gas of any kind and yet the value of $e/m$ will remain unaltered.*** This constant value, when we measure $e/m$ in the c.g.s. system of magnetic units, is equal to about $1.7 \times 10^7$. ***If we compare this with the value of the ratio of the mass to the charge of electricity carried by any system previously known, we find that it is of quite a different order of magnitude.*** Before the **cathode rays** were investigated, the charged atom of hydrogen met with in the electrolysis of liquids was the system which had the greatest known value of $e/m$, and in this case the value is only $10^4$, hence ***for the corpuscle in the cathode rays the value of $e/m$ is 1,700 times the value of the corresponding quantity for the charged hydrogen atom.*** This discrepancy must arise in one or other of two ways: ***either the mass of the corpuscle must be very small compared with that of the atom of hydrogen***, which until quite recently was the smallest mass recognized in physics, ***or else the charge on the corpuscle must be very much greater than that on the hydrogen atom.*** Now it has been shown by a method which I shall shortly describe, that the electric charge is practically the same in the two cases; hence we are driven to the conclusion that ***the mass of the corpuscle is only about 1/1,700 of that of the hydrogen atom.*** Thus, ***the atom is not the ultimate limit to the subdivision of matter***; we may go further and get to the ***corpuscle***, and at this stage the ***corpuscle*** is the same from whatever source it may be derived.

**Corpuscles very widely distributed**

It is not only from what may be regarded as a somewhat artificial and sophisticated source, viz. *cathode rays*, that we can obtain *corpuscles*. When once they had been discovered, *it was found that they are of very general occurrence*. They are given out by metals when raised to a red heat; indeed, any substance when heated gives out *corpuscles* to some extent. We can detect the emission of them from some substances, such as rubidium and the alloy of sodium and potassium, even when they are cold; and it is perhaps allowable to suppose that there is some emission by all substances, though our instruments are not at present sufficiently delicate to detect it unless it is unusually large.

*Corpuscles* are also given out by metals and other bodies, but especially by the alkali metals, when these are exposed to light.

They are being continually given out in large quantities and with very great velocities by *radioactive substances* such as uranium and radium; they are produced in large quantities when salts are put into flames, and there is good reason to suppose that *corpuscles* reach us from the sun.

The *corpuscle* is thus very widely distributed, but *wherever it is found, it preserves its individuality, e/m being always equal to a certain constant value*. The *corpuscle* appears to form a part of all kinds of matter under the most diverse conditions; it seems natural therefore to regard it as *one of the bricks of which atoms are built up*.

**Magnitude of the electric charge carried by the corpuscle**

I shall now return to the proof that *the very large value of e/m for the corpuscle, as compared with that for the atom of hydrogen, is due to the smallness of m the mass, and not to the greatness of e the charge*. We can do this by actually measuring the value of *e*, availing ourselves for this purpose of a discovery by C. T. R. Wilson, that a charged particle acts as a nucleus round which water vapor condenses and forms drops of water. If we have air saturated with water vapor and cool it, so that it would be super-saturated if there were no deposition of moisture, we know that if any dust is present, the particles of dust act as nuclei round which the water condenses and we get the familiar phenomena of fog and rain. If the air is quite dust free, we can, however, cool it very considerably without any deposition of moisture taking place. If there is no dust, C. T. R. Wilson has shown that the cloud does not form until the temperature has been lowered to such a point that the supersaturation is about eightfold. When however this temperature is reached, a thick fog forms even in dust-free air.

***When charged particles are present in the gas, Wilson showed that a much smaller amount of cooling is sufficient to produce the fog***, a fourfold super saturation being all that is required when the charged particles are those which occur in a gas when it is in a state in which it conducts electricity. Each of the charged particles becomes the center round which a drop of water forms; the drops form a cloud, and thus the charged particles, however small to begin with, now become visible and can be observed.

The effect of the ***charged particles*** on the formation of a cloud can be shown very distinctly by the following experiment: A vessel which is in contact with water is saturated with moisture at the temperature of the room. This vessel is in communication with a cylinder in which a large piston slides up and down. The piston to begin with is at the top of its travel; by suddenly exhausting the air from below the piston, the pressure of the air above it will force it down with great rapidity, and the air in the vessel will expand very quickly. When, however, air expands, it gets cool; thus, the air in the vessel previously saturated is now super-saturated. If there is no dust present, no deposition of moisture will take place, unless the air is cooled to such a low temperature that the amount of moisture required to saturate it is only about 1/8 of that actually present.

Now the amount of cooling, and therefore of supersaturation, depends upon the travel of the piston; the greater the travel the greater the cooling. Suppose the travel is regulated so that the supersaturation is less than eightfold and greater than fourfold. We now free the air from dust by forming cloud after cloud in the dusty air; as the clouds fall, they carry the dust down with them, just as in nature the air is cleared by showers. We find at last that when we make the expansion no cloud is visible.

The gas is now made in a conducting state by bringing a little radium near the vessel; this fills the gas with large quantities of both positively and negatively electrified particles. On making the expansion now an exceedingly dense cloud is formed. That this is due to the electrification in the gas can be shown by the following experiment:

Along the inside walls of the vessel, we have two vertical insulated plates which can be electrified. If these plates are charged, they will drag the electrified particles out of the gas as fast as they are formed, so that in this way we can get rid of, or at any rate largely reduce, the number of electrified particles in the gas. If the expansion is now made with the plates charged before bringing up the radium, there is only a very small cloud formed.

We can use the drops to find the charge on the particles, for when we know the travel of the piston, we can deduce the amount of supersaturation, and hence the amount of water deposited when the cloud forms. The water is deposited in the form of a number of small drops all of the same size; thus, the number of drops will be the volume of the water deposited divided by the volume of one of the drops. Hence, if we find the volume of one of the drops, we can find the number of drops which are formed round the charged particles. If the particles are not too numerous, each will have a drop round it, and we can thus find the number of electrified particles.

From the rate at which the drops slowly fall we can determine their size. In consequence of the viscosity or friction of the air small bodies do not fall with a constantly accelerated velocity, but soon reach a speed which remains uniform for the rest of the fall; the smaller the body the slower this speed. Sir George Stokes has shown that $v$, the speed at which a drop of rain falls, is given by the formula

$$v = 2/9 \text{ x } ga^2/\mu$$

where $a$ is the radius of the drop, g the acceleration due to gravity, and $\mu$ the coefficient of viscosity of the air.

If we substitute the values g and $\mu$, we get

$$v = 1.28 \text{ x } 10^6 . a^2.$$

Hence if we measure $v$ we can determine $a$, the radius of the drop.

We can in this way find the volume of a drop, and may therefore, as explained above, calculate the number of drops and therefore the number of electrified particles.

It is a simple matter to find by electrical methods the total quantity of electricity on these particles; and hence, as we know the number of particles, we can deduce at once the charge on each particle.

This was the method by which I first determined *the charge on the particle*; H. A. Wilson has since used a simpler method founded on the following principles: C. T. R. Wilson has shown that the drops of water condense more easily on negatively electrified particles than on positively electrified ones. Thus, by adjusting the expansion, it is possible to get drops of water round the negative particles and not round the positive; with this expansion, therefore, all the drops are negatively electrified. The size of these drops and therefore their weight can, as before, be determined by measuring the speed at which they fall under gravity. Suppose now, that we hold above the drops a positively electrified body: then, since the drops are negatively electrified, they will be attracted towards the positive electricity, and thus the downward force on the drops will be diminished and they will not fall so rapidly as they did when free from electrical attraction. If we adjust the electrical attraction so that the upward force on each drop is equal to the weight of the drop, the drops will not fall at all, but will, like Mahornet's coffin, remain suspended between heaven and earth. If then we adjust the electrical force until the drops are in equilibrium and neither fall nor rise, we know that the upward force on each drop is equal to the weight of the drop, which we have already determined by measuring the rate of fall when the drop was not exposed to any electrical force. If $X$ is the electrical force, $e$ the charge on the drop, and $w$ its weight, we have, when there is equilibrium,

$$X e = w.$$

Since $X$ can easily be measured and $w$ is known, we can use this relation to determine $e$, the charge on the drop. The value of $e$, found by these methods, is $3.0 \times 10^{-10}$ electrostatic units, or $10^{-20}$ electromagnetic units. This value is the same as that of the charge carried by a hydrogen atom in the electrolysis of dilute solutions, an approximate value of which has been long known.

It might be objected that the charge measured in the preceding experiments is the charge on a molecule or collection of molecules of the gas, and not the charge on a *corpuscle*.

This objection does not, however, apply to another form in which I tried the experiment, where the charges on the *corpuscles* were got, not by exposing the gas to the effects of radium, but by allowing ultraviolet light to fall on a metal plate in contact with the gas. *In this case, as experiments made in a very high vacuum show, the electrification, which is entirely negative, escapes from the metal in the form of corpuscles*. When a gas is present, the *corpuscles* strike against the molecules of the gas and stick to them.

Thus, though it is the molecules which are charged, the charge on a molecule is equal to the charge on a *corpuscle*, and when we determine the charge on the molecules by the methods I have just described, we determine the charge carried by the *corpuscle*.

The value of the charge when the electrification is produced by ultraviolet light is the same as when the electrification is produced by radium.

We have just seen that $e$, the charge on the *corpuscle*, is in electromagnetic units equal to $10^{-20}$, and we have previously found that $e/m$, $m$ being the mass of a corpuscle, is equal to $1.7 \times 10^7$, hence $m = 6 \times 10^{-28}$ grammes.

We can realize more easily what this means if we express the mass of the *corpuscle* in terms of the mass of the atom of hydrogen.

We have seen that for the ***corpuscle*** $e/m = 1.7 \times 10^7$. If $E$ is the charge carried by an atom of hydrogen in the electrolysis of dilute solutions, and $M$ is the mass of the hydrogen atom, $E/M = 10^4$; hence $e/m = 1,700\ E/M$.

We have already stated that the value of $e$ found by the preceding methods agrees well with the value of $E$ which has long been approximately known. Townsend has used a method in which the value of $e/E$ is directly measured, and has shown in this way also that $e$ equal to $E$. Hence, since $e/m = 1,700\ E/M$, we have $M = 1,700\ m$, i.e. ***the mass of a corpuscle is only about 1/1,700 part of the mass of the hydrogen atom***.

In all known cases in which negative electricity occurs in gases at very low pressures, it occurs in the form of ***corpuscles***, small bodies with an invariable charge and mass. The case is entirely different with positive electricity.

[  499  ]

LVIII. *Collision of α Particles with Light Atoms. By C. G. Darwin, M.A., Reader in Mathematical Physics, University of Manchester* [*].

1. IN order to account for the fact that α particles are sometimes deflected through large angles, Rutherford † put forward the hypothesis that the positive electricity in an atom carries its mass and is concentrated in an excessively small region. The experiments of Geiger and Marsden ‡ fully confirmed this theory, and they succeeded in calculating the nuclear charge for a few substances. They found it to be about half the atomic weight multiplied by the electronic charge. Rutherford's calculations apply to the case where an α particle collides with a heavy atom, and, for simplicity, he assumes that the collision does not set the atom in motion. His work requires modification to include the fact that the nucleus of the atom is set in motion, and it is this question that is treated in the present paper.

2. We first require the result of a single collision. In finding this we may neglect the effects of the electrons in the atom on account of their small mass. Let $EM$ be the charge and mass of the α particle, $em$ of the nucleus. Let $V$ be the initial velocity of the α particle. The nucleus starts from rest. There are four unknown quantities to be determined by the collision, the velocity $v$ and deflexion $\phi$ of the α particle, and the velocity $u$ and direction of motion of the nucleus. The latter is in the plane determined by the initial and final motions of the α particle. Let it be at angle $\theta$ from the initial direction. For convenience $\theta$ is measured in the opposite direction from $\phi$. When one of these quantities is known the other three can be determined by the principles of momentum and energy. To find the frequency of a given set of values of $v$ &c. we must calculate the orbits.

The equations of momentum and energy are :—

$$\left.\begin{array}{l} MV = Mv \cos \phi + mu \cos \theta, \\ 0 = Mv \sin \phi - mu \sin \theta, \\ MV^2 = Mv^2 + mu^2. \end{array}\right\} \quad . \quad . \quad . \quad . \quad (1)$$

* Communicated by Sir Ernest Rutherford, F.R.S.
† Rutherford, Phil. Mag. vol. xxi. p. 669 (1911).
‡ Geiger and Marsden, Phil. Mag.

# Bohr, N. H. D. (October 7, 1885 – November 18, 1962)

Bohr was a Danish physicist who made foundational contributions to understanding atomic structure and quantum theory, for which he received the Nobel Prize in Physics in 1922. Bohr was also a philosopher and a promoter of scientific research.

*Bohr developed the Bohr model of the atom, in which he proposed that energy levels of electrons are discrete and that the electrons revolve in stable orbits around the atomic nucleus but can jump from one energy level (or orbit) to another.* Although the Bohr model has been supplanted by other models, its underlying principles remain valid. He conceived the principle of complementarity: that items could be separately analyzed in terms of contradictory properties, like behaving as a wave or a stream of particles. The notion of complementarity dominated Bohr's thinking in both science and philosophy.

Bohr was born in Copenhagen, Denmark, on 7 October 1885, the second of three children of Christian Bohr, a professor of physiology at the University of Copenhagen, and Ellen Bohr (née Adler), who was the daughter of David B. Adler from the wealthy Danish Jewish Adler banking family. He had an elder sister, Jenny, and a younger brother Harald. Jenny became a teacher, while Harald became a mathematician and footballer who played for the Danish national team at the 1908 Summer Olympics in London. Niels was a passionate footballer as well, and the two brothers played several matches for the Copenhagen-based Akademisk Boldklub (Academic Football Club), with Niels as goalkeeper.

Bohr was educated at Gammelholm Latin School, starting when he was seven. In 1903, Bohr enrolled as an undergraduate at Copenhagen University. His major was physics, which he studied under Professor Christian Christiansen, the university's only professor of physics at that time. He also studied astronomy and mathematics under Professor Thorvald Thiele, and philosophy under Professor Harald Høffding, a friend of his father.

In 1905 a gold medal competition was sponsored by the Royal Danish Academy of Sciences and Letters to investigate a method for measuring the surface tension of liquids that had been proposed by Lord Rayleigh in 1879. This involved measuring the frequency of oscillation of the radius of a water jet. Bohr conducted a series of experiments using his father's laboratory in the university; the university itself had no physics laboratory. To complete his experiments, he had to make his own glassware, creating test tubes with the required elliptical cross-sections. He went beyond the original task, incorporating improvements into both Rayleigh's theory and his method, by taking into account the viscosity of the water, and by working with finite amplitudes instead of just infinitesimal ones. His essay, which he submitted at the last minute, won the prize. He later submitted an improved version of the paper to the Royal Society in London for publication in the Philosophical Transactions of the Royal Society.

Harald became the first of the two Bohr brothers to earn a master's degree, which he earned for mathematics in April 1909. Niels took another nine months to earn his on the electron theory of metals, a topic assigned by his supervisor, Christiansen. Bohr subsequently elaborated his master's thesis into his much-larger Doctor of Philosophy (dr. phil.) thesis. He surveyed the literature on the subject, settling on a model postulated by Paul Drude and elaborated by Hendrik Lorentz, in which the electrons in a metal are considered to behave like a gas. Bohr extended Lorentz's model, but was still unable to account for phenomena like the Hall effect, and concluded that electron theory could not fully explain the magnetic properties of metals. The thesis was accepted in April 1911, and Bohr conducted his formal defence on 13 May. Harald had received his doctorate the previous year. Bohr's thesis was groundbreaking, but attracted little interest outside Scandinavia because it was written in Danish, a Copenhagen University requirement at the time. In 1921, the Dutch physicist Hendrika Johanna van Leeuwen would independently derive a theorem in Bohr's thesis that is today known as the Bohr–Van Leeuwen theorem.

In 1910, Bohr met Margrethe Nørlund, the sister of the mathematician Niels Erik Nørlund. Bohr resigned his membership in the Church of Denmark on 16 April 1912, and he and Margrethe were married in a civil ceremony at the town hall in Slagelse on 1 August. Years later, his brother Harald similarly left the church before getting married. Bohr and Margrethe had six sons. The oldest, Christian, died in a boating accident in 1934, and another, Harald, died from childhood meningitis. Aage Bohr became a successful physicist, and in 1975 was awarded the Nobel Prize in physics, like his father. Hans became a physician; Erik, a chemical engineer; and Ernest, a lawyer. Like his uncle Harald, Ernest Bohr became an Olympic athlete, playing field hockey for Denmark at the 1948 Summer Olympics in London

In September 1911, Bohr, supported by a fellowship from the Carlsberg Foundation, travelled to England, where most of the theoretical work on the structure of atoms and molecules was being done. He met J. J. Thomson of the Cavendish Laboratory and Trinity College, Cambridge. He attended lectures on electromagnetism given by James Jeans and Joseph Larmor, and did some research on cathode rays, but failed to impress Thomson. He had more success with younger physicists like the Australian William Lawrence Bragg, and New Zealand's Ernest Rutherford, whose 1911 small central nucleus Rutherford model of the atom had challenged Thomson's 1904 plum pudding model. Bohr received an invitation from Rutherford to conduct post-doctoral work at Victoria University of Manchester, where Bohr met George de Hevesy and Charles Galton Darwin (whom Bohr referred to as "the grandson of the real Darwin").

Bohr returned to Denmark in July 1912 for his wedding, and travelled around England and Scotland on his honeymoon. On his return, he became a *privatdocent* at the University of Copenhagen, giving lectures on thermodynamics, and, after being promoted to a *docent* in 1913, began teaching medical students.

In 1913, in an attempt to develop certain outlines of a *theory of line–spectra* based on a suitable application of the fundamental ideas introduced by Planck in his *theory of temperature-radiation* to *the theory of the nucleus atom* of Rutherford, Bohr demonstrated that it is possible in this way to obtain a sample interpretation of some of the main laws governing the line-spectra of the elements, and especially to obtain a deduction of the well-known Balmer formula for the hydrogen spectrum.

The first of his three papers, which later became famous as "the trilogy", was given before the Physical Society of Copenhagen in December 1913. ***He adapted Rutherford's nuclear structure to Planck's quantum theory and so created his model of the atom.*** [Bohr, N. (1913). *On the Constitution of Atoms and Molecules, Part I. Phil. Mag.*, 26, 151, 1-24; https://doi.org/10.1080/14786441308634955; Bohr, N. (1913). *On the Constitution of Atoms and Molecules, Part II. Systems Containing Only a Single Nucleus. Phil. Mag.*, 26, 153, 476-502; https://doi.org/10.1080/14786441308634993; Bohr, N. (1913). *On the Constitution of Atoms and Molecules, Part III Systems containing several nuclei. Phil. Mag.*, 26, 155, 857-75; https://doi.org/10.1080/14786441308635031. See below.]

Planetary models of atoms were not new, but Bohr's treatment was. Taking the 1912 paper by Darwin* on the role of electrons in the interaction of alpha particles with a nucleus as his starting point, he advanced the theory of electrons travelling in orbits of quantized "stationary states" around the atom's nucleus in order to stabilize the atom.

[* Darwin, C. G. (1912). A theory of the absorption and scattering of the alpha rays. *Phil. Mag.*, 23, 138, 901-20; https://doi.org/10.1080/14786440608637291.]

***Bohr's great achievement was the synthesis of Rutherford's atom model with Planck's quantum hypothesis.*** He supposed the atom to consist of a nucleus with a positive charge Ze and Z electrons with charge – e each, moving according to the laws of classical mechanics. He introduced the idea that an electron could drop from a higher-energy orbit to a lower one, in the process emitting a quantum of discrete energy. This became a basis for what is now known as the *old quantum theory*. From a set of assumptions concerning the stationary state of an atom and the frequency of the radiation emitted or absorbed when the atom passes from one such state to another, he showed that it is possible to obtain a simple interpretation of the main laws governing the line spectra of the elements, and especially to deduce the Balmer formula for the hydrogen spectrum.

[In 1885, Johann Balmer had come up with his Balmer series to describe the visible spectral lines of a hydrogen atom:

$$1/\lambda = R_H \left(1/2^2 - 1/n^2\right) \text{ for } n = 3, 4, 5, \ldots$$

where $\lambda$ is the wavelength of the absorbed or emitted light and $R_H$ is the Rydberg constant. Balmer's formula was corroborated by the discovery of additional spectral lines, but for thirty years, no one could explain why it worked.]

In the first paper of his trilogy, Bohr was able to derive it from his model:

$$R_Z = 2\pi^2 m_e Z^2 e^4/h^3$$

where $m_e$ is the electron's mass, e is its charge, h is Planck's constant and Z is the atom's atomic number (1 for hydrogen).

Many older physicists, like Thomson, Rayleigh and Hendrik Lorentz, did not like the trilogy, but the younger generation, including Rutherford, David Hilbert, Albert Einstein, Enrico Fermi, Max Born and Arnold Sommerfeld saw it as a breakthrough. Its acceptance was entirely due to its ability to explain phenomena which stymied other models, and to predict results that were subsequently verified by experiments.

Bohr did not enjoy teaching medical students. In 1914, he decided to return to Manchester, where Rutherford had offered him a job as a reader in place of Darwin, whose tenure had expired. Bohr accepted. He took a leave of absence from the University of Copenhagen, which he started by taking a holiday in Tyrol with his brother and aunt. There, he visited the University of Göttingen and the Ludwig Maximilian University of Munich, where he met Sommerfeld and conducted seminars on the trilogy. The First World War broke out while they were in Tyrol, greatly complicating the trip back to Denmark and Bohr's subsequent voyage with Margrethe to England, where he arrived in October 1914.

In 1916, after two years in Manchester, Bohr was appointed to the Chair of Theoretical Physics at the University of Copenhagen, a position created especially for him. In April 1917 Bohr began a campaign to establish an Institute of Theoretical Physics. He gained the support of the Danish government and the Carlsberg Foundation, and sizeable contributions were also made by industry and private donors, many of them Jewish. Legislation establishing the institute was passed in November 1918. Now known as the Niels Bohr Institute, it opened on 3 March 1921, with Bohr as its director. His family moved into an apartment on the first floor.

Bohr's institute served as a focal point for researchers into quantum mechanics and related subjects in the 1920s and 1930s, when most of the world's best known theoretical physicists spent some time in his company. Bohr became widely appreciated as their congenial host and eminent colleague.

The Bohr model worked well for hydrogen and ionized single electron Helium which impressed Einstein, but could not explain more complex elements. By 1919, Bohr was moving away from the idea that electrons orbited the nucleus and developed heuristics to describe them. The rare-earth elements posed a particular classification problem for chemists, because they were so chemically similar.

Bohr founded the Institute of Theoretical Physics at the University of Copenhagen, now known as the Niels Bohr Institute, which opened in 1920. Bohr mentored and collaborated with physicists including Hans Kramers, Oskar Klein, George de Hevesy, and Werner Heisenberg. He predicted the existence of a new zirconium-like element, which was named hafnium, after the Latin name for Copenhagen, where it was discovered. Later, the element bohrium was named after him.

*It was not until his 1921 paper that he showed that the chemical properties of each element were largely determined by the number of electrons in the outer orbits of its atoms*. [Bohr, N. (1921). *The structure of the atom and the physical and chemical properties of the elements*. (Based on an address given before a joint meeting of the Physical and Chemical Societies of Copenhagen on the 18th of October 1921.) *Fysisk Tidsskrift*, xix. p. 153, 1921. Translation in Bohr, N. (1922). *The Theory of Spectra and Atomic Constitution*. Cambridge University Press.]

An important development came in 1924 with Wolfgang Pauli's discovery of the *Pauli exclusion principle*, which put Bohr's models on a firm theoretical footing.

*In 1922 Bohr was awarded the Nobel Prize in Physics "for his services in the investigation of the structure of atoms and of the radiation emanating from them"*. The award thus recognized both the Trilogy and his early leading work in the emerging field of quantum mechanics. *For his Nobel lecture, Bohr gave his audience a comprehensive survey of what was then known about the structure of the atom, including the correspondence principle, which he had formulated*.

This states that the behavior of systems described by quantum theory reproduces classical physics in the limit of large quantum numbers.

The discovery of Compton scattering by Arthur Holly Compton in 1923 convinced most physicists that light was composed of photons, and that energy and momentum were conserved in collisions between electrons and photons. In 1924, Bohr, Kramers and John C. Slater, an American physicist working at the Institute in Copenhagen, proposed the Bohr–Kramers–Slater theory (BKS). It was more a program than a full physical theory, as the ideas it developed were not worked out quantitatively. BKS theory became the final attempt at understanding the interaction of matter and electromagnetic radiation on the basis of the *old quantum theory*, in which quantum phenomena were treated by imposing quantum restrictions on a classical wave description of the electromagnetic field.

Modelling atomic behavior under incident electromagnetic radiation using "virtual oscillators" at the absorption and emission frequencies, rather than the (different) apparent frequencies of the Bohr orbits, led Max Born, Werner Heisenberg and Kramers to explore different mathematical models. They led to the development of *matrix mechanics*, the first form of modern quantum mechanics. The BKS theory also generated discussion of, and renewed attention to, difficulties in the foundations of the old quantum theory. The most provocative element of BKS – that momentum and energy would not necessarily be conserved in each interaction, but only statistically – was soon shown to be in conflict with experiments conducted by Walther Bothe and Hans Geiger. In light of these results, Bohr informed Darwin that "there is nothing else to do than to give our revolutionary efforts as honorable a funeral as possible".

Heisenberg first came to Copenhagen in 1924, then returned to Göttingen in June 1925, shortly thereafter developing the mathematical foundations of quantum mechanics. When he showed his results to Max Born in Göttingen, Born realized that they could best be expressed using matrices.

The introduction of spin by George Uhlenbeck and Samuel Goudsmit in November 1925 was a milestone. The next month, Bohr travelled to Leiden to attend celebrations of the 50th anniversary of Hendrick Lorentz receiving his doctorate. When his train stopped in Hamburg, he was met by Wolfgang Pauli and Otto Stern, who asked for his opinion of the spin theory. Bohr pointed out that he had concerns about the interaction between electrons and magnetic fields. When he arrived in Leiden, Paul Ehrenfest and Albert Einstein informed Bohr that Einstein had resolved this problem using relativity. Bohr then had Uhlenbeck and Goudsmit incorporate this into their paper.

*This work attracted the attention of the British physicist Paul Dirac, who came to Copenhagen for six months in September 1926.* Austrian physicist Erwin Schrödinger also visited in 1926. His attempt at explaining quantum physics in classical terms using wave mechanics impressed Bohr, who believed it contributed "so much to mathematical clarity and simplicity that it represents a gigantic advance over all previous forms of quantum mechanics".

When Kramers left the institute in 1926 to take up a chair as professor of theoretical physics at the Utrecht University, Bohr arranged for Heisenberg to return and take Kramers' place as a lektor at the University of Copenhagen. Heisenberg worked in Copenhagen as a university lecturer and assistant to Bohr from 1926 to 1927.

Bohr became convinced that light behaved like both waves and particles and, in 1927, experiments confirmed the de Broglie hypothesis that matter (like electrons) also behaved like waves. He conceived the philosophical principle of complementarity: that items could have apparently mutually exclusive properties, such as being a wave or a stream of particles, depending on the experimental framework.

*In February 1927, Heisenberg developed the first version of the uncertainty principle*, presenting it using a thought experiment where an electron was observed through a gamma-ray microscope. Bohr was dissatisfied with Heisenberg's argument, since it required only that a measurement disturb properties that already existed, rather than the more radical idea that the electron's properties could not be discussed at all apart from the context they were measured in. In a paper presented at the Volta Conference at Como in September 1927,

Bohr emphasized that Heisenberg's uncertainty relations could be derived from classical considerations about the resolving power of optical instruments. Understanding the true meaning of complementarity would, Bohr believed, require "closer investigation". Einstein preferred the determinism of classical physics over the probabilistic new quantum physics to which he himself had contributed.

In 1914 Carl Jacobsen, the heir to Carlsberg breweries, bequeathed his mansion to be used for life by the Dane who had made the most prominent contribution to science, literature or the arts, as an honorary residence. Harald Høffding had been the first occupant, and upon his death in July 1931, the Royal Danish Academy of Sciences and Letters gave Bohr occupancy. He and his family moved there in 1932. He was elected president of the Academy on 17 March 1939.

During the 1930s, Bohr helped refugees from Nazism. After Denmark was occupied by the Germans, he had a famous meeting with Heisenberg, who had become the head of the German nuclear weapon project. In September 1943 word reached Bohr that he was about to be arrested by the Germans, and he fled to Sweden. From there, he was flown to Britain, where he joined the British Tube Alloys nuclear weapons project, and was part of the British mission to the Manhattan Project. After the war, Bohr called for international cooperation on nuclear energy. He was involved with the establishment of CERN and the Research Establishment Risø of the Danish Atomic Energy Commission and became the first chairman of the Nordic Institute for Theoretical Physics in 1957.

By 1929 the phenomenon of beta decay prompted Bohr to again suggest that the law of conservation of energy be abandoned, but Enrico Fermi's hypothetical neutrino and the subsequent 1932 discovery of the neutron provided another explanation. This prompted Bohr to create a new theory of the compound nucleus in 1936, which explained how neutrons could be captured by the nucleus. In this model, the nucleus could be deformed like a drop of liquid. He worked on this with a new collaborator, the Danish physicist Fritz Kalckar, who died suddenly in 1938.

The discovery of nuclear fission by Otto Hahn in December 1938 (and its theoretical explanation by Lise Meitner) generated intense interest among physicists. Bohr brought the news to the United States where he opened the Fifth Washington Conference on Theoretical Physics with Fermi on 26 January 1939. When Bohr told George Placzek that this resolved all the mysteries of transuranic elements, Placzek told him that one remained: the neutron capture energies of uranium did not match those of its decay. Bohr thought about it for a few minutes and then announced to Placzek, Léon Rosenfeld and John Wheeler that "I have understood everything." Based on his liquid drop model of the nucleus, Bohr concluded that it was the uranium-235 isotope and not the more abundant uranium-238 that was primarily responsible for fission with thermal neutrons.

Bohr was elected president of the Royal Danish Academy of Sciences and Letters on 17 March 1939.

In April 1940, John R. Dunning demonstrated that Bohr was correct. In the meantime, Bohr and Wheeler developed a theoretical treatment which they published in a September 1939 paper on "The Mechanism of Nuclear Fission".

The rise of Nazism in Germany prompted many scholars to flee their countries, either because they were Jewish or because they were political opponents of the Nazi regime. Bohr offered the refugees temporary jobs at the institute, provided them with financial support, arranged for them to be awarded fellowships from the Rockefeller Foundation, and ultimately found them places at institutions around the world. In April 1940, early in the Second World War, Nazi Germany invaded and occupied Denmark. Bohr kept the Institute running, but all the foreign scholars departed.

Bohr was aware of the possibility of using uranium-235 to construct an atomic bomb, referring to it in lectures in Britain and Denmark shortly before and after the war started, but he did not believe that it was technically feasible to extract a sufficient quantity of uranium-235. In September 1941, Heisenberg, who had become head of the German nuclear energy project, visited Bohr in Copenhagen. During this meeting the two men took a private moment outside, the content of which has caused much speculation, as both gave differing accounts.

According to Heisenberg, he began to address nuclear energy, morality and the war, to which Bohr seems to have reacted by terminating the conversation abruptly while not giving Heisenberg hints about his own opinions. In 1957, Heisenberg wrote to Robert Jungk, who was then working on the book *Brighter than a Thousand Suns: A Personal History of the Atomic Scientists*. Heisenberg explained that he had visited Copenhagen to communicate to Bohr the views of several German scientists, that production of a nuclear weapon was possible with great efforts, and this raised enormous responsibilities on the world's scientists on both sides. When Bohr saw Jungk's depiction in the Danish translation of the book, he drafted (but never sent) a letter to Heisenberg, stating that he never understood the purpose of Heisenberg's visit, was shocked by Heisenberg's opinion that Germany would win the war, and that atomic weapons could be decisive.

In September 1943, word reached Bohr and his brother Harald that the Nazis considered their family to be Jewish, since their mother was Jewish, and that they were therefore in danger of being arrested. The Danish resistance helped Bohr and his wife escape by sea to Sweden on 29 September.

When the news of Bohr's escape reached Britain, Lord Cherwell sent a telegram to Bohr asking him to come to Britain. Bohr arrived in Scotland on 6 October in a de Havilland Mosquito operated by the British Overseas Airways Corporation (BOAC). The Mosquitos were unarmed high-speed bomber aircraft that had been converted to carry small, valuable cargoes or important passengers. By flying at high speed and high altitude, they could cross German-occupied Norway, and yet avoid German fighters. Bohr was warmly received by James Chadwick and Sir John Anderson, but for security reasons Bohr was kept out of sight. He was given an apartment at St James's Palace and an office with the British Tube Alloys nuclear weapons development team. Bohr was astonished at the amount of progress that had been made.

Chadwick arranged for Bohr to visit the United States as a Tube Alloys consultant. On 8 December 1943, Bohr arrived in Washington, D.C., where he met with the director of the Manhattan Project, Brigadier General Leslie R. Groves, Jr. He visited Einstein and Pauli at the Institute for Advanced Study in Princeton, New Jersey, and went to Los Alamos in New Mexico, where the nuclear weapons were being designed.

Bohr did not remain at Los Alamos, but paid a series of extended visits over the course of the next two years. Robert Oppenheimer credited Bohr with acting "as a scientific father figure to the younger men", most notably Richard Feynman. Bohr is quoted as saying, "They didn't need my help in making the atom bomb." Oppenheimer gave Bohr credit for an important contribution to the work on modulated neutron initiators. "This device remained a stubborn puzzle," Oppenheimer noted, "but in early February 1945 Niels Bohr clarified what had to be done."

Oppenheimer suggested that Bohr visit President Franklin D. Roosevelt to convince him that the Manhattan Project should be shared with the Soviets in the hope of speeding up its results. Bohr's friend, Supreme Court Justice Felix Frankfurter, informed President Roosevelt about Bohr's opinions, and a meeting between them took place on August 26, 1944. Roosevelt suggested that Bohr return to the United Kingdom to try to win British approval. When Churchill and Roosevelt met at Hyde Park on September 19, 1944, they rejected the idea of informing the world about the project, and the aide-mémoire of their conversation contained a rider that "enquiries should be made regarding the activities of Professor Bohr and steps taken to ensure that he is responsible for no leakage of information, particularly to the Russians".

With the war now ended, Bohr returned to Copenhagen on August 25, 1945, and was re-elected President of the Royal Danish Academy of Arts and Sciences on September 21.

Bohr died of heart failure at his home in Carlsberg on November 18, 1962.

## Bohr, N. (1913). On the Constitution of Atoms and Molecules, Part I. – Binding of Electrons by Positive Nuclei.

[*Phil. Mag.*, 26, 151, 1-24; https://doi.org/10.1080/14786441308634955.]

Received July, 1913.

Bohr took the next step by adapting Rutherford's theory of atomic structure to Planck's quantum hypothesis creating his model of the atom. *He supposed the atom to consist of a nucleus with a positive charge Ze and Z electrons with charge – e each, moving according to the laws of classical mechanics. He introduced the idea that an electron could drop from a higher-energy orbit to a lower one, in the process emitting a quantum of discrete energy*. This became a basis for what is now known as the *old quantum theory*. From a set of assumptions concerning the stationary state of an atom and the frequency of the radiation emitted or absorbed when the atom passes from one such state to another, he showed that it is possible to obtain a simple interpretation of the main laws governing the line spectra of the elements, and especially to deduce the Balmer formula for the hydrogen spectrum. Addresses mechanism of the binding of electrons by a positive nucleus in relation to Planck's theory.

### Introduction

In order to explain the results of experiments on scattering of α rays by matter Prof. Rutherford[1] has given a theory of the *structure of atoms.*

[1] Rutherford, E. (1911). The scattering of α and β particles by matter and the structure of the atom. *Phil. Mag.*, 6, 21, 669-88; https://www.chemteam.info/Chem-History/ Rutherford-1911.html.

According to this theory, *the atom consists of a positively charged nucleus surrounded by a system of electrons kept together by attractive forces from the nucleus; the total negative charge of the electrons is equal to the positive charge of the nucleus*. Further, the *nucleus* is assumed to be the seat of the essential part of the mass of the atom, and to have linear dimensions exceedingly small compared with the linear dimensions of the whole atom. The number of *electrons* in an atom is deduced to be approximately equal to half the atomic weight. Great interest is to be attributed to this atom-model; for, as Rutherford has shown, the assumption of the existence of nuclei, as those in question, seems to be necessary in order to account for the results of the experiments on large angle scattering of the α rays.[2]

[2] See also Geiger, H. & Marsden, E. (April, 1913). The Laws of Deflexion of a Particles through Large Angles. *Phil. Mag.*, 6, 25, 148, 604-23; https://www.chemteam.info/Chem-History/GeigerMarsden-1913/GeigerMarsden-1913.html.

In an attempt to explain some of the properties of matter on the basis of this atom-model we meet, however, with difficulties of a serious nature arising from the ***apparent instability of the system of electrons***: difficulties purposely avoided in atom-models previously considered, for instance, in the one proposed by Sir. J. J. Thomson[3].

[3] Thomson, J. J. (March, 1904). *On the Structure of the Atom: an Investigation of the Stability and Periods of Oscillation of a number of Corpuscles arranged at equal intervals around the Circumference of a Circle; with Application of the Results to the Theory of Atomic Structure.* Phil. Mag., 6, 7, 39, 237-265; http://dx.doi.org/10.1080/1478644040 9463107.

According to the theory of the latter the atom consists of a sphere of uniform positive electrification, inside which the electrons move in circular orbits.

The principal ***difference between the atom-models proposed by Thomson and Rutherford*** consists in the circumstance that the forces acting on the electrons in the atom-model of Thomson allow of certain configurations and motion of the electrons for which the system is in a stable equilibrium; such configurations, however, apparently do not exist for the second atom-model. The nature of the difference in question will perhaps be most clearly seen by noticing that among the quantities characterizing the first atom a quantity appears – the radius of the positive sphere – of dimensions of a length and of the same order of magnitude as the linear extension of the atom, while such a length does not appear among the quantities characterizing the second atom, viz. the charges and masses of the electrons and the positive nucleus; nor can it be determined solely by help of the latter quantities.

The way of considering a problem of this kind has, however, undergone essential alterations in recent years owing to the development of the theory of the energy radiation, and the direct affirmation of the new assumptions introduced in this theory, found by experiments on very different phenomena such as specific heats, photoelectric effect, Rontgen-rays, &c. ***The result of the discussion of these questions seems to be a general acknowledgment of the inadequacy of the classical electrodynamics in describing the behavior of system of atomic size.***[4]

[4] See for instance (1912). *La théorie du rayonnement et les quanta.* (Radiation theory and quanta.) Rapports et discussions de la réunion tenue à Bruxelles, du 30 octobre au 3 novembre 1911, sous les auspices de M.E. Solvay. Paris, 1912.

Whatever the alteration in the laws of motion of the electrons may be, ***it seems necessary to introduce in the laws in question a quantity foreign to the classical electrodynamics, i.e., Planck's constant, or as it often is called the elementary quantum of action***. By the introduction of this quantity the question of the stable configuration of the electrons in the atoms is essentially changed, as this constant is of such dimensions and magnitude that it, together with the mass and charge of the particles, can determine a length of the order of magnitude required.

***This paper is an attempt to show that the application of the above ideas to Rutherford's atom-model affords a basis for a theory of the constitution of atoms***. It will further be shown that from this theory we are led to a theory of the ***constitution of molecules***. In the present ***first part of the paper the mechanism of the binding of electrons by a positive nucleus is discussed in relation to Planck's theory***. It will be shown that it is possible from the point of view taken to account in a simple way for the law of the line spectrum of hydrogen. Further, reasons are given for a principal hypothesis on which the considerations contained in the following parts are based.

I wish here to express my thinks to Prof. Rutherford for his kind and encouraging interest in this work.

## Part I. – Binding of Electrons by Positive Nuclei.

### § 1. General Considerations

The inadequacy of the classical electrodynamics in accounting for the properties of atoms from an atom-model as Rutherford's, will appear very clearly if we consider a simple system consisting of a positively charged nucleus of very small dimensions and an electron describing closed orbits around it. For simplicity, let us assume that the mass of the electron is negligibly small in comparison with that of the nucleus, and further, that the velocity of the electron is small compared with that of light.

Let us at first assume that there is no energy radiation. In this case the electron will describe stationary elliptical orbits. The frequency of revolution $\omega$ and the major-axis of the orbit $2a$ will depend on the amount of energy W which must be transferred to the system in order to remove the electron to an infinitely great distance apart from the nucleus. Denoting the charge of the electron and of the nucleus by – e and E respectively and the mass of the electron by m, we thus get

$$\omega = \sqrt{2}/\pi \cdot W^{3/2}/eE\sqrt{m}, \qquad 2a = eE/W. \qquad (1)$$

Further, it can easily be shown that the mean value of the kinetic energy of the electron taken for a whole revolution is equal to W. We see that if the value of W is not given, there will be no values of $\omega$ and $a$ characteristic for the system in question.

***Let us now, however, take the effect of the energy radiation into account***, calculated in the ordinary way from the acceleration of the electron. In this case the electron will no longer describe stationary orbits. W will continuously increase, and the electron will approach the nucleus describing orbits of smaller and smaller dimensions, and with greater and greater frequency; the electron on the average gaining in kinetic energy at the same time as the whole system loses energy. This process will go on until the dimensions of the orbit are the same order of magnitude as the dimensions of the electron or those of the nucleus. A simple calculation shows that the energy radiated out during the process considered will be enormously great compared with that radiated out by ordinary molecular processes.

It is obvious that the behavior of such a system will be very different from that of an atomic system occurring in nature. In the first place, the actual atoms in their permanent state to have absolutely fixed dimensions and frequencies. Further, if we consider any process, the result seems always to be that after a certain amount of energy characteristic for the systems in question is radiated out, the system will again settle down in a stable state of equilibrium, in which the distance apart of the particles are of the same order of magnitude as before the process.

Now *the essential point in Planck's theory of radiation is that the energy radiation from an atomic system does not take place in the continuous way assumed in the ordinary electrodynamics, but that it, on the contrary, takes place in distinctly separated emissions*, the amount of energy radiated out from an atomic vibrator of frequency $v$ in a single emission being equal to $\tau h v$, where $\tau$ is an entire number, and h is a universal constant.[5]

[5] See for instance, Planck, M. (1910). *Zur Theorie der Wärmestrahlung.* (On the theory of thermal radiation.) *Ann. Phys.*, 336, 4, 758-68; https://doi.org/10.1002/andp.191033 60406; (1912). *Über die Begründung des Gesetzes der schwarzen Strahlung.* (On the Justification of the Law of Black Radiation.) *Ann. Phys.*, 342, 4, 642-56; https://doi.org/10.1002/andp.19123420403; (1911). *Eine neue Strahlungshypothese.* (A new radiation hypothesis.) *Verh. Phys. Ges.*, 13, 138-48.

Returning to the simple case of an electron and a positive nucleus considered above, let us assume that the electron at the beginning of the interaction with the nucleus was at a great distance apart from the nucleus, and had no sensible velocity relative to the latter. Let us further assume that the electron after interaction has taken place has settled down in a stationary orbit around the nucleus. We shall, for reasons referred to later, assume that the orbit in question is circular: this assumption will, however, make no alteration in the calculations for system containing only a single electron.

Let as now assume that, during the binding of the electron, a homogeneous radiation is emitted of a frequency $v$, equal to half the frequency of revolution of the electron in its final orbit; then from Planck's theory, we might expect that the amount of energy emitted by the process considered is equal to $\tau h v$, where h is Planck's constant an entire number. If we assume that the radiation emitted is homogeneous, the second assumption concerning the frequency of the radiation suggests itself, since the frequency of revolution of the electron at the beginning of the emission is 0. The question, however, of the rigorous validity of both assumptions, and also of the application made of Planck's theory, will be more closely discussed in § 3.

Putting

$$W = \tau h\omega/2, \tag{2}$$

we get by help of the formula (1)

$$W = 2\pi^2 me^2 E^2/\tau^2 h^2, \quad \omega = 4\pi^2 me^2 E^2/\tau^3 h^3, \quad 2a = \tau^2 h^2/2\pi^2 meE. \tag{3}$$

If in these expressions we give $\tau$ different values, we get a series of values for W, $\omega$, and $a$ corresponding to a series of configurations of the system. According to the above considerations, we are led to assume that these configurations will correspond to states of the system in which there is no radiation of energy; states which consequently will be stationary as long as the system is not disturbed from outside. We see that the value of W is greatest if $\tau$ has its smallest value 1. This case will therefore correspond to the most stable of the system, i.e., will correspond to the binding of the electron for the breaking up of which the greatest amount of energy is required.

Putting in the above expressions $\tau = 1$ and $E = e$, and introducing the experimental values

$$e = 4.7 \cdot 10^{-10}, \qquad e/m = 5.31 \cdot 10^{17}, \qquad h = 6.5 \cdot 10^{-27},$$

we get

$$2a = 1.1 \cdot 10^{-8} \text{ cm}, \quad \omega = 6.2 \cdot 10^{15} \text{ 1/sec}, \quad W/e = 13 \text{ volt}.$$

*We see that these values are of the same order of magnitude as the linear dimensions of the atoms, the optical frequencies, and the ionization-potentials*. The general importance of Planck's theory for the discussion of the behavior of atomic system was originally pointed out by Einstein.[6]

[6] Einstein, A. (1905). *Über einen die Erzeugung und Verwandlung des Lichtes betreffenden heuristischen Gesichtspunkt*. (On a Heuristic Point of View Concerning the Production and Transformation of Light.) *Ann. Physik*, 4, 17, 132-48; Einstein, A. (1906). *Theorie der Lichterzeugung und Lichtabsorption*. (On the Theory of Light Production and Light Absorption.) *Ann. Physik*, 4, 20, 199-206; Einstein, A. (1907). *Theorie der Strahlung und die Theorie der Spezifischen Warme*. (Theory of radiation and the theory of specific heat.) *Ann. Physik*, 4, 22, 180-90.

…

It will now be attempted to show that the difficulties in question disappear if we consider the problems from the point of view taken in this paper. Before proceeding it may be useful to restate briefly the ideas characterizing the calculations on p. 5. The principal assumptions used are:

(1) That the dynamical equilibrium of the systems in the stationary states can be discussed by help of the ordinary mechanics, while the passing of the systems between different stationary states cannot be treated on that basis.

(2) That the latter is followed by the emission of a homogeneous radiation, for which the relation between the frequency and the amount of energy emitted is the one given by Planck's theory.

The first assumption seems to present itself; for it is known that the ordinary mechanism cannot have an absolute validity, but will only hold in calculations of certain mean values of the motion of the electrons. On the other hand, in the calculations of the dynamical equilibrium in a stationary state in which there is no relative displacement of the particles, we need not distinguish between the actual motions and their mean values. *The second assumption is in obvious conflict with the ordinary ideas of electrodynamics, but appears to be necessary in order to account for experimental facts.*

In the calculations on page 5 we have further made use of the more special assumptions, viz., that the different stationary states correspond to the emission of a different number of Planck's energy-quanta, and that the frequency of the radiation emitted during the passing of the system from a state in which no energy is yet radiated out to one of the stationary states, is equal to half the frequency of revolution of the electron in the latter state. We can, however (see § 3), also arrive at the expressions (3) for the stationary states by using assumptions of somewhat different from. We shall, therefore, postpone the discussion of the special assumptions, and first show how by the help of the above principal assumptions, and of the expressions (3) for the stationary states, we can account for the line-spectrum of hydrogen.

## § 2. Emission of Line-spectra

*Spectrum of Hydrogen. – General evidence indicates that an atom of hydrogen consists simply of a single electron rotating round a positive nucleus of charge e.* [9]

> [9] See for instance Bohr, N. (1913). *Phil. Mag.*, 25, 24. The conclusion drawn in the paper cited in strongly supported by the fact that hydrogen, in the experiments on positive rays of Sir. J. J. Thomson, is the only element which never occurs with a positive charge corresponding to the loss of more than one electron (compare Bohr, N. (1912). *Phil. Mag.*, 24, 672).

The reformation of a hydrogen atom, when the electron has been removed to great distances away from the nucleus – e.g. by the effect of electrical discharge in a vacuum tube – will accordingly correspond to the binding of an electron by a positive nucleus considered on p. 5. If in (3) we put E = e, we get for the total amount of energy radiated out by the formation of one of the stationary states,

$$W_\tau = 2\pi^2 m e^4 / \tau^2 h^2.$$

The amount of energy emitted by the passing of the system from a state corresponding to $\tau = \tau_1$ to one corresponding to $\tau = \tau_2$, is consequently

$$W_{\tau 2} - W_{\tau 1} = 2\pi^2 m e^4 / h^2 \cdot (1/\tau_2^2 - 1/\tau_1^2).$$

If now we suppose that the radiation is question is homogeneous, and that the amount of energy emitted is equal to $h\nu$, where $\nu$ is the frequency of the radiation, we get

$$W_{\tau 2} - W_{\tau 1} = h\nu$$

and from this

$$\nu = 2\pi^2 m e^4 / h^3 \cdot (1/\tau_2^2 - 1/\tau_1^2). \tag{4}$$

***We see that this expression accounts for the law connecting the lines in the spectrum of hydrogen***. If we put $\tau_2 = 2$ and let $\tau_1$ vary, we get the ordinary Balmer series. If we put $\tau_3 = 3$, we get the series in the ultra-red observed by Paschen[10] and previously suspected by Ritz.

[10] Paschen, F. (1908). *Ann. Phys.*, 27, 565.

If we put $\tau_2 = 1$ and $\tau = 4, 5, \ldots$ , we get series respectively in the extreme ultraviolet and the extreme ultra-red, which are not observed, but the existence of which may be expected.

The agreement in question is quantitative as well as qualitative. Putting
$$e = 4.7 \cdot 10^{-10}, \qquad e/m = 5.31 \cdot 10^{17} \qquad \text{and } h = 6.5 \cdot 10^{-27},$$
we get
$$2\pi^2 m e^4 / h^3 = 3.1 \cdot 10^{15}.$$

The observed value for the factor outside the bracket in the formula (4) is $3.290 \cdot 10^{15}$. ***The agreement between the theoretical and observed values is inside the uncertainty due to experimental errors in the constants entering in the expression for the theoretical value.*** We shall in § 3 return to consider the possible importance of the agreement in question.

It may be remarked that the fact, that it has not been possibly to observe more than 12 lines of the Balmer series in experiments with vacuum tubes, while 33 lines are observed in the spectra of some celestial bodies, is just what we should expect from the above theory. According to the equation (3) the diameter of the orbit of the electron in the different stationary states is proportional to $\tau^2$. For $\tau = 12$ the diameter is equal to $1.6 \cdot 10^{-6}$ cm, or equal to mean distance between the molecules in a gas at a pressure of about 7 mm mercury; for $\tau = 33$ the diameter is equal to $1.2 \cdot 10^{-5}$ cm, corresponding to the mean distance of the molecules at a pressure of about 0.02 mm mercury. According to the theory the necessary condition for the appearance of a great number of lines is therefore a very small density of the gas; for simultaneously to obtain an intensity sufficient for observation the space filled with the gas must be very great. If the theory is right, we may therefore never expect to be able in experiments with vacuum tubes to observe the lines corresponding to high numbers of the Balmer series of the emission spectrum of hydrogen; it might, however, be possible to observe the lines by investigation of the absorption spectrum of this gas. (see § 4).

It will be observed that we in the above way do not obtain other series of lines, generally ascribed to hydrogen; for instance, the series first observed by Pickering[11] in the spectrum of the star ζ Puppis, and the set of series recently found by Fowler[12] by experiments with vacuum tubes containing a mixture of hydrogen and helium.

[11] Pickering, E. C. (1897). *Astrophys. J.* IV., p. 369 (1896); v. p. 92.
[12] Fowler, A. (Dec, 1912). *Month. Not. Roy. Astr. Soc.*, LXXIII.

We shall, however, see that, by help of the above theory, ***we can account naturally for these series of lines if we ascribe them to helium.***

A neutral atom of the latter element consists, according to Rutherford's theory, of a positive nucleus of charge 2e and two electrons. Now considering the binding of a single electron by a helium nucleus, we get putting E = 2e in the expressions (3) on page 5, and proceeding in in exactly the same way as above,

$$\nu = 8\pi^2 me^4/h^3 \cdot (1/\tau_2^2 - 1/\tau_1^2) = 2\pi^2 me^4/h^3 \cdot \{1/(\tau_2/2)^2 - 1/(\tau_1/2^2)\}.$$

If we in this formula put $\tau_1 = 1$ or $\tau_2 = 2$, we get series of lines in the extreme ultra-violet. If we put $\tau_2 = 3$, and let $\tau_1$ vary, we get a series which includes 2 of the series observed by Fowler, and denoted by him as the first and second principal series of the hydrogen spectrum. If we put $\tau_2 = 4$, we get the series observed by Pickering in the spectrum of ζ Puppis. Every second of the lines in this series is identical with a line in the Balmer series of the hydrogen spectrum; the presence of hydrogen in the star in question may therefore account for the fact that these lines are of a greater intensity than the rest of the lines in the series. The series is also observed in the experiments of Fowler, and denoted in his paper as the Sharp series of the hydrogen spectrum. If we finally in the above formula put $\tau_2 = 5$, 6, …, we get series, the strong lines of which are to be expected in the ultra-red.

The reason why the spectrum considered is not observed in ordinary helium tubes may be that in such tubes the ionization of helium is not so complete in the star considered or in the experiments of Fowler, where a strong discharge was sent through a mixture of hydrogen and helium. ***The condition for the appearance of the spectrum is, according to the above theory, that helium atoms are present in a state in which they have lost both their electrons.*** Now we must assume that the amount of energy to be used in removing the second electron from a helium atom is much greater than that to be used in removing the first. Further, it is known from experiments on positive rays, that hydrogen atoms can acquire a negative charge; therefore, the presence of hydrogen in the experiments of Fowler may affect that more electrons are removed from some of the helium atoms than would be the case if only helium were present.

***Spectra of other substances***. — in case of systems containing more electrons we must – in conformity with the result of experiments – expect more complicated laws for the line-spectra than those considered. I shall try to show that the point of view taken above allows, at any rate, a certain understanding of the laws observed. According to Rydberg's theory — with the generalization given by Ritz[13] –

[13] Ritz, W. (1908). *Über ein neues Gesetz der Serienspektren. Phys. Zeitschr.*, 9, 521-9; Ritz, W. (1908). *On a new law of series spectra. Astrophys. J.* 28, 237–243. [The *Ritz combination principle* was crucial in making sense of the regularities in the line spectra of atoms. It was a key principle that guided Bohr in constructing a quantum theory of line spectra. Observations of spectral lines revealed that pairs of line frequencies combine (add) to give the frequency of another line in the spectrum. The Ritz combination rule is $v(nk) + v(km) = v(nm)$, which follows from Eqs. (26) and (27). As a universal, exact law of spectroscopy, the *Ritz rule* provided a powerful tool to analyze spectra and to discover new lines. Given the measured frequencies $v_1$ and $v_2$ of two known lines in a spectrum, the Ritz rule told spectroscopists to look for new lines at the frequencies $v_1 + v_2$ or $v_1 - v_2$."]

the frequency corresponding to the lines of the spectrum of an element can be expressed by

$$v = F_\tau (\tau_1) - F_s(\tau_2),$$

where $\tau_1$ and $\tau_2$ are entire numbers, and $F_1$, $F_2$, $F_3$, … are functions of $\tau$ which approximately are equal to $K/(\tau + a_1)^2$ , $K/(\tau + a_2)^2$ , … K is a universal constant, equal to the factor outside the bracket in the formula (4) for the spectrum of hydrogen. The different series appear if we put $\tau_1$ or $\tau_2$ equal to a fixed number and let the other vary.

***The circumstance that the frequency can be written as a difference between two functions of entire numbers suggests an origin of the lines in the spectra in question similar to the one that we have assumed for hydrogen***; i.e. that the lines correspond to a radiation emitted during the passing of the system between two different stationary states. For system containing more than one electron the detailed discussion may be very complicated, as there will be many different configurations of the electrons which can be taken into consideration as stationary states. This may account for the difference sets of series in the line spectra emitted from the substances in question. Here I shall only try to show how, by help of the theory, it can be simply explained that the constant K entering in Rydberg's formula is the same for all substances. Let us assume that the spectrum in question corresponds to the radiation emitted during the binding of an electron; and let us further assume that the system including the electron considered is neutral. The force on the electron, when at a great distance apart the nucleus and the electrons previously bound, will be very nearly the same as the above case of the binding of an electron by a hydrogen nucleus. The energy corresponding to one of the stationary states will therefore for $\tau$ great be very nearly equal to that given by the expression (3) on p. 5, if we put $E = e$. For $\tau$ great we consequently get

$$\lim \, [\tau^2 \cdot F_1(\tau)] = \lim[\tau^2 \cdot F_2(\tau)] = \ldots = 2\pi^2 me^4/h^3,$$

in conformity with Rydberg's theory.

…

As mentioned in the introduction, ***the above hypothesis will be used in a following communication as a basis for a theory of the constitution of atoms and molecules***. It will be shown that it leads to results which seem to be in conformity with experiments on a number of different phenomena.

The foundation of the hypothesis has been sought entirely in its relation with Planck's theory of radiation; by help of considerations given later it will be attempted to throw some further light on the foundation of it from another point of view.

April 5, 1913.

## Bohr, N. (1913). On the Constitution of Atoms and Molecules, Part II. Systems Containing Only a Single Nucleus[1].

[*Phil. Mag.*, 26, 153, 476-502; https://doi.org/ 10.1080/14786441308634993.]

Received July, 1913.

[1]Part I was published in (1913). *Phil. Mag.*, 26, 151, 1-24.

Application of Bohr's model of the atom to systems containing a single nucleus.

### § 1 General Assumptions

Following the theory of Rutherford, we shall assume that the atoms of the elements consist of a positively charged nucleus surrounded by a cluster of electrons. The nucleus is the seat of the essential part of the mass of the atom, and has linear dimensions exceedingly small compared with the distance apart of the electrons in the surrounding cluster.

As in the previous paper, we shall assume that the cluster of electrons is formed by the successive binding by the nucleus of electrons initially nearly at rest, energy at the same time being radiated away. This will go on until, when the total negative charge on the bound electrons is numerically equal to the positive charge on the nucleus, the system will be neutral and no longer able to exert sensible forces on electrons at distances from the nucleus great in comparison with the dimensions of the orbits of the bound electrons. We may regard the formation of helium from α rays as an observed example of a process of this kind, an α particle on this view being identical with the nucleus of a helium atom.

***On account of the small dimensions of the nucleus, its internal structure will not be of sensible influence on the constitution of the cluster of electrons, and consequently will have no effect on the ordinary physical and chemical properties of the atom***. The latter properties on this theory will depend entirely on the total charge and mass of the nucleus; the internal structure of the nucleus will be of influence only on the phenomena of radioactivity.

From the result of experiments on large-angle scattering of α-rays, Rutherford[2] found an electric charge on the nucleus corresponding per atom to a number of electrons approximately equal to half the atomic weight.

[2] Compare also Geiger, H. & Marsden, E. (April, 1913). *The Laws of Deflection of a Particles through Large Angles. Phil. Mag.*, 6, 25, 148, 604-23; https://www.chemteam. info/Chem-History/GeigerMarsden-1913/GeigerMarsden-1913.html.

This result seems to be in agreement with the number of electrons per atom calculated from experiments on scattering of Rontgen radiation.[3]

[3] Compare Barkla, C.G. (1911). *Note on the energy of scattered X-radiation. Phil. Mag.*, 6, 21, 125, 648-52; https://doi.org/10.1080/14786440508637077.

The total experimental evidence supports the hypothesis[4] that ***the actual number of electrons in a neutral atom with a few exceptions is equal to the number which indicated the position of the corresponding element in the series of element arranged in order of increasing atomic weight***.

[4] Compare Broek, A. V. D. (1913). *Zeit. Phys.*, 14, 32.

For example, on this view, the atom of oxygen which is the eighth element of the series has eight electrons and a nucleus carrying eight unit charges.

We shall assume that the electrons are arranged at equal angular intervals in coaxial rings rotating round the nucleus. In order to determine the frequency and dimensions of the rings we shall use the main hypothesis of the first paper, viz.; that ***in the permanent state of an atom the angular momentum of every electron round the center of its orbit is equal to the universal value h/2π,*** where h is Planck's constant. We shall take as a condition of stability, that the total energy of the system in the configuration in question is less than in any neighboring configuration satisfying the same condition of the angular momentum of the electrons.

If the charge on the nucleus and the number of electrons in the different rings is known, the condition in regard to the angular momentum of the electrons will, as shown in § 2, completely determine the configuration of the system. i.e., the frequency of revolution and the linear dimensions of the rings. Corresponding to different distributions of the electrons in the rings, however, there will, in general, be more than one configuration which will satisfy the condition of the angular momentum together with the condition of stability.

In § 3 and § 4 it will be shown that, on the general view of the formation of the atoms, we are led to indications of the arrangement of the electrons in the rings which are consistent with those suggested by the chemical properties of the corresponding element.

In § 5 will be shown that it is possible from the theory to calculate the momentum velocity of cathode rays necessary to produce the characteristic Rontgen radiation from the element, and that this is in approximate agreement with the experimental values.

In § 6 the phenomena of radioactivity will be briefly considered in relation of the theory.

## § 2 Configuration and Stability of the System

Let us consider an electron of charge e and mass m which moves in a circular orbit of radius a with a velocity $v$ small compared with the velocity of light. Let us denote the radial force acting on the electrons by $e^2/a^2F$; F will in general be dependent on $a$. The condition of dynamical equilibrium gives

$$m v^2/a = e^2/a^2 F.$$

Introducing the condition of universal constancy of the angular momentum of the electron, we have

$$m v a = h/2\pi.$$

From these two conditions we now get

$$a = h^2/4\pi^2 e^2 m \cdot F^{-1} \quad \text{and} \quad v = 2\pi e^2/h \cdot F; \tag{1}$$

and for the frequency of revolution $\omega$ consequently

$$\omega = 4\pi^2 e^2 m/h^2 \cdot F^2. \tag{2}$$

If F is known, the dimensions and frequency of the corresponding orbit are simply determined by (1) and (2). For a ring of $n$ electrons rotating round a nucleus of charge $ne$ we have (comp. Part I., 20)

$$F = N - s_n, \text{ where } s_n = \frac{1}{4} \cdot \Sigma_{s=1}^{s=n-1} \operatorname{cosec} s\pi/n.$$

...

If system of rings rotating round a nucleus in a single plane is stable for small displacements of the electrons perpendicular to this plane, there will in general be no stable configurations of the rings, satisfying the condition of the constancy of the angular momentum of the electrons, in which all the rings are not situated in the plane. An exception occurs in the special case of two rings containing equal numbers of electrons; in this case there may be a stable configuration in which the two rings have equal radii and rotate in parallel planes at equal distances from the nucleus, the electrons in the one ring being situated just opposite the intervals between the electrons in the other ring. The latter configuration, however, is unstable if the configuration in which all the electrons in the two rings are arranged in a single ring is stable.

### § 3 Constitution of Atoms containing very few Electrons

At stated in § 1, the condition of the universal constancy of the angular momentum of the electrons, together with the condition of stability, is in most cases not sufficient to determine completely the constitution of the system. On the general view of formation of atoms, however, and by making use of the knowledge of the properties of the corresponding elements, it will be attempted, in this section and the next, to obtain indications of what configurations of the electrons may be expected to occur in the atoms. In these considerations we shall assume that the number of electrons in the atom is equal to the number which indicates the position of the corresponding element in the series of elements arranged in order of increasing atomic weight. Exceptions to this rule will be supposed to occur only at such places in the series where deviation from the periodic law of the chemical properties of the elements are observed. In order to show clearly the principles used we shall first consider with some detail those atoms containing very few electrons.

…

## § 4 Atoms containing a greater numbers of Electrons

From the examples discussed in the former section it will appear that the problem of the arrangement of the electrons in the atoms is intimately connected with the question of the confluence of two rings of electrons rotating round a nucleus outside each other, and satisfying the condition of the universal constancy of the angular momentum. apart from the necessary conditions of stability for displacements of the electrons perpendicular to the plane of the orbits, the present theory gives very little information on this problem. It seems, however, possible by the help of simple considerations to throw some light on the question.

…

On the same lines, the presence of the group of the rare earths indicates that for still greater values of N another gradual alteration of the innermost rings will take place. Since, however, for elements of higher atomic weight than those of this group, the laws connection the vibration of the chemical properties with the atomic weight are similar to these between the elements of low atomic weight, we may conclude that the configuration of the innermost electrons will be again repeated. ***The theory, however, is not sufficiently complete to give a definite answer to such problems.***

## § 5 Characteristic Rontgen Radiation

According to the theory of emission of radiation given in Part I., the ordinary line-spectrum of an element is emitted during the reformation of an atom when one or more of the electrons in the other rings are removed. In analogy it may be supposed that the characteristic Rontgen radiation is sent out during the setting down of the system if electrons in inner rings are removed by some agency, e.g. by impact of cathode particles. This view of the origin of the characteristic Rontgen radiation has been proposed by Sir. J. J. Thomson. Without any special assumption in regard to the constitution of the radiation, we can from this view determine the minimum velocity of the cathode rays necessary to produce the characteristic Rontgen radiation of a special type by calculating the energy necessary to remove one of the electrons from the different rings. Even if we know the numbers of electrons in the rings, a rigorous calculation of this momentum energy might still be complicated, and the result largely dependent on the assumptions used; for, as mentioned in Part I., p. 19, ***the calculation cannot be performed entirely on the basis of the ordinary mechanics*** …

…

## § 6 Radioactive Phenomena

According to the present theory the cluster of electrons surrounding the nucleus is formed with emission of energy, and the configuration is determined by the condition that the energy emitted is a maximum. The stability involved by these assumptions seems to be in agreement with the general properties of matter. It is, however, in striking opposition to the phenomena of radioactivity, and according to the theory the origin of the latter phenomena may therefore be sought elsewhere than in the electronic distribution round the nucleus.

A necessary consequence of Rutherford's theory of the structure of atoms is that the α-particles have their origin in the nucleus. On the present theory it seems also necessary that the nucleus is the seat of the expulsion of the high-speed β-particles. In the first place, the spontaneous expulsion of a β-particle from the cluster of electrons surrounding the nucleus would be something quite foreign to the assumed properties of the system. further, the expulsion of an α-particle can hardly be expected to produce a lasting effect on the stability of the cluster of electrons. The effect of the expulsion will be of two different kinds. Partly the particle may collide with the bound electrons during its passing through the atom. This effect will be analogous to that produced by bombardment of atoms of other substances by α-rays and cannot be expected to give rise to a subsequent expulsion of β-rays. Partly the expulsion of the particle will involve an alteration in the configuration of the bound electrons, since the charge remaining on the nucleus is different from the original.

…

The question of the origin of β-particles may also be considered from another point of view, based on a consideration of the chemical and physical properties of the radioactive substances. As is well known, several of these substances have very similar chemical properties and have hitherto resisted every attempt to separate them by chemical means. There is also some evidence that the substances in question show the same line-spectrum.[14]

[14] See Russel, A.S. & Rossi, R. (1912). *An Investigation of the Spectrum of Ionium. Proc. Roy. Soc. A.*, 87, 598, 478-84; https://www.jstor.org/stable/93182.

It has been suggested by several writers that the substances are different only in radio-active properties and atomic weight but identical in all other physical and chemical respects. according to the theory, this would mean that the charge on the nucleus, as well as the configuration of the surrounding electrons, was identical in some of the elements, the only difference being the mass and the internal condition of the nucleus. From the considerations of § 4 this assumption is already strongly suggested by the fact that the number of radioactive substances is greater than the number of places at our disposal in the periodic system. If, however, the assumption is right, the fact that two apparently identical elements emit β-particles of different velocities, shows that the β-rays as well as the α-rays have their origin in the nucleus.

…

In escaping from the nucleus, the β-rays may be expected to collide with the bound electrons in the inner rings. This will give rise to an emission of a characteristic radiation of the same type as the characteristic Rontgen radiation emitted from elements of lower atomic weight by impact of cathode-rays. The assumption that the emission of γ-rays is due to collisions of β-rays with bound electrons is proposed by Rutherford[16] in order to account for the numerous groups of homogeneous β-rays expelled from certain radioactive substances.

[16] Rutherford, E. (1912). *The Origin of Beta and Gamma Rays from Radioactive Substances. Phil. Mag.*, 6, 24, 453-62; https://doi.org/10.1080/14786441008637351; Rutherford, E. (1912). *On the Energy of the Group of Beta Rays from Radium. Phil. Mag.*, 6, 24, 893-4. …

# PART III   The proton.

## Rutherford, E. 1st Baron Rutherford of Nelson (August 30, 1871 – October 19, 1937)

Rutherford was a New Zealand physicist and chemist who was a pioneering researcher in both atomic and nuclear physics. He has been described as "the father of nuclear physics" and "the greatest experimentalist since Michael Faraday." *In 1908, he was awarded the Nobel Prize in Chemistry "for his investigations into the disintegration of the elements, and the chemistry of radioactive substances."* He was the first Oceanian Nobel laureate, and the first to perform Nobel-awarded work in Canada.

Ernest Rutherford was born on 30 August 1871 in Brightwater, New Zealand, the fourth of twelve children of James Rutherford, an immigrant farmer and mechanic from Perth, Scotland, and Martha Thompson, a schoolteacher from Hornchurch, England.

When Rutherford was age 5, he moved to Foxhill, New Zealand, and attended Foxhill School. At 11 in 1883, the Rutherford family moved to Havelock, a town in the Marlborough Sounds. The move was made to be closer to the flax mill Rutherford's father developed. Ernest studied at Havelock School.

In 1887, on his second attempt, he won a scholarship to study at Nelson College. On his first examination attempt, he had the highest mark of anyone from Nelson. When he was awarded the scholarship, he had received 580 out of 600 possible marks. After being awarded the scholarship, Havelock School presented him with a five-volume set of books titled The Peoples of the World. He studied at Nelson College between 1887 and 1889, and was head boy in 1889. He also played in the school's rugby team. He was offered a cadetship in government service, but he declined as he still had 15 months of college remaining.

In 1889, after his second attempt, he won a scholarship to study at Canterbury College, University of New Zealand, between 1890 and 1894. He participated in its debating society and the Science Society. At Canterbury, he was awarded a complex B.A. in Latin, English and Mathematics in 1892, a M.A. in Mathematics and Physical Science in 1893, and a B.Sc. in Chemistry and Geology in 1894.

In 1895 he was awarded an 1851 Research Fellowship from the Royal Commission for the Exhibition of 1851, to travel to England for postgraduate study in the Cavendish Laboratory at the University of Cambridge. In 1897, he was awarded a B.A. Research Degree and the Coutts-Trotter Studentship from Trinity College, Cambridge.

When Rutherford began his studies at Cambridge, he was among the first 'aliens' (those without a Cambridge degree) allowed to do research at the university, and was additionally honored to study under J. J. Thomson.

Under Thomson's leadership, Rutherford worked on the conductive effects of X-rays on gases, which led to the ***discovery of the electron***, the results first presented by Thomson in 1897. Hearing of Henri Becquerel's experience with uranium, Rutherford started to explore its radioactivity, discovering two types that differed from X-rays in their penetrating power.

In 1898, Rutherford accepted the Macdonald Chair of Physics at McGill University in Montreal, Canada, on Thomson's recommendation.

Continuing his research in Canada, ***in 1899 he coined the terms "alpha ray" and "beta ray" to describe these two distinct types of radiation***. [Rutherford, E.* (January, 1899). *Uranium Radiation and the Electrical Conduction Produced by It. Phil. Mag.*, 5, xlvii, pp. 109-163; https://www.chemteam.info/Chem-History/Rutherford-Alpha&Beta.html. See below.]

In 1900, in Christchurch, Rutherford married Mary Georgina Newton (1876–1954), to whom he had been engaged before leaving New Zealand. They had one daughter, Eileen Mary (1901–1930), who married the physicist Ralph Fowler, and died during the birth of her fourth child. Rutherford's hobbies included golf and motoring.

From 1900 to 1903, he was joined at McGill by the young chemist Frederick Soddy (Nobel Prize in Chemistry, 1921) for whom he set the problem of identifying the noble gas emitted by the radioactive element thorium, a substance which was itself radioactive and would coat other substances. Once he had eliminated all the normal chemical reactions, Soddy suggested that it must be one of the inert gases, which they named thoron. This substance was later found to be 220Rn, an isotope of radon. They also found another substance they called Thorium X, later identified as 224Rn, and continued to find traces of helium. They also worked with samples of "Uranium X" (protactinium), from William Crookes, and radium, from Marie Curie. Rutherford further investigated thoron in conjunction with R.B. Owens and found that a sample of radioactive material of any size invariably took the same amount of time for half the sample to decay (in this case, 11 1/2 minutes), a phenomenon for which he coined the term "half-life".

Rutherford and Soddy published their paper *Law of Radioactive Change* to account for all their experiments. [Rutherford, E. & Soddy, F. (1903) *Radioactive Change. Phil. Mag.*, 5, 576-91. Reproduced in Romer, A., *The Discovery of Radioactivity and Transmutation.* (Dover, 1964), 152-66]. Until then, atoms were assumed to be the indestructible basis of all matter; and although Curie had suggested that radioactivity was an atomic phenomenon, the idea of the atoms of radioactive substances breaking up was a radically new idea. Rutherford and Soddy demonstrated that radioactivity involved the spontaneous disintegration of atoms into other, as yet, unidentified matter.

In 1903, Rutherford considered a type of radiation, discovered (but not named) by French chemist Paul Villard in 1900, as an emission from radium, and realized that this observation must represent something different from his own *alpha and beta rays*, due to its very much greater penetrating power. Rutherford therefore gave this third type of radiation the name of *gamma ray*. All three of Rutherford's terms are in standard use today – other types of radioactive decay have since been discovered, but Rutherford's three types are among the most common. In 1904, Rutherford suggested that radioactivity provides a source of energy sufficient to explain the existence of the Sun for the many millions of years required for the slow biological evolution on Earth proposed by biologists such as Charles Darwin. The physicist Lord Kelvin had argued earlier for a much younger Earth, based on the insufficiency of known energy sources, but Rutherford pointed out, at a lecture attended by Kelvin, that radioactivity could solve this problem. In 1907, he returned to Britain to take the Langworthy Professorship at the Victoria University of Manchester.

In Manchester, Rutherford continued his work with *alpha radiation*. In conjunction with Hans Geiger, he developed zinc sulfide scintillation screens and ionization chambers to count alpha particles. By dividing the total charge accumulated on the screen by the number counted, Rutherford determined that the charge on the alpha particle was two. In late 1907, Ernest Rutherford and Thomas Royds allowed alphas to penetrate a very thin window into an evacuated tube. As they sparked the tube into discharge, the spectrum obtained from it changed, as the alphas accumulated in the tube. Eventually, the clear spectrum of helium gas appeared, proving that alphas were at least ionized helium atoms, and probably helium nuclei.

***In 1908, Rutherford was awarded the Nobel Prize in Chemistry "for his investigations into the disintegration of the elements, and the chemistry of radioactive substances."*** (Nobel Lecture, December 11, 1908, *The Chemical Nature of the Alpha Particles from Radioactive Substances*; https://www.nobelprize.org/prizes/chemistry/1908/rutherford/ lecture/.)

Rutherford continued to make ground-breaking discoveries long after receiving the Nobel prize in 1908. In 1911, Rutherford conducted groundbreaking experiments, particularly the *gold foil experiment*, which led to the formulation of his *nuclear model of the atom*.

**The Gold Foil Experiment**

The gold foil experiment, also known as the Geiger-Marsden experiment, was conducted between 1908 and 1913 by scientists Hans Geiger and Ernest Marsden under the supervision of Ernest Rutherford at the University of Manchester. At that time, the prevailing atomic model was J.J. Thomson's "plum pudding model," which suggested that atoms were composed of a positively charged "pudding" with negatively charged electrons embedded within it. Under his direction in 1909, Hans Geiger and Ernest Marsden, performed the Geiger–Marsden experiment, which demonstrated the nuclear nature of atoms by measuring the deflection of alpha particles passing through a very thin sheet of gold foil. They expected the alpha particles to pass through with minimal deflection, based on the plum pudding model. However, *Rutherford was inspired to ask Geiger and Marsden in this experiment to look for alpha particles with very high deflection angles*, which was not expected according to any theory of matter at that time. *Such deflection angles, although rare, were found.* They observed that while most particles passed through, a small fraction were deflected at large angles, and some even bounced back. This unexpected result indicated that the atom must have a small, dense, positively charged core, which they termed the **nucleus**.

Reflecting on these results in one of his last lectures, Rutherford was quoted as saying: "It was quite the most incredible event that has ever happened to me in my life. It was almost as incredible as if you fired a 15-inch shell at a piece of tissue paper and it came back and hit you." *It was Rutherford's interpretation of this data that led him to propose the nucleus, a very small, charged region containing much of the atom's mass.*

In 1910 Rutherford, with Geiger and mathematician Harry Bateman published their classic paper describing the first analysis of the distribution in time of radioactive emission, a distribution now called the Poisson distribution.

**Rutherford's Nuclear Model (1911)**

His *Nuclear Model*, proposed in 1911, revolutionized atomic theory *by introducing the concept of a dense nucleus at the center of the atom, surrounded by orbiting electrons.* [Rutherford, E. (May 1911). LXXIX. *The scattering of α and β particles by matter and the structure of the atom. Phil. Mag.* 6, 21, 125, 669–88; http://dx.doi.org/10.1080/14786440508637080. See below.] This model emerged as a significant advancement over the previous Thomson "plum pudding" model, which suggested that atoms were composed of a diffuse cloud of positive charge with electrons embedded within it.

*[Key Features of the Nuclear Model*

1.      The nucleus is a tiny, dense core at the center of the atom, containing most of the atom's mass and all of its positive charge. It is significantly smaller than the atom itself, with a size ratio comparable to a marble in a football stadium.

2.      Electrons, which are negatively charged, orbit the nucleus at a relatively large distance, occupying most of the atom's volume. This arrangement is similar to planets orbiting the sun, leading to the model being sometimes referred to as the planetary model.

3.      The majority of the atom is composed of empty space, which explains why most alpha particles passed through the gold foil without deflection.

4.      The atom is electrically neutral, meaning the number of protons in the nucleus equals the number of electrons orbiting it.]

*Rutherford's nuclear model was a pivotal step in the development of atomic theory*. It laid the groundwork for future models, particularly the Bohr model, which introduced quantized electron orbits. Rutherford's findings challenged the existing theories and opened new avenues for research in nuclear physics and chemistry, ultimately leading to a deeper understanding of atomic structure. *Rutherford's Nuclear Model fundamentally changed the way scientists understood the atom, emphasizing the importance of the nucleus and the arrangement of electrons*, and it remains a cornerstone of modern atomic theory.

Together with Thomas Royds, Rutherford is credited with proving that *alpha radiation is composed of helium nuclei. In 1911, he theorized that atoms have their charge concentrated in a very small nucleus*. He arrived at this theory through his discovery and interpretation of Rutherford scattering during the gold foil experiment performed by Hans Geiger and Ernest Marsden.

In 1912, he invited *Niels Bohr* to join his lab, leading to the *Bohr model of the atom* (which postulated that electrons moved in specific orbits about the compact nucleus). *Bohr adapted Rutherford's nuclear structure to be consistent with Max Planck's quantum hypothesis*.

*The resulting Bohr model was the basis for quantum mechanical atomic physics of Heisenberg which remains valid today*. [Bohr, N. (1913). *On the Constitution of Atoms and Molecules, Part I. – Binding of Electrons by Positive Nuclei. Phil. Mag.*, 26, 151, 1-24; https://doi.org/10.1080/ 14786441308634955. Bohr took the next step by adapting Rutherford's theory of atomic structure to Planck's quantum hypothesis creating his model of the atom. See below.]

**Discovery of the proton**

Together with H.G. Moseley, Rutherford developed the atomic numbering system in 1913. Rutherford and Moseley's experiments used cathode rays to bombard various elements with streams of electrons and observed that each element responded in a consistent and distinct manner. Their research was the first to assert that each element could be defined by the properties of its inner structures – an observation that later led to the discovery of the atomic nucleus. This research led Rutherford to theorize that the hydrogen atom (at the time the least massive entity known to bear a positive charge) was a sort of "positive electron" – a component of every atomic element. [Rutherford, E. (March, 1914). *The Structure of the Atom. Phil. Mag.,* 6, 27, p. 488–98; see below. The present paper and the accompanying paper by Mr. C. Darwin deal with certain points in connection with the "nucleus" theory of the atom which were purposely omitted in my first communications on that subject (Phil. Mag., May 1911).]

During World War I, Rutherford worked on a top-secret project to solve the practical problems of submarine detection. Both Rutherford and Paul Langevin suggested the use of piezoelectricity, and Rutherford successfully developed a device which measured its output. The use of piezoelectricity then became essential to the development of ultrasound as it is known today.

In 1917, he performed the first artificially induced nuclear reaction by conducting experiments in which nitrogen nuclei were bombarded with **alpha particles**. These experiments led him to discover the emission of a subatomic particle that he initially called the "hydrogen atom", but later (more precisely) renamed the **proton**. He is also credited with developing the atomic numbering system alongside Henry Moseley. His other achievements include advancing the fields of radio communications and ultrasound technology.

It was not until 1919 that Rutherford expanded upon his *theory of the "positive electron"* with a series of experiments beginning shortly before the end of his time at Manchester. He found that nitrogen, and other light elements, ejected a proton, which he called a "hydrogen atom," when hit with α (alpha) particles. In particular, he showed that particles ejected by alpha particles colliding with hydrogen have unit charge and 1/4 the momentum of alpha particles. [Rutherford, E. (June, 1919*). Collision of α particles with light atoms. I-IV. Phil. Mag.,* 37, 537-87. See below.]

This result showed Rutherford that hydrogen nuclei were a part of nitrogen nuclei (and by inference, probably other nuclei as well). Such a construction had been suspected for many years, on the basis of atomic weights that were integral multiples of that of hydrogen; see

Prout's hypothesis. Hydrogen was known to be the lightest element, and its nuclei presumably the lightest nuclei.

In Rutherford's four-part article on the "*Collision of α-particles with light atoms*" [(July, 1919). *Nature*, 103, 415–418 https://doi.org/10.1038/103415a0] he reported two additional fundamental and far-reaching discoveries. First, he showed that at high angles the scattering of alpha particles from hydrogen differed from the theoretical results he himself published in 1911. These were the first results to probe the interactions that hold a nucleus together. Second, he showed that α-particles colliding with nitrogen nuclei would react rather than simply bounce off. One product of the reaction was the proton; the other product was shown by Patrick Blackett, Rutherford's colleague and former student, to be oxygen.

$$^{14}N + \alpha \rightarrow {}^{17}O + p.$$

Rutherford therefore recognized "that the nucleus may increase rather than diminish in mass as the result of collisions in which the proton is expelled".

Rutherford returned to the Cavendish Laboratory in 1919, succeeding J. J. Thomson as Cavendish Professor of Physics and Director of the Cavendish Laboratory, a position he held until his death in 1937. In the same year, the first controlled experiment to split the nucleus was performed by John Cockcroft and Ernest Walton, working under his direction. During his tenure, Nobel prizes were awarded to James Chadwick for discovering the *neutron* (in 1932), John Cockcroft and Ernest Walton for an experiment that was to be known as "splitting the atom" using a particle accelerator, and Edward Appleton for demonstrating the existence of the ionosphere.

**Development of proton and neutron theory**

In 1919–1920, Rutherford continued his research on the "hydrogen atom" to confirm that alpha particles break down nitrogen nuclei and to affirm the nature of the products.

*Rutherford decided that a hydrogen nucleus was possibly a fundamental building block of all nuclei*, and also possibly a new fundamental particle as well, since nothing was known to be lighter than that nucleus, thus, confirming and extending the work of Wilhelm Wien, who in 1898 discovered the proton in streams of ionized gas. In a 1919 paper [Rutherford, E. (1919). "*Collision of α particles with light atoms*". *Philosophical Magazine*. 37: 571], Rutherford reported the apparent discovery of a new doubly charged particle of mass 3, denoted the X++. The X++ particle was later determined to have mass 4 and to be just a low-energy alpha particle. Nevertheless, Rutherford had conjectured the existence of the *deuteron, a +1 charged particle of mass 2, and a neutral particle of mass*

*1.* The former is the nucleus of ***deuterium***, discovered in 1931 by Harold Urey. The mass of the hypothetical neutral particle would be little different from that of the proton. Rutherford determined that such a zero-charge particle would be difficult to detect by available techniques.

In 1920, Rutherford gave a Bakerian lecture at the Royal Society entitled the "*Nuclear Constitution of Atoms*" [see below], a summary of recent experiments on atomic nuclei and conclusions as to the structure of atomic nuclei. This suggested the likely existence of two new particles: one consisting of a ***hydrogen nucleus***, and another of one consisted of a ***hydrogen nucleus and a closely bound electron***.

By 1920, the existence of electrons within the atomic nucleus was widely assumed. It was assumed the nucleus consisted of ***hydrogen nuclei*** in number equal to the atomic mass number. But, since each hydrogen nucleus had charge +1 e, the nucleus required a smaller number of "internal electrons" each of charge −1 e to give the nucleus its correct total charge.

In 1920 Rutherford postulated the ***hydrogen nucleus*** to be a new particle, which he dubbed the ***proton***, establishing a distinct name for the smallest known positively-charged particle of matter (that can exist independently anyway). [Rutherford, E. (July, 1920). *Nuclear Constitution of Atoms. Roy. Soc. Proc. A*, 97, 686, 374–400; https://doi.org/10.1098/rspa. 1920.0040. Bakerian lecture at the Royal Society. See below.]

The ***alpha particle*** was known to be very stable, and it was assumed to retain its identity within the nucleus. The alpha particle was presumed to consist of four protons and two closely bound electrons to give it +2 charge and mass 4. Such a model was consistent with the scattering of alpha particles from heavy nuclei, as well as the charge and mass of the many isotopes that had been identified. There were other motivations for the ***proton-electron model***. As noted by Rutherford at the time, "We have strong reason for believing that the nuclei of atoms contain electrons as well as positively charged bodies ...", namely, it was known that beta radiation was electrons emitted from the nucleus.

In 1921, while working with Niels Bohr, Rutherford theorized about the existence of ***neutrons***, which could somehow compensate for the repelling effect of the positive charges of protons by causing an attractive nuclear force and thus keep the nuclei from flying apart, due to the repulsion between protons. William Harkins, an American chemist, named the uncharged particle the ***neutron***. About that same time the word ***proton*** was adopted for the ***hydrogen nucleus***.

Rutherford's discoveries included the concept of radioactive half-life, the radioactive element radon, and the differentiation and naming of **alpha and beta radiation**. [*Uranium Radiation and the Electrical Conduction Produced by It*. by E. Rutherford, M.A., B.SC. formerly 1851 Science Scholar, Coutts Trotter Student, Trinity College, Cambridge; Phil. Mag. (January, 1899), 5, xlvii, pp. 109-163. Communicated by Professor J. J. Thomson, F.R.S.]

These experiments show that the uranium radiation is complex, and that there are present at least two distinct types of radiation--one that is very readily absorbed, which will be termed for convenience the α radiation, and the other of a more penetrative character, which will be termed the β radiation.

The character of the β radiation seems to be independent of the nature of the filter through which it has passed. It was found that radiation of the same intensity and of the same penetrative power was obtained by cutting off the α radiation by thin sheets of aluminum, tinfoil, or paper. The β radiation passes through all the substances tried with far greater facility than the α radiation. For example, a plate of thin cover glass placed over the uranium reduced the rate of leak to one-thirtieth of its value; the β radiation, however, passed through it with hardly any loss of intensity.

**In 1932, Rutherford's theory of neutrons was proved by his associate James Chadwick,** who recognized **neutrons** immediately when they were produced by other scientists and later himself, in bombarding beryllium with alpha particles. [Chadwick, J. (February, 1932). *Possible Existence of a Neutron. Nature*, 129, 312; https://doi.org/10.1038/129312a0. See below.]  In 1935, Chadwick was awarded the Nobel Prize in Physics for this discovery.

For some time before his death, Rutherford had a small hernia, which he neglected to have repaired, and it eventually became strangulated, rendering him violently ill. He had an emergency operation in London, but died in Cambridge four days later, on October 19, 1937, at the age of 66, of what physicians termed "intestinal paralysis." After cremation at Golders Green Crematorium, he was given the high honor of burial in Westminster Abbey, near Isaac Newton, Charles Darwin, and other illustrious British scientists.
In honor of his scientific advancements, Rutherford was recognized as a baron of the United Kingdom. After his death in 1937, he was buried in Westminster Abbey near Charles Darwin and Isaac Newton. The chemical element rutherfordium ($^{104}$Rf) was named after him in 1997.

## Rutherford, E.* (January, 1899). Uranium Radiation and the Electrical Conduction Produced by It.

*Phil. Mag.*, 5, xlvii, pp. 109-163; https://www.chemteam.info/Chem-History/Rutherford-Alpha&Beta.html.

* Cambridge M.A., B.SC. formerly 1851 Science Scholar, Coutts Trotter Student, Trinity College. This paper was published by Rutherford, his student, only two and a half years after J.J. Thomson published his paper, *On the cathode rays*.

Communicated by Professor J. J. Thomson, F.R.S.

The remarkable radiation emitted by uranium and its compounds has been studied by its discoverer, Becquerel, and the results of his investigations on the nature and properties of the radiation have been given in a series of papers in the *Comptes Rendus*. *

* (1896), *Comptes Rendus,* pp. 420, 501, 559, 689, 762, 1086; (1897), pp. 438, 800.

He showed that the radiation, continuously emitted from uranium compounds, has the power of passing through considerable thicknesses of metals and other opaque substances; it has the power of acting on a photographic plate and of discharging positive and negative electrification to an equal degree. The gas through which the radiation passes is made a temporary conductor of electricity and preserves its power of discharging electrification for a short time after the source of radiation has been removed.

The results of Becquerel showed that **Röntgen** and **uranium** radiations were very similar in their power of penetrating solid bodies and producing conduction in a gas exposed to them; but there was an essential difference between the two types of radiation. He found that **uranium radiation could be refracted and polarized, while no definite results showing polarization or refraction have been obtained for Röntgen radiation**. It is the object of the present paper to investigate in more detail the nature of uranium radiation and the electrical conduction produced. As most of the results obtained have been interpreted on the **ionization theory of gases** which was introduced to explain the electrical conduction produced by Röntgen radiation, a brief account is given of the theory and the results to which it leads.

In the course of the investigation, the following subjects have been considered:

§ 1. Comparison of methods of investigation.
§ 2. Refraction and polarization of uranium radiation.
§ 3. Theory of ionization of gases.

## § 1. Comparison of Methods of Investigation

The properties of uranium radiation may be investigated by two methods, one depending on the action on a photographic plate and the other on the discharge of electrification. The photographic method is very slow and tedious, and admits of only the roughest measurements. Two- or three-days' exposure to the radiation is generally required to produce any marked effect on the photographic plate. In addition, when we are dealing with very slight photographic action, the fogging of the plate, during the long exposures required, by the vapors of substances* is liable to obscure the results.

    * Russell, (1897). *Proc. Roy. Soc.*

On the other hand, the method of testing the electrical discharge caused by the radiation is much more rapid than the photographic method, and also admits of fairly accurate quantitative determinations.

The question of polarization and refraction of the radiation can, however, only be tested by the photographic method. The electrical experiment (explained in § 2) to test refraction is not very satisfactory.

## § 2. Polarization and Refraction

The almost identical effects produced in gases by uranium and Röntgen radiation (which will be described later) led me to consider the question whether the two types of radiation did not behave the same in other respects.

In order to test this, experiments were tried to see if uranium radiation could be polarized or refracted. Becquerel** had found evidence of polarization and refraction, but in repeating experiments similar to those tried by him, I have been unable to find any evidence of either.

       ** (1896), *Comptes Rendus,* p. 559.

A large number of photographs by the radiation have been taken under various conditions, but in no case have I been able to observe any effect on the photographic plate which showed the presence of polarization or refraction.

In order to avoid fogging of the plate during the long exposures required, by the vapors of substances, lead was employed as far as possible in the neighborhood of the plate, as its effect on the film is very slight.

A brief account will now be given of the experiments on refraction and polarization.

***Refraction.*** A thick lead plate was taken and a long narrow slit cut through it; this was placed over a uniform layer of uranium oxide; the arrangement was then equivalent to a line source of radiation and a slit. Thin prisms of glass, aluminum, and paraffin-wax were fixed at intervals on the lead plate with their edges just covering the slit. A photographic plate was supported 5 mm. from the slit. The plate was left for a week in a dark box. On developing a dark line was observed on the plate. This line was not appreciably broadened or displaced above the prisms. Different sizes of slits gave equally negative results. If there was any appreciable refraction, we should expect the image of the slit to be displaced from the line of the slit.

Becquerel* examined the opacity of glass for uranium radiation in the solid and also in a finely-powdered state by the method of electric leakage, and found that, if anything, the transparency of the glass for the radiation was greater in the finely divided than in the solid state.

       * (1896), *Comptes Rendus,* p. 559.

I have repeated this experiment and obtained the same result. As Becquerel stated, it is difficult to reconcile this result with the presence of refraction.

*Polarization.* An arrangement very similar to that used by Becquerel was employed. A deep hole was cut in a thick lead plate and partly filled with uranium oxide. A small tourmaline covered the opening. Another small tourmaline was cut in two and placed on top of the first, so that in one half of the opening the tourmalines were crossed and in the other half uncrossed. The tourmalines were very good optically. The photographic plate was supported 1 to 3 mm. above the tourmalines. The plate was exposed four days, and on developing a black circle showed up on the plate, but in not one of the photographs could the slightest difference in the intensity be observed. Becquerel* stated that in his experiment the two halves were unequally darkened and concluded from this result that the radiation was doubly refracted by tourmaline, and that the two rays were unequally absorbed.

## § 3. Theory of Ionization

To explain the conductivity of a gas exposed to Röntgen radiation, the theory** has been put forward that the rays in passing through the gas produce positively and negatively charged particles in the gas, and that the number produced per second depends on the intensity of the radiation and the pressure.

** Thomson, J. J. & Rutherford, E. (November 1896), *Phil. Mag.*

These carriers are assumed to be so small that they will move with a uniform velocity through a gas under a constant potential gradient. The term ion was given to them from analogy with electrolytic conduction, but in using the term it is not assumed that the ion is necessarily of atomic dimensions; it may be a multiple or submultiple of the atom.

Suppose we have a gas between two plates exposed to the radiation and that the plates are kept at a constant difference of potential. A certain number of ions will be produced per second by the radiation and the number produced will in general depend on the pressure of the gas. Under the electric field the positive ions travel towards the negative plate and the negative ions towards the other plate, and consequently a current will pass through the gas. Some of the ions will also recombine, the rate of recombination being proportional to the square of the number present. The current passing through the gas for a given intensity of radiation will depend on the difference of potential between the plates, but when the potential difference is greater than a certain value the current will reach a maximum. When this is the case all the ions are removed by the electric field before they can recombine.

The positive and negative ions will be partially separated by the electric field, and an excess of ions of one sign may be blown away, so that a charged gas will be obtained. If the ions are not uniformly distributed between the plates, the potential gradient will be disturbed by the movement of the ions.

If energy is absorbed in producing ions, we should expect the absorption to be proportional to the number of ions produced and thus depend on the pressure. If this theory be applied to uranium radiation we should expect to obtain the following results:

(1) Charged carriers produced through the volume of the gas.
(2) Ionization proportional to the intensity of the radiation and the pressure.
(3) Absorption of radiation proportional to pressure.
(4) Existence of saturation current.
(5) Rate of recombination of the ions proportional to the square of the number present.
(6) Partial separation of positive and negative ions.
(7) Disturbance of potential gradient under certain conditions between two plates exposed to the radiation.

The experiments now to be described sufficiently indicate that the theory does form a satisfactory explanation of the electrical conductivity produced by uranium radiation.

In all experiments to follow, the results are independent of the sign of the charged plate, unless the contrary is expressly stated.

## § 4. Complex Nature of Uranium Radiation

Before entering on the general phenomena of the conduction produced by uranium radiation, an account will be given of some experiments to decide whether the same radiation is emitted by uranium and its compounds and whether the radiation is homogeneous. Röntgen and others have observed that the x-rays are in general of a complex nature, including rays of wide differences in their power of penetrating solid bodies. The penetrating power is also dependent to a large extent on the stage of exhaustion of the Crookes tube.

In order to test the complexity of the radiation, an electrical method was employed. The general arrangement is shown in Fig. 1.

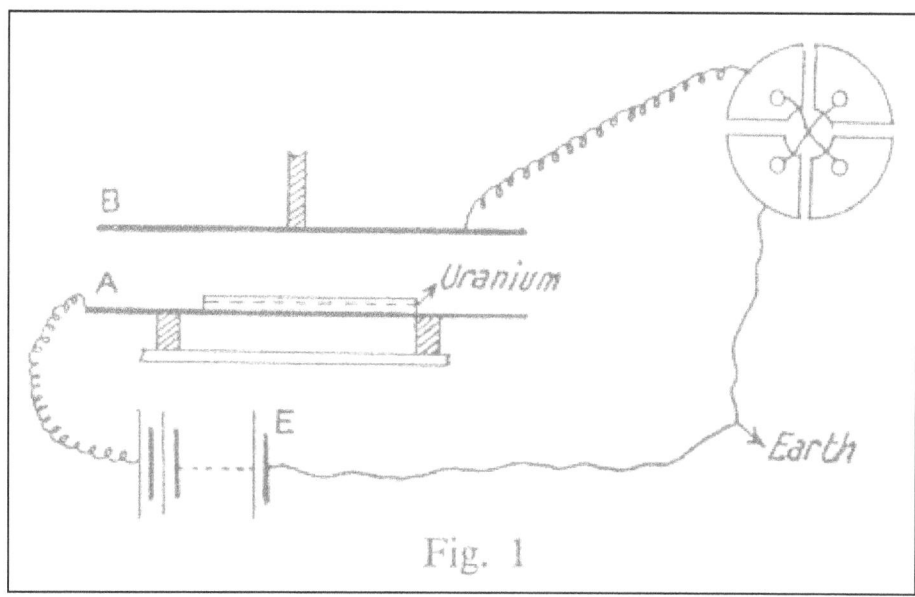

Fig. 1

The metallic uranium or compound of uranium to be employed was powdered and spread uniformly over the center of a horizontal zinc plate A, 20 cm square. A zinc plate B, 20 cm square, was fixed parallel to A and 4 cm from it. Both plates were insulated. A was connected to one pole of a battery of 50 volts, the other pole of which was to earth; B was connected to one pair of quadrants of an electrometer, the other pair of which was connected to earth.

Under the influence of the uranium radiation there was a rate of leak between the two plates A and B. The rate of movement of the electrometer needle, when the motion was steady, was taken as a measure of the current through the gas.

Successive layers of thin metal foil were then placed over the uranium compound and the rate of leak determined for each additional sheet. The table (p. 174) shows the results obtained for thin Dutch metal.

In the third column the ratio of the rates of leak for each additional thickness of metal leaf is given. Where two thicknesses were added at once, the square root of the observed ratio is taken, for three thicknesses the cube root. The table shows that for the first ten thicknesses of metal the rate of leak diminished approximately in a geometrical progression as the thickness of the metal increased in arithmetical progression.

THICKNESS OF METAL LEAF 0.00008 CM.
LAYER OF URANIUM OXIDE ON PLATE

| Number of Layers | Leak per min. in scale divisions | Ratio for each layer |
|---|---|---|
| 0 | 91 | |
| 1 | 77 | 0.85 |
| 2 | 60 | 0.78 |
| 3 | 49 | 0.82 |
| 4 | 42 | 0.86 |
| 5 | 33 | 0.79 |
| 6 | 24.7 | 0.75 |
| 8 | 15.4 | 0.79 |
| 10 | 9.1 | 0.77 |
| 13 | 5.8 | 0.86 |

It will be shown later (§ 8) that the rate of leak between two plates for a saturating voltage is proportional to the intensity of the radiation after passing through the metal. The voltage of 50 employed was not sufficient to saturate the gas, but it was found that the comparative rates of leak under similar conditions for 50 and 200 volts between the plates were nearly the same. When we are dealing with very small rates of leak, it is advisable to employ as small a voltage as possible, in order that any small changes in the voltage of the battery should not appreciably affect the result. For this reason, the voltage of 50 was used, and the comparative rates of leak obtained are very approximately the same as for saturating electromotive forces.

Since the rate of leak diminishes in a geometrical progression with the thickness of metal, we see from the above statement that the intensity of the radiation falls off in a geometrical progression, i.e. according to an ordinary absorption law. This shows that the part of the radiation considered is approximately homogeneous. With increase of the number of layers the absorption commences to diminish. This is shown more clearly by using uranium oxide with layers of thin aluminum leaf (see table, p. 175).

It will be observed that for the first three layers of aluminum foil, the intensity of the radiation falls off according to the ordinary absorption law, and that, after the fourth thickness, the intensity of the radiation is only slightly diminished by adding another eight layers.

THICKNESS OF ALUMINIUM FOIL 0.0005 CM.

| Number of layers of Aluminum foil | Leak per minute in scale divisions | Ratio |
|---|---|---|
| 0 | 182 | |
| 1 | 77 | 0.42 |
| 2 | 33 | 0.43 |
| 3 | 14.6 | 0.44 |
| 4 | 9.4 | 0.65 |
| 12 | 7 | |

The aluminum foil in this case was about 0.0005 cm thick, so that after the passage of the radiation through 0.002 cm of aluminum the intensity of the radiation is reduced to about one-twentieth of its value. The addition of a thickness of 0.001 cm of aluminum has only a small effect in cutting down the rate of leak. The intensity is, however, again reduced to about half of its value after passing through an additional thickness of 0.05 cm, which corresponds to one hundred sheets of aluminum foil.

These experiments show that the uranium radiation is complex, and that there are present at least two distinct types of radiation--one that is very readily absorbed, which will be termed for convenience the *α radiation*, and the other of a more penetrative character, which will be termed the *β radiation*.

The character of the β radiation seems to be independent of the nature of the filter through which it has passed. It was found that radiation of the same intensity and of the same penetrative power was obtained by cutting off the α radiation by thin sheets of aluminum, tinfoil, or paper. The β radiation passes through all the substances tried with far greater facility than the α radiation. For example, a plate of thin cover glass placed over the uranium reduced the rate of leak to one-thirtieth of its value; the β radiation, however, passed through it with hardly any loss of intensity.

Some experiments with different thicknesses of aluminum seem to show, as far as the results go, that the β radiation is of an approximately homogeneous character. The following table gives some of the results obtained for the β radiation from uranium oxide:

β RADIATION

| Thickness of Alumimum | Rate of Leak |
|---|---|
| 0.005 | 1 |
| 0.028 | 0.68 |
| 0.051 | 0.48 |
| 0.09 | 0.25 |

The rate of leak is taken as unity after the α radiation has been absorbed by passing through ten layers of aluminum foil. The intensity of the radiation diminishes with the thickness of metal traversed according to the ordinary absorption law. It must be remembered that when we are dealing with the β radiation alone, the rate of leak is, in general, only a few per cent of the leak due to the α radiation, so that the investigation of the homogeneity of the β radiation cannot be carried out with the same accuracy as for the α radiation. As far, however, as the experiments have gone, the results seem to point to the conclusion that the β radiation is approximately homogeneous, although it is possible that other types of radiation of either small intensity or very great penetrating power may be present.

## § 5. Radiation emitted by different Compounds of Uranium

All the compounds of uranium examined gave out the two types of radiation, and the penetrating power of the radiation for both the α and β is the same for all compounds.

The following table shows the results obtained for some of the uranium compounds.

THICKNESS OF ALUMINIUM FOIL 0.0005 cm.

| Number of layers of Aluminum foil | Proportionate Rate of Leak | | | |
|---|---|---|---|---|
| | Uranium metal | Uranium Nitrate | Uranium Oxide | Uranium Potassium Sulphate |
| 0 | 1 | 1 | 1 | 1 |
| 1 | 0.51 | 0.43 | 0.42 | 0.42 |
| 2 | 0.35 | 0.28 | 0.18 | 0.27 |
| 3 | - | 0.17 | 0.08 | 0.17 |
| 4 | - | 0.15 | 0.05 | 0.12 |
| 5 | 0.15 | - | - | - |
| 12 | - | 0.125 | 0.04 | 0.11 |

Fig. 2 shows graphically some of the results obtained for the various uranium compounds. The ordinates represent rates of leak, and the abscissae thicknesses of aluminum through which the radiation has passed.

The different compounds of uranium gave different rates of leak, but, for convenience of comparison, the rate of leak due to the uncovered salt is taken as unity.

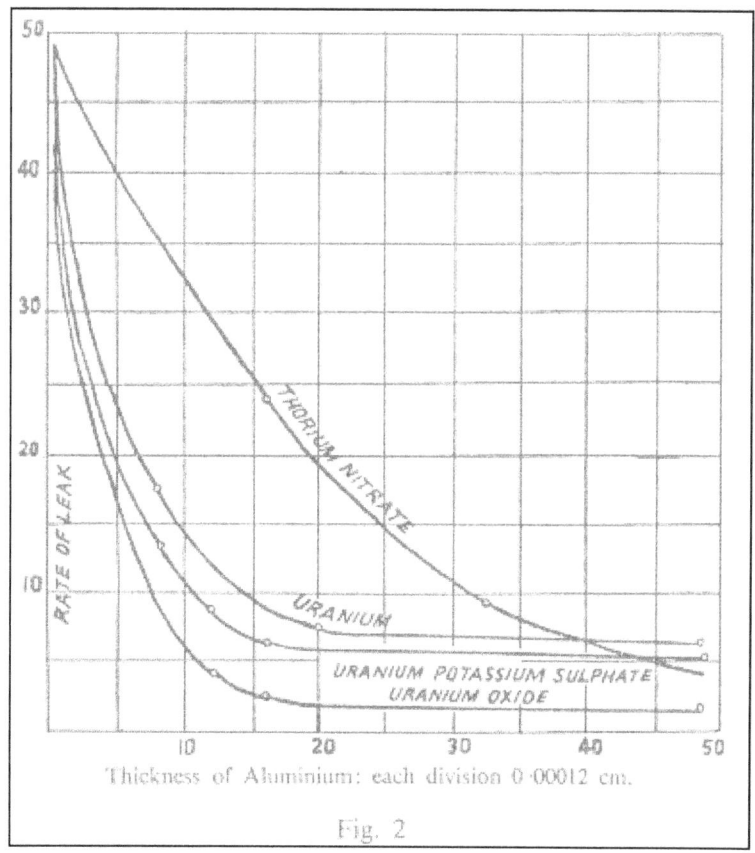

Thickness of Aluminium: each division 0·00012 cm.

Fig. 2

It will be seen that the rate of decrease is approximately the same for the first layer of metal, and that the rate of decrease becomes much slower after four thicknesses of foil.

The rate of leak due to the β radiation is a different proportion of the total amount in each case. The uranium metal was used in the form of powder, and a smaller area of it was used than in the other cases. For the experiments on uranium oxide a thin layer of fine powder was employed, and we see, in that case, that the β radiation bears a much smaller proportion to the total than for the other compounds. When a thick layer of the oxide was used there was, however, an increase in the ratio, as the following table shows:

| Number of layers of Aluminum foil | Rate of Leak | |
| --- | --- | --- |
| | Thin layer of Uranium Oxide | Thick layer of Uranium Oxide |
| 0 | 1 | 1 |
| 1 | 0.42 | 5 |
| 2 | 0.18 | - |
| 4 | 0.05 | 0.12 |
| 8 | - | 0.113 |
| 12 | 0.04 | - |
| 18 | - | 0.11 |

The amount of the α radiation depends chiefly on the surface of the uranium compound, while the β radiation depends also on the thickness of the layer. The increase of the rate of leak due to the β radiation with the thickness of the layer indicates that the β radiation can pass through a considerable thickness of the uranium compound. Experiments showed that the leak due to the α radiation did not increase much with the thickness of the layer. I did not, however, have enough uranium salt to test the variation of the rate of leak due to the β radiation for thick layers.

The rate of leak from a given weight of uranium or uranium compound depends largely on the amount of surface. The greater the surface, the greater the rate of leak. A small crystal of uranium nitrate was dissolved in water, and the water then evaporated so as to deposit a thin layer of the salt over the bottom of the dish. This gave quite a large leakage. The leakage in such a case is due chiefly to the β radiation.

Since the rate of leak due to any uranium compound depends largely on its amount of surface, it is difficult to compare the quantity of radiation given out by equal amounts of different salts: for the result will depend greatly on the state of division of the compound. It is possible that the apparently very powerful radiation obtained from pitchblende by Curie* may be partly due to the very fine state of division of the substance rather than to the presence of a new and powerful radiating substance.

* (July, 1898), *Comptes Rendus,* p. 175.

The rate of leak due to the β radiation is, as a rule, small compared with that produced by the α radiation. It is difficult, however, to compare the relative intensities of the two kinds. The α radiation is strongly absorbed by gases (§ 8), while the β radiation is only slightly so. It will be shown later (§ 8) that the absorption of the radiation by the gas is approximately proportional to the number of ions produced. If, therefore, the β radiation is only slightly absorbed by the gas, the number of ions produced by it is small, i.e. the rate of leak is small. The comparative rates of leak due to the α and β radiations is thus dependent on the relative absorption of the radiations by the gas as well as on the relative intensity.

The photographic actions of the α and β radiations have also been compared. A thin uniform layer of uranium oxide was sprinkled over a glass plate; one half of the plate was covered by a piece of aluminum of sufficient thickness to practically absorb the α radiation. The photographic plate was fixed about 4 mm from the uranium surface. The plate was exposed 48 hours, and, on developing, it was found that the darkening of the two halves was not greatly different. On the one half of the plate the action was due to the β radiation alone, and on the other due to the α and β radiations together. Except when the photographic plate is close to the uranium surface, the photographic action is due principally to the β radiation.

## § 6. Transparency of Substances to the two Types of Radiation

If the intensity of the radiation in traversing a substance diminishes according to the ordinary absorption law, the ratio $r$ of the intensity of the radiation after passing through a distance $d$ of the substance to the intensity when the substance is removed is given by

$$r = e^{-\lambda d}$$

where $\lambda$ is the coefficient of absorption and $e = 2.7$.

In the following table a few values of $\lambda$ are given for the α and β radiations, assuming in each case that the radiation is simple and that the intensity falls off according to the above law:

| Substance | $\lambda$ for the α radiation | $\lambda$ for the β radiation |
|---|---|---|
| Dutch metal | 2700 | - |
| Aluminium | 1600 | 15 |
| Tinfoil | 2650 | 108 |
| Copper | - | 49 |
| Silver | - | 97 |
| Platinum | - | 240 |
| Glass | - | 5.6 |

The above results show what a great difference there is in the power of penetration of the two types of radiation. The transparency of aluminum for the β radiation is over one hundred times as great as for the α radiation. The opacity of the metals aluminum, copper, silver, platinum for the β radiation follows the same order as their atomic weights. Aluminum is the most transparent of the metals used, but glass is more transparent than aluminum for the β radiation. Platinum has an opacity sixteen times as great as aluminum. For the α radiation, aluminum is more transparent than Dutch metal or tinfoil.

For a thickness of aluminum 0.09 cm the intensity of the β radiation was reduced to 0.25 of its value; for a thickness of copper 0.03 cm. the intensity was reduced to 0.23 of its value. These results are not in agreement with some given by Becquerel, * who found copper was more transparent than aluminum for uranium radiation.

  * (1896), *Comptes Rendus,* p.763.

The β radiation has a penetrating power of about the same order as the radiation given out by an average x-ray bulb. Its power of penetration is, however, much less than for the rays from a 'hard' bulb. The α radiation, on the other hand, is far more easily absorbed than rays from an ordinary bulb, but is very similar in its penetrating power to the secondary radiation** sent out when x-rays fall upon a metal surface.

  ** (1898), Perrin, *Comptes Rendus,* cxxiv, p. 455; Sagnac, *Comptes Rendus.*

It is possible that the α radiation is a secondary radiation set up at the surface of the uranium by the passage of the β radiation through the uranium, in exactly the same way as a diffuse radiation is produced at the surface of a metal by the passage of Röntgen rays through it. There is not, however, sufficient evidence at present to decide the question.

[§ 7 to § 18. have been deleted.]

## § 19 General Remarks

The cause and origin of the radiation continuously emitted by uranium and its salts still remain a mystery. All the results that have been obtained point to the conclusion that uranium gives out types of radiation which, as regards their effects on gases, are similar to Röntgen rays and the secondary radiation emitted by metals when Röntgen rays fall upon them. If there is no polarization or refraction the similarity is complete. J. J. Thomson* has suggested that the regrouping of the constituents of the atom may give rise to electrical effects such as are produced in the ionization of a gas.

  * (1898). *Proc. Camb. Phil. Soc.,* ix, pt. viii, p. 397.

Röntgen's* and Wiedemann's** results seem to show that in the process of ionization a radiation is emitted which has similar properties to easily absorbed Röntgen radiation.

* (1898). *Wied. Annal.,* lxiv.
** (1895). *Zeit. f. Electrochemie,* ii., p. 159.

The energy spent in producing uranium radiation is probably extremely small, so that the radiation could continue for long intervals of time without much diminution of internal energy of the uranium. The effect of the temperature of the uranium on the amount of radiation given out has been tried. An arrangement similar to that in § 11 was employed. The radiation was completely absorbed in the gas. The vessel was heated up to about 200° C; but not much difference in the rate of discharge was observed. The results of such experiments are very difficult to interpret, as the variation of ionization with temperature is not known.

I have been unable to observe the presence of any secondary radiation produced when uranium radiation falls upon a metal. Such a radiation is probably produced, but its effects are too small for measurement.

In conclusion, I desire to express my best thanks to Professor J. J. Thomson for his kindly interest and encouragement during the course of this investigation.

Cavendish Laboratory

September 1, 1898

## Rutherford, E. (December, 1908). The Chemical Nature of the Alpha Particles from Radioactive Substances

Nobel Lecture, December 11, 1908;
https://www.nobelprize.org/prizes/chemistry/1908/rutherford/lecture/.)

In 1908, Rutherford was awarded the Nobel Prize in Chemistry "*for his investigations into the disintegration of the elements, and the chemistry of radioactive substances*." The discovery of radioactivity in 1896 led to a series of more in-depth investigations. In 1899 Ernest Rutherford demonstrated that there were at least two distinct types of radiation: *alpha radiation* and *beta radiation*. He discovered that radioactive preparations gave rise to the formation of gases. Working with Frederick Soddy, Rutherford advanced the hypothesis that helium gas could be formed from radioactive substances. In 1902 they formulated a revolutionary theory: that elements could disintegrate and be transformed into other elements. [Ernest Rutherford – Facts. NobelPrize.org.; https://www.nobelprize.org/ prizes/chemistry/1908/rutherford/facts/.]

The study of the properties of the α-rays has played a notable part in the development of radioactivity and has been instrumental in bringing to light a number of facts and relationships of the first importance. With increase of experimental knowledge there has been a growing recognition that a large part of radioactive phenomena is intimately connected with the expulsion of the α-particles. In this lecture an attempt will be made to give a brief historical account of the development of our knowledge of the α-rays and to trace the long and arduous path trodden by the experimenter in the attempts to solve the difficult question of the chemical nature of the α-particles. α-rays were first observed in 1899 as a special type of radiation and during the last six years there has been a persistent attack on this great problem, which has finally yielded to the assault when the resources of the attack seemed almost exhausted.

Shortly after his discovery of the radiating power of uranium by the photographic method, Becquerel showed that the radiation from uranium like the Röntgen-rays possessed the property of discharging an electrified body. [*Röntgen rays, commonly known as X-rays, are a form of high-energy electromagnetic radiation* discovered by Wilhelm Conrad Röntgen in 1895.] In a detailed investigation of this property, I examined the effect on the rate of discharge by placing successive layers of thin aluminum foil over the surface of a layer of uranium oxide and was led to the conclusion that *two types of radiation of very different penetrating power were present*. The conclusions at that period were summed up as follows:

"These experiments show that the uranium, radiation is complex and that there are present *at least two distinct types of radiation – one that is very readily absorbed, which will be termed for convenience the a-radiation, and the other of a more penetrative character, which will be termed the b-radiation."*[1]

[1] Rutherford, E. (January, 1899). *Uranium Radiation and the Electrical Conduction Produced by It. Phil. Mag.*, 5, 47, 109-163, p. 116; https://www.chemteam. info/Chem-History/Rutherford-Alpha&Beta.html.

When other radioactive substances were discovered, it was seen that the types of radiation present were analogous to the β– and α-rays of uranium and when a still more penetrating type of radiation from radium was discovered by Villard, the term γ-rays was applied to them. The names thus given soon came into general use as a convenient nomenclature for the three distinct types of radiation emitted from uranium, radium, thorium, and actinium. On account of their insignificant penetrating power, the α-rays were at first considered of little importance and attention was mainly directed to the more penetrating β-rays. With the advent of active preparations of radium, Giesel in 1899 showed that the β-rays from this substance were easily deflected by a magnetic field in the same direction as a stream of cathode rays and consequently appeared to be a stream of projected particles carrying a negative charge. *The proof of the identity of the β-particles with the electrons constituting the cathode rays was completed in 1900* by Becquerel, who showed that the β-particles from radium had about the same small mass as the electrons and *were projected at a speed comparable with the velocity of light*. Time does not allow me to enter into the later work of Kaufmann and others on this subject, which has greatly extended our knowledge of the constitution and mass of *electrons*.

In the meantime, further investigation had disclosed that the α-particles produced most of the ionization observed in the neighborhood of an unscreened radioactive substance, and that most of the energy radiated was in the form of α-rays. It was calculated by Rutherford and McClung in 1901 that one gram of radium radiated a large amount of energy in the form of α-rays.

The increasing recognition of the importance of the α-rays in radioactive phenomena led to attempts to determine the nature of this easily absorbed type of radiation. Strutt (Lord Rayleigh) in 1901 and Sir William Crookes in 1902 suggested that they might possibly prove to be *projected particles carrying a positive charge*. I independently arrived at the same conclusion from consideration of a variety of evidence. If this were the case, the α-rays should be deflected by a magnetic field. Preliminary work showed that the deflection was very slight if it occurred at all. Experiments were continued at intervals over a period of two years and it was not until 1902, when a preparation of radium of activity 19,000 was available, that I was able to show conclusively that the particles were deflected by a magnetic field, though in a very minute degree compared with the β-rays. This showed that *the α-rays consisted of projected charged particles while the direction of deflection indicated that each particle carried a positive charge*. The α-particles were shown to be deflected also by an electric field and from the magnitude of the deflection, it was deduced that the velocity of the swiftest particles was about $2.5 \times 10^9$ cm per second, or one-twelfth the velocity of light, while the value of e/m – the ratio of the charge carried by the particle to its mass – was found to be 5,000 electromagnetic units. Now it is known from the data of the electrolysis of water that the value of e/m for the hydrogen atom is 9,650. If the α-particle carried the same positive charge as the unit fundamental charge of the hydrogen atom, it was seen that *the mass of the α-particle was about twice that of the hydrogen atom*. On account of the complexity of the rays it was recognized that the results were only approximate, but the experiments indicated clearly that the α-particle was atomic in mass and might prove ultimately to be either a hydrogen or a helium atom or the atom of some unknown element of light atomic weight. These experiments were repeated by Des Coudres in 1903 with similar results, while Becquerel showed the deflection of the α-rays in a magnetic field by the photographic method.

This proof that the α-particles consisted of actual charged atoms of matter projected with an enormous velocity at once threw a flood of light on radioactive processes, in particular upon another important series of investigations which were being contemporaneously carried on in the Laboratory at Montreal in conjunction with Mr. F. Soddy. Had time permitted, it would have been of interest to consider in some detail the nature of these researches which placed on a firm foundation the now generally accepted "*transformation theory*" *of radioactivity*. From a close examination of the substances thorium, radium, and uranium, Rutherford and Soddy had reached the conclusion that radioactive bodies were in a state of transformation, as a result of which a number of new substances were produced entirely distinct in chemical and physical character from the parent element. From the independence of the rate of transformation of chemical and physical agencies, it was recognized that *the transformation was atomic and not molecular in character*. Each of these new bodies was shown to lose its radioactive properties according to a definite law. Even before the discovery of the material nature of the α-rays, it had been considered probable that the radiation from any particular substance accompanied the breaking up of its atoms. The proof that the α-particle was an ejected atom of matter at once strengthened this conclusion and at the same time gave a more concrete and definite representation of the processes occurring in radioactive matter. The point of view reached by us at that time is clearly seen from the following quotation, which with little alteration holds good today. "The results obtained so far point to the conclusion that the beginning of the succession of chemical changes taking place in radioactive bodies is due to the emission of the α-rays, i.e. the projection of a heavy charged mass from the atom. The portion left behind is unstable, undergoing further chemical changes which are again accompanied by the emission of α-rays, and in some cases also of β-rays.

The power possessed by the radioactive bodies of apparently spontaneously projecting large masses with enormous velocities supports the view that *the atoms of these substances are made up, in part at least, of rapidly rotating or oscillating systems of heavy charged bodies, large compared with the electron*. The sudden escape of these masses from their orbit may be due either to the action of internal forces or external forces of which we have at present no knowledge."[2]

[2] Rutherford, E. & Soddy, F. (1903), *Phil. Mag.*, 5, 106.

Consider for a moment the explanation of the changes in *radium*. A minute fraction of the radium atoms is supposed each second to become unstable, breaking up with explosive violence. A fragment of the atom – and α-particle – is ejected at a high speed, and the residue of the atom, which has a lighter weight than before, becomes an atom of a new substance, the radium emanation. The atoms of this substance are far more unstable than those of radium and explode again with the expulsion of an α-particle. As a result, the atom of *radium A* makes its appearance and the process of disintegration thus started continues through a long series of stages.

I can only refer in passing here to the large amount of work done by various experimenters in analyzing the long series of transformations of radium and thorium and actinium; the linking up of radium with uranium and the discovery by Boltwood of the long looked-for and elusive parent of radium, viz. ionium. This phase of the subject is of unusual interest and importance but has only an indirect bearing on the subject of my lecture. It has been shown that the great majority of the transition elements produced by the transformation of uranium and thorium break up with the expulsion of α-particles. A few, however, throw off only β-particles, while some are "rayless", i.e. undergo transformation without the expulsion of high-speed α– and β-particles. It is necessary to suppose that in these latter cases the atoms break up with the expulsion of α-particles at a speed too low to be detected, or, as is more probable, undergo a process of atomic rearrangement without the expulsion of material particles of atomic dimensions.

Another striking property of radium was soon seen to be connected with the expulsion of α-particles. In 1903 P. Curie and Laborde showed that *radium was a self-heating substance* and was always above the temperature of the surrounding air. It seemed probable from the beginning that the effect must be the result of the heating effect due to the impact of the α-particles on the radium. Consider for a moment a pellet of radium enclosed in a tube. The α-particles are shot out in great numbers equally from all parts of the radium and in consequence of their slight penetrating power are all stopped in the radium itself or by the walls of the tube. The energy of motion of the α-particles is converted into heat. On this view *the radium is subject to a fierce and unceasing bombardment by its own particles and is heated by its own radiation*. This was confirmed by the work of Rutherford and Barnes in 1903, who showed that three quarters of the heating effect of radium was not directly due to the radium but to its product, the emanation, and that each of the different substances produced in radium gave out heat in proportion to the energy of the $\alpha$-particles expelled from it. These experiments brought clearly to light *the enormous energy, compared with the weight of matter involved, which was emitted during the transformation of the emanation*. It can readily be calculated that one kilogram of the radium-emanation and its products would initially emit energy at the rate of 14,000 horse-power, and during its life would give off energy corresponding to about 80,000 horse-power for one day.

It was thus clear that the heating effect of radium was mainly a secondary phenomenon resulting from the bombardment by its own $\alpha$-particles. It was evident also that all the radioactive substances must emit heat in proportion to the number and energy of the $\alpha$-particles expelled per second.

We must now consider another discovery of the first importance. In discussing the consequences of the disintegration theory, Rutherford and Soddy drew attention to the fact that any stable substances produced during the transformation of the radio-elements should be present in quantity in the radioactive minerals, where the processes of transformation have been taking place for ages. This suggestion was first put forward in 1902.[3]

[3] Rutherford, E. & Soddy, F. (1902). *Phil. Mag.*, 4, 582.

"In the light of these results and the view that has already been put forward of the nature of radioactivity, the speculation naturally arises whether the presence of helium in minerals and its invariable association with uranium and thorium, may not be connected with their radioactivity, and again[4].

[4] Rutherford, E. & Soddy, F. (1903). *Phil. Mag.*, 5, 453.

"It is therefore to be expected that if any of the unknown ultimate products of the changes of a radioactive element are gaseous, they would be found occluded, possibly in considerable quantities, in the natural minerals containing that element. This lends support to the suggestion already put forwards, that possibly helium is an ultimate product of the disintegration of one of the radioactive elements, since it is only found in radioactive minerals."

It was at the same time recognized that it was quite possible that the $\alpha$-particle itself might prove to be a helium atom. As only weak preparations were then available, it did not seem feasible at that time to test whether helium was produced from radium. About a year later, thanks to Dr. Giesel of Braunschweig, preparations of pure radium bromide were made available to experimenters. Using 30 milligrams of Giesel's preparation, Sir William Ramsay and Soddy in 1903 were able to show conclusively that helium was present in radium some months old and that the emanation produced helium. This discovery was of the greatest interest and importance, for it brought to light that in addition to a series of transition elements, radium also gave rise in its transformation to a stable form of matter.

A fundamental question immediately arose as to the position of helium in the scheme of transformations of radium. Was the helium the end or final product of transformation of radium or did it arise at some other stage or stages? In a letter to Nature[5] I pointed out that probably helium was derived from the α-particles fired out by the α-ray products of radium and made an approximate estimate of the rate of production of helium by radium.

[5] Rutherford, E. (Aug. 20, 1903). Letter in *Nature*, 69.

It was calculated that the amount of helium produced per gram of radium should lie between 20 and 200 cubic millimeters per year and probably nearer the latter estimate. The data available for calculation at that time were imperfect, but it is of interest to note that the rate of production of helium recently found by Sir James Dewar, in 1908, viz. 134 cubic millimeters per year, is not far from the value calculated as most probable at that time.

These estimates of the rate of production of helium were later modified as new and more accurate data became available. In 1905, I measured the charge carried by the α-particles from a thin film of radium. Assuming that each α-particle carried the ionic charge measured by J.J. Thomson, I showed that $6.2 \times 10^{10}$ α-particles were expelled per second per gram of radium itself and four times this number when radium was in equilibrium with its three α-ray products. The rate of production of helium calculated on these data was 240 cubic millimeters per gram per year.

In the meantime, by the admirable researches of Bragg and Kleeman in 1904, our knowledge of the character of the absorption of the α-particles by matter had been much extended. It had long been known that the absorption of α-particles by matter was different in many respects from that of the β-rays. Bragg showed that these differences arose from the fact that the α-particle, on account of its great energy of motion, was not deflected from its path like the β-particle, but travelled in nearly a straight line, ionizing the molecules in its path. From a thin film of matter of one kind, the α-particles were all projected at the same speed and lost their power of producing ionization suddenly, after traversing a certain definite distance of air. The velocity of the α-particles in this view were reduced by their passage through matter by equal amounts. These conclusions of Bragg were confirmed by experiments I made by the photographic method. As a source of rays, a thin film of radium C, deposited from the radium-emanation on a thin wire, was used. By examining the deflection of the rays in a magnetic field, it was found that the rays were homogeneous and were expelled from the surface of the wire at an identical speed. By passing the rays through a screen of mica or aluminum, it was found *that the velocity of all the α-particles were reduced by the same amount* and the issuing beam was still homogeneous.

A remarkable result was noted. *All α-particles apparently lost their characteristic properties of ionization, phosphorescence and photographic action, at exactly the same point while they were still moving at a speed of about 9,000 kilometres per second*. At this critical speed, the α-particle suddenly vanishes from our ken and can no longer be followed by the methods of observation at our command.

The use of a homogeneous source of α-rays like radium C at once suggested itself as affording a basis for a more accurate determination of the value of e/m for the α-particle and for seeing whether the value was consistent with the view that the α-particle was a charged atom of helium. In the course of a long series of experiments, I proved that *the α-particles, whether expelled from radium, thorium or actinium, were identical in mass* and must consist of the same kind of matter.

The velocity of expulsion of the α-particles from different kinds of active matter varied over comparatively narrow limits but the value of e/m was constant and equal to 5,070. This value was not very different from the one originally found. A difficulty at once arose in interpreting this result. We have seen that the value of e/m for the hydrogen atom is 9,650. If the α-particle carried the same positive charge as the hydrogen atom, the value of e/m for the α-particle would indicate that *its mass was twice that of the hydrogen atom*, i.e. equal to the mass of a hydrogen molecule. It seemed very improbable that hydrogen should be ejected in a molecular and not an atomic state as a result of the atomic explosion. If, however, the α-particle carried a charge equal to twice that of the hydrogen atom, the mass of the α-particle would work out at nearly four, i.e. a mass nearly equal to that of the atom of helium.

*I suggested that, in all probability, the β-particle was a helium atom which carried two unit charges*. On this view, every radioactive substance which emitted α-particles must give rise to helium. This at once offered an explanation of the fact observed by Debierne that actinium as well as radium produced helium. It was pointed out that the presence of a double charge of helium-atom was not altogether improbable for reasons to be given later (p. 138).

While the evidence as a whole strongly supported the view that the α-particle was a helium atom, it was found exceedingly difficult to obtain a decisive experimental proof of the relation. *If it could be shown experimentally that the α-particle did in reality carry two unit charges, the proof of the relation would be greatly strengthened*. For this purpose, *an electrical method was devised by Rutherford and Geiger for counting directly the α-particles expelled from a radioactive substance*. The ionization produced in a gas by a single α-particle is exceedingly small and would be difficult to detect electrically except by a very refined method. Recourse was had to an automatic method of magnifying the ionization produced by an α-particle. For this purpose, it was arranged that the α-particles should be fired through a small opening into a vessel containing air or other gas at a low pressure, exposed to an electric field near the sparking value. Under these conditions the ions produced by the passage of the α-particle through the gas generate a large number of fresh ions by collision. In this way it was found possible to magnify the electrical effect due to an α-particle several thousand times. The entrance of an α-particle into the testing vessel was then indicated by a sudden deflection of the electrometer needle. This method was developed into an accurate method of counting the number of α-particles fired in a known time through the small aperture of the testing vessel. From this was deduced the total number of α-particles expelled per second from any thin film of radioactive matter. *In this way it was shown that 3.4 x 10$^{10}$ α-particles are expelled per second from one gram of radium itself and from each of its α-ray products in equilibrium with it.*

The correctness of this method was indicated by another, quite distinct method of counting. Sir William Crookes and Elster and Geitel had shown that the α-particles falling on a screen of phosphorescent zinc sulphide produced a number of scintillations. Using specially prepared screens, Rutherford and Geiger counted the number of these scintillations per second with the aid of a microscope. It was found that, within the limit of experimental error, the number of scintillations per second on a screen agreed with the number of α-particles impinging on it, counted by the electrical method. It was thus clear that each α-particle produced a visible scintillation on the screen, and that either the electrical or the optical method could be used for counting the α-particles. Apart from the purpose for which these experiments were made, the results are of great interest and importance, for it is the first time that it has been found possible to detect a single atom of matter by its electrical and optical effect. This is of course only possible because of the great velocity of the α-particle.

*Knowing the number of α-particles expelled from radium from the counting experiment, the charge carried by each α-particle was determined by measuring the total positive charge carried by all the α-particles expelled*. It was found that each α-particle carried a positive charge of 9.3 x 10$^{-10}$ electrostatic units. From a consideration of the experimental evidence of the charge carried by the ions in gases, *it was concluded that the α-particle did carry two unit charges*, and that the unit charge carried by the hydrogen atom was equal to 4.65 x 10$^{-10}$ units. From a comparison of the known value of e/m for the α-particle with that of the hydrogen atom, *it follows that an α-particle is a projected atom of helium carrying two charges*, or, to express it in another way, *the α-particle, after its charge is neutralized, is a helium atom*.

The data obtained from the counting experiments allow us to calculate simply the magnitude of a number of important radioactive quantities. It was found that the calculated values of the life of radium, of the volume of the emanation, and of the heating effect of radium were in excellent agreement with the values found experimentally. A test of the correctness of these methods of calculation was forthcoming shortly after the publication of these results. Rutherford and Geiger calculated, on the assumption that the α-particle was a helium atom, that one gram of radium in equilibrium should produce a volume of 158 cubic millimeters of helium per year. Sir James Dewar in 1908 carried out a long experimental investigation on the rate of production of helium by radium, and showed that one gram of radium in equilibrium produced about 134 cubic millimeters per year. Considering the difficulty of the investigation, the agreement between the experimental and calculated values is very good and is strong evidence in support of the identity of the α-particle with a helium atom.

While the whole train of evidence we have considered indicates with little room for doubt that the α-particle is a projected helium atom, there was still wanting a decisive and incontrovertible proof of the relationship. It might be argued, for example, that the helium atom appeared as a result of the disintegration of the radium atom in the same way as the atom of the emanation and had no direct connection with the α-particle. If one helium atom were liberated at the same time that an α-particle was expelled, experiment and calculation might still agree and yet the α-particle might be an atom of hydrogen or of some unknown substance.

In order to remove this possible objection, it is necessary to show that the α-particles, collected quite independently of the active matter from which they are expelled, give rise to helium. With this purpose in view some experiments were recently (1908) made by Rutherford and Royds. A large quantity of emanation was forced into a glass tube which had walls so thin that the α-particles were fired right through them, though the walls were impervious to the emanation itself. The α-particles were projected into the glass walls of an outer sealed vessel and were gradually released into the exhausted space between the emanation tube and the outer vessel. After some days a bright spectrum of helium was observed in the outer vessel. There is, however, one objection to this experiment. It might be possible that the helium observed had diffused through the thin glass walls from the emanation. This objection was removed by showing that no trace of helium appeared, when the emanation was replaced by a larger volume of helium itself. We may thus confidently conclude *that the α-particles themselves give rise to helium, and are atoms of helium*.

Further experiments showed that when the α-particles were fired through the glass walls into a thin sheet of lead or tin, helium could always be obtained from the metals after a few hours' bombardment.

Considering the evidence together, *we conclude that the α-particle is a projected atom of helium, which has, or in some way during its flight acquires, two unit charges of positive electricity*. It is somewhat unexpected that the atom of a monatomic gas like helium should carry a double charge. It must not however be forgotten that *the α-particle is released at a high speed as a result of an intense atomic explosion, and plunges through the molecules of matter in its path*. Such conditions are exceptionably favorable to the release of loosely attached electrons from the atomic system. *If the α-particle can lose two electrons in this way, the double positive charge is explained*.

We have seen that there is every reason to believe that the α-particles, so freely expelled from the great majority of radioactive substances, are identical in mass and constitution and must consist of atoms of helium. We are consequently driven to the conclusion that the atoms of the primary radioactive elements like uranium and thorium must be built up in part at least of atoms of helium. These atoms are released at definite stages of the transformations at a rate independent of control by laboratory forces. There is good reason to believe that in the majority of cases, a single helium atom is expelled during the atomic explosion. This is certainly the case for radium itself and its series of products. On the other hand, Bronson has drawn attention to certain cases, viz. the emanations of actinium and of thorium, where apparently two and three atoms of helium respectively are expelled at one time. No doubt these exceptions will receive careful investigation in the future. It is of interest to note that uranium itself appears to expel two α-particles for one from each of its products. Knowing the number of atoms of helium expelled from the atom of each product, we can at once calculate the atomic weights of the products. For example, in the uranium-ionium-radium series, uranium expels two α-particles and each of the six following α-ray products one, i.e. eight in all. Taking the atomic weight of uranium as 238.5, the atomic weight of ionium should be 230.5, of radium 226.5, of the emanation 222.5, and so on. It is of interest to note that the atomic weight of radium deduced in this way is in close agreement with the latest experimental values. The atomic weight of the end-product of radium, resulting from the transformation of radium F (polonium) should be $238.5 - 8 \times 4 = 206.5$, or a value close to that for lead. Long ago, Boltwood suggested from examination of analyses of old uranium minerals, that lead was in all probability a transformation product of the uranium-radium series. The coincidence of numbers is certainly striking, but a direct proof of the production of lead from radium will be required before this conclusion can be considered as definitely established.

It is very remarkable that a chemically inert element like helium should play such a prominent part in the constitution of the atomic systems of uranium and thorium and radium. It may well be that this property of helium of forming complex atoms is in some way connected with its inability to enter into ordinary chemical combinations. It must not be forgotten that uranium and thorium and each of their transformation products must be regarded as distinct chemical elements in the ordinary sense. They differ from ordinary elements in the comparative instability of their atomic systems. The atoms break up spontaneously with great violence, expelling in many cases an atom of helium at a high speed. All the evidence is against the view that uranium or thorium or radium can be regarded as an ordinary molecular compound of helium with some known or unknown element, which breaks up into helium. The character of the radioactive transformations and their independence of temperature and other agencies have no analogy in ordinary chemical changes.

Apart from their radioactivity and high atomic weight, uranium, thorium, and radium show no especially distinctive chemical behavior. Radium for example is closely allied in general chemical properties to barium. It is consequently not unreasonable to suppose that other elements may be built up in part of helium, although the absence of radioactivity may prevent us from obtaining any definite proof. On this view, it may prove significant that the atomic weights of many elements differ by four – the atomic weight of helium-or a multiple of four. Time is too limited to discuss in greater detail these and other interesting questions which have been raised by the proof of the chemical nature of the α-particle. *

* The lecture was illustrated with lantern slides and experiments on the radium emanation.

## Rutherford, E. (May, 1911). LXXIX. The scattering of α and β particles by matter and the structure of the atom.[1]

*Phil. Mag. 6,* 21, 125, 669–88; http://dx.doi.org/10.1080/14786440508637080.

[1] Communicated by the author. A brief account of this paper was communicated to the Manchester Literary and Philosophical Society in February, 1911.

Received April, 1911.

[*Alpha particles are helium-4 nuclei, consisting of two protons and two neutrons, and carry a positive charge of +2e.* They have a mass of approximately 6.6464835 x $10^{-27}$ kg and travel at speeds up to about one-tenth of the speed of light. They are deflected by electric and magnetic fields and have a range of about 3-4 cm in air.

*Beta particles are high-energy electrons. They carry a negative charge of -1e* and can reach speeds up to 50% the speed of light. Beta particles are easily deflected by magnetic and electric fields and have a greater range in air, able to travel several meters.

*Gamma rays are electromagnetic radiation.* They show no deflection in the electric field and travel in a straight line. Gamma rays are highly penetrating.]

§ 1 It is well known that the α and β particles suffer deflections from their rectilinear paths by encounters with atoms of matter. This scattering is far more marked for the β than for the α particle on account of the much smaller momentum and energy of the former particle. There seems to be no doubt that such swiftly moving particles pass through the atoms in their path, and that the deflections observed are due to the strong electric field traversed within the atomic system. It has generally been supposed that the scattering of a pencil of α or β rays in passing through a thin plate of matter is the result of a multitude of small scattering by the atoms of matter traversed. The observations, however, of Geiger and Marsden[2] on the scattering of α rays indicate that some of the α particles must suffer a deflection of more than a right angle at a single encounter.

[2] (1909), *Proc. Roy. Soc.,* LXXXII, p. 495.

They found, for example, that a small fraction of the incident α particles, about 1 in 20,000 turned through an average angle of 90 degrees in passing through a layer of gold–foil about .00004 cm. thick, which was equivalent in stopping power of the α particle to 1.6 millimeters of air. Geiger[3] showed later that the most probable angle of deflections for a pencil of α particles traversing a gold–foil of this thickness was about 0.87°.

[3] (1910), *Proc. Roy. Soc.*, LXXXIII. p. 492.

A simple calculation based on the theory of probability shows that the chance of an α particle being deflected through 90 degrees is vanishingly small. In addition, it will be seen later that the distribution of the α particles for various angles of large deflection does not follow the probability law to be expected if such deflections are made up of a large number of small deviations. It seems reasonable to suppose that the deflection through a large angle is due to a single atomic encounter, for the chance of a second encounter of a kind to produce a large deflection must in most cases be exceedingly small. A simple calculation shows that the atom must be a seat of an intense electric field in order to produce such a large deflection at a single encounter.

Recently Sir J.J. Thomson[4] has put forward a theory to explain the scattering of electrified particles in passing through small thickness of matter.

[4] (1910), *Camb. Lit. & Phil. Soc.*, XV. pt. 5.

The atom is supposed to consist of a number N of negatively charged ***corpuscles***, accompanied by an equal quantity of positive electricity uniformly distributed throughout a sphere. The deflection of a negatively electrified particle in passing through the atom is ascribed to two causes– (1) the repulsion of the ***corpuscles'*** distribution through the atom, and (2) the attraction of the positive electricity in the atom. The deflection of the particle in passing through the atom is supposed to be small, while the average deflection after a large number $m$ of encounters was taken as $\sqrt{m} . \theta$, where $\theta$ is the average deflection due to a single atom. It was shown that the number N of the electrons within the atom could be deduced from observations of the scattering of electrified particles. The accuracy of this theory of compound scattering was examined experimentally by Crowther[5] in a later paper.

[5] Crowther, (1910), *Proc. Roy. Soc.*, LXXXIV. p.226.

His result apparently confirmed the main conclusions of the theory, and he deduced, on the assumption that the positive electricity was continuous, that the number of electrons in an atom was about three times its atomic weight.

The theory of Sir J.J. Thomson is based on the assumption that the scattering due to a single encounter is small, and the particular structure assumed for the atom does not admit of a very large deflection of an α particle in traversing a single atom, unless it be supposed that the diameter of the sphere of positive electricity is minute compared with the diameter of the sphere of influence of the atom.

Since the α and β particles traverse the atom, it should be possible from a close study of the nature of the deflection to form some idea of the constitution of the atom to produce the effects observed. In fact, the scattering of high–speed charged particles by the atoms of matter is one of the most promising methods of attack of this problem. The development of the scintillation method of counting single α particles affords unusual advantages of investigation, and the researches of H. Geiger by this method have already added much to our knowledge of the scattering of α rays by matter.

§ 2 We shall first examine theoretically the single encounters [6] with an atom of simple structure, which is able to produce large deflections of an α particle, and then compare the deductions from the theory with the experimental data available.

> [6] The deviation of a particle throughout a considerable angle from an encounter with a single atom will in this paper be called *"single" scattering*. The deviation of a particle resulting from a multitude of small deviations will be termed *"compound" scattering*.

Consider an atom which contains a charge ± $N_e$ at its center surrounded by a sphere of electrification containing a charge ∓ $N_e$ supposed uniformly distributed throughout a sphere of radius is the fundamental unit of charge, which in this paper is taken as 4.65 × $10^{-10}$ E.S. unit. We shall suppose that for distances less than $10^{-12}$ cm the central charge and also the charge on the α particle may be supposed to be concentrated at a point. It will be shown that the main deductions from the theory are independent of whether the central charge is supposed to be positive or negative. For convenience, the sign will be assumed to be positive. The question of the stability of the atom proposed need not be considered at this stage, for this will obviously depend upon the minute structure of the atom, and on the motion of the constituent charged parts.

In order to from some idea of the forces required to deflect an α particle through a large angle, consider an atom containing a positive charge $N_e$ at its center, and surrounded by a distribution of negative electricity $N_e$ uniformly distributed within a sphere of radius R. The electric force X and the potential V at a distance r from the center of an atom for a point inside the atom, are given by

$$X = N_e \left( 1/r^2 - r/R^3 \right)$$
$$V = N_e \left( 1/r - 3/2R + r^2/R^3 \right).$$

Suppose an α particle of mass m velocity u and charge E shot directly towards the center of the atom. It will be brought to rest at a distance b from the center given by

$$1/2mu^2 = N_eE \, (1/b - 3/2R + b^2/R^3).$$

It will be seen that b is an important quantity in later calculations. Assuming that the central charge is 100e. it can be calculated that the value of b for an α particle of velocity $2.09 \times 10^9$ cm. per second is about $3.4 \times 10^{-12}$ cm. In this calculation b is supposed to be very small compared with R. Since R is supposed to be of the order of the radius of the atom, viz. $10^{-8}$ cm, it is obvious that the α particle before being turned back penetrates so close to the central charge, that the field due to the uniform distribution of negative electricity may be neglected. In general, a simple calculation shows that for all deflections greater than a degree, we may without sensible error suppose the deflection due to the field of the central charge alone. Possible single deviations due to the negative electricity, if distributed in the form of corpuscles, are not taken into account at this stage of the theory. It will be shown later that its effect is in general small compared with that due to the central field.

Fig. 1

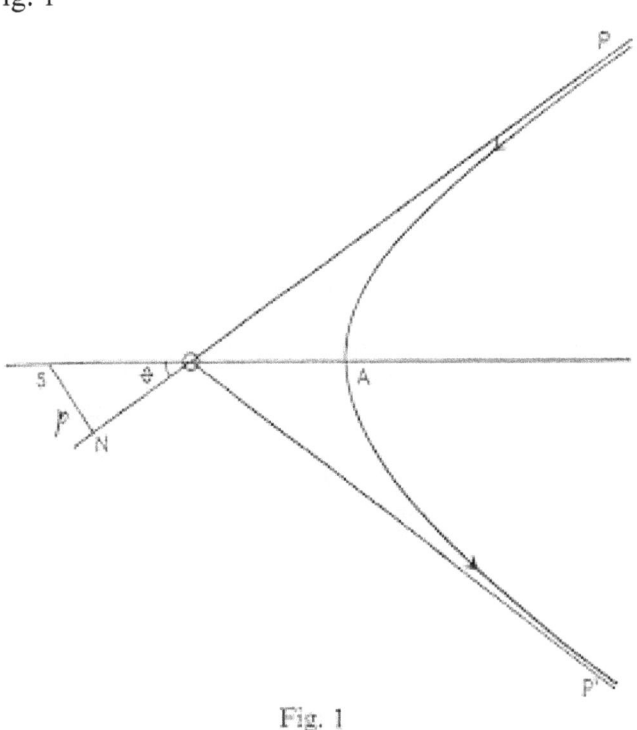

Fig. 1

Consider the passage of a positive electrified particle close to the center of an atom. Supposing that the velocity of the particle is not appreciably charged by its passage through the atom, the path of the particle under the influence of a repulsive force varying inversely

145

as the square of the distance will be a hyperbola with the center of the atom S as the external focus. Suppose the particle to enter the atom in the direction PO [Fig. 1], and that the direction on motion on escaping the atom is OP'. OP and OP' make equal angles with the line SA, where A is the apse of the hyperbola. p = SN = perpendicular distance from center in direction of initial motion of particle. Let angle POA= $\theta$. Let V = velocity of particle on entering the atom, v its velocity at A, then from consideration of *angular momentum*.

$$pV = SA \cdot v$$

From *conservation of energy*

$$1/2mV^2 = 1/2mv^2 - N_eE/SA,$$
$$v^2 = V^2 (1 - b/SA).$$

Since the eccentricity is sec $\theta$,

$$SA = SO + OA = p \csc \theta (1 + \cos \theta) = 2 \cot \theta/2,$$
$$P^2 = SA (SA - b) = p \cot \theta /2(p \cot \theta /2 - b)$$
$$b = 2p \cot \theta.$$

The angle of deviation $\phi$ of the particle is $\pi - 2\theta$ and

$$\cot \phi/2 = 2p/b \ast \tag{1}$$

*A simple consideration shows that the deflection is unaltered if the forces are attractive instead of repulsive

This gives the angle of deviation of the particle in terms of b, and the perpendicular distance of the direction of projection from the center of the atom.

For illustration, the angle of deviation $\phi$ for different values of p/b are shown in the following table: –

| p/b .... | 10 | 5 | 2 | 1 | 0.5 | 0.25 | 0.125 |
|---|---|---|---|---|---|---|---|
| $\phi$ ...... | $5^0$-7 | $11^04$ | $28^0$ | $53^0$ | $90^0$ | $127^0$ | $152^0$ |

## § 3 Probability of single deflection through any angle.

...

146

## § 7 General Considerations

In comparing the theory outlined in this paper with the experimental result, it has been supposed that the atom consists of a central charge supposed concentrated at a point, and that the large single deflections of the α and β particles are mainly due to their passage through the strong central field. The effect of the equal and opposite compensating charge supposed distributed uniformly throughout a sphere has been neglected. Some of the evidence in support of these assumptions will now be briefly considered. For concreteness, consider the passage of a high speed α particle through an atom having a positive central charge Ne, and surrounded by a compensating charge of N electrons. Remembering that the mass, momentum, and kinetic energy of the α particle are very large compared with the corresponding values for an electron in rapid motion, it does not seem possible from dynamic considerations that an α particles can be deflected through a large angle by a close approach to an electron, even if the latter be in rapid motion and constrained by strong electrical forces. It seems reasonable to suppose that the chance of single deflections through a large angle due to this cause, if not zero, must be exceedingly small compared with that due to the central charge.

It is of interest to examine how far the experimental evidence throws light on the question of the extent of the distribution of the central charge. Suppose, for example, the central charge to be composed of N unit charges distributed over such a volume that the large single deflections are mainly due to the constituent and not to the external field produced by the distribution. It has shown (§ 3) that the fraction of the α particles scattered through a large angle is proportional to $(N_eE)^2$, where $N_e$ is the central charge concentrated at a point and E the charge on the deflected particle. If, however, this charge is distributed in single units, the fraction of the α particles scattered through a given angle is proportional to $N_e^2$ instead of $N^2e^2$. In this calculation, the influence of mass of the constituent particles has been neglected, and account has only been taken of its electric field. Since it has been shown that the value of the central point charge for gold must be about 100, the value of the distributed charge requires to produce the same proportion of single deflections through a large angle should be at least 10,000. Under these conditions the mass of the constituent particle would be small compared with that of the α particle, and the difficulty arises of the production of large single deflections at all. In addition, with such a large distributed charge, the effect of compound scattering is relatively more important than that of single scattering. For example, the probable small angle of deflection of a pencil of α particles passing through a thin gold–foil would be much greater than that experimentally observed by Geiger (§ b-c). The large and small angle scattering could not then be explained by the assumption of a central charge of the same value. Considering the evidence as a whole, it seems simplest to suppose that the atom contains a central charge distributed through a very small volume, and that the large single deflections are due to the central charge as a whole, and not to its constituents. At the same time, the experimental evidence is not precise enough to negate the possibility that a small fraction of the positive charge may be carried by satellites extending some distance from the center. Evidence on this point could be obtained by examining whether the same central charge is required to explain the single deflections of α and β particles; for the α particle must approach much closer to the center of the atom than the β particle of average speed to suffer the same large deflection.

The general data available indicate that the value of this central charge for different atoms is approximately proportional to their atomic weights, at any rate for atoms heavier than aluminum. It will be of great interest to examine experimentally whether such a simple relation holds also for the lighter atoms. In cases where the mass of the deflecting atom (for example, hydrogen, helium, lithium) is not very different from that of the α particle, the general theory of single scattering will require modification, for it is necessary to take into account the movements of the atom itself (see § 4). It is interest to note that Nagaoka[12] has mathematically considered the properties of a "Saturnian" atom which he supposed to consist of a central attracting mass surrounded by rings of rotating electrons.

[12] Nagaoka, (1904), *Phil. Mag.*, VII. p. 445.

He showed that such a system was stable if the attractive force was large. From the point of view considered in this paper, the chance of large deflection would practically be unaltered, whether the atom is considered to be a disk or a sphere. It may be remarked that the approximate value found for the central charge of the atom of gold (100e) is about that to be expected if the atom of gold consisted of 49 atoms of helium, each carrying a charge 2e. This may be only a coincidence, but it is certainly suggestive in view of the expulsion of helium atoms carrying two unit charges from radioactive matter.

The deductions from the theory so far considered are independent of the sign of the central charge, and it has not so far been found possible to obtain definite evidence to determine whether it be positive or negative. It may be possible to settle the question of sign by consideration of the difference of the laws of absorption of the β particle to be expected on the two hypotheses, for the effect of radiation in reducing the velocity of the β particle should be far more marked with a possible than with a negative center. If the central charge be positive, it is easily seen that a positively charge mass, if released from the center of a heavy atom, would acquire a great velocity in moving through the electric field. It may be possible in this way to account for the high velocity of expulsion of α particles without supposing that they are initially in rapid motion within the atom.

Further consideration of the application of this theory to these and other questions will be reserved for a later paper [see below], when the main deductions of the theory have been tested experimentally. Experiments in this direction are already in progress by Geiger and Marsden.

University of Manchester
April 1911.

The preceding paper by Rutherford sets forth his theory of the scattering of α particles by atoms composed of a small, centrally located, positively charged nucleus surrounded by a sphere of equal but uniformly distributed negative charge whose effect on the scattering of the particles is negligible. The orders of magnitude envisioned were, roughly, for the nuclear radius about $10^{-12}$ cm, and for the whole atom about $10^{-8}$ cm. If one imagines that discrete negative electrons were present instead of a distribution of negative charge, then the nuclear atom would be mostly empty space.

**Rutherford, E. (March, 1914). The Structure of the Atom. ***

*Phil. Mag.,* 6, 27, p. 488–98.

      * Communicated by the Author.

The present paper and the accompanying paper by Mr. C. Darwin [see below] deal with certain points in connection with the "nucleus" theory of the atom which were purposely omitted in my first communications on that subject (*Phil. Mag.*, May 1911).

A brief account is given of the later investigation which have been made to test the theory and of the deductions which can be drawn from them. At the same time a brief statement is given of recent observations on the passage of α particles through hydrogen, which throw important light on the dimensions of the nucleus.

**Scattering of α particles**

In my previous paper (*loc. cit.*) I pointed out the importance of the study of the passage of the high speed α and β particles through matter as a means of throwing light on the internal structure of the atom. Attention was drawn to the remarkable fact, first observed by Geiger and Marsden**, that a small fraction of the swift α particles from radioactive substances were able to be deflected through an angle of more than 90° as the results of an encounter with a single atom.

      ** (1909), *Proc. Roy. Soc. A.*, lxxxii. p. 495.

It was shown that the type of atom devised by Lord Kelvin and worked out in great detail by Sir J. J. Thomson was unable to produce such large deflections unless the diameter of the positive sphere was exceedingly small. In order to account for this large angle scattering of α particles, I supposed that the atom consisted of a positively charged nucleus of small dimensions in which practically all the mass of the atom was concentrated. The nucleus was supposed to be surrounded by a distribution of electrons to make the atom electrically neutral, and extending to distances from the nucleus comparable with the ordinary accepted radius of the atom. Some of the swift α particles passes through the atoms in their path and entered the intense electric field in the neighborhood of the nucleus and were deflected from their rectilinear path. In order to suffer a deflection of more than a few degrees, the α particle has to pass very close to the nucleus, and it was assumed that the field of force in this region was not appreciably affected by the external electronic distribution. Supposing that the forces between the nucleus and the α particle are repulsive and follow the law of inverse squares the α particle describes a hyperbolic orbit round the nucleus and its deflection can be simply calculated.

It was deduced from this theory that the number of α particles falling normally on unit area of a surface and making an angle φ with the direction of the incident rays is proportional to

(1) $\operatorname{cosec}^4 \varphi/2$ or $1/\varphi^4$ if φ be small;

(2) the number of atoms per unit volume of the scattering material;

(3) thickness of scattering material $t$ provided this is small;

(4) square of the nucleus charge N$e$;

(5) and is inversely proportional to $(mu^2)^2$, where $m$ is the mass of the α particle and $u$ is velocity.

From the data of scattering on α particles previously given by Geiger[*], it was deduced that the value of the nucleus charge was equal to about half the atomic weight multiplied by the electronic charge.

[*] (1910), *Proc. Roy. Soc A.*, lxxxiii. p. 492.

Experiments were begun by Geiger and Marsden[**] to test the laws of theory.

[**] Geiger & Marsden, (1913), *Phil. Mag.*, xxv. p. 604.

The general experimental method employed by them consisted in allowing a narrow pencil of a particles to fall normally on a thin film of matter, and observing by the scintillation method the number scattered through different angles. This was a very difficult and laborious piece of work involving the counting of many thousands of particles. They found that their results were in very close accord with the theory. When the thickness of the scattering film was very small the amount of scattering was directly proportional to the thickness and varied inversely as the fourth power of the velocity of the incident α particles. A special study was made of the number of α particles scattered through angles varying between 5 and 150. Although over this range the number decreased in the ratio 200,000 to 1, the relation between number and angle agreed with the theory within the limit of experimental error. They found that the scattering of different atoms of matter was approximately proportional to the square of the atomic weight, showing that the charge on the nucleus was nearly proportional to the atomic weight. By determining the number of α particles scattering from thin films of gold, they concluded that the nucleus charge was equal to about half the atomic weight multiplied by the electronic charge. On the account of the difficulties of this experiment, the actual number could not be considered correct within more than 20 percent.

The experimental results of Geiger and Marsden were thus in complete according with the predictions of the theory, and indicated the essential correctness of this hypothesis of the structure of the atom.

In determining the magnitude of single scattering, I assumed in my previous paper, for simplicity of calculation, that the atom was at rest during an encounter with an α particle. In an accompanying paper, Mr. C. Darwin has worked out the relations to be expected when account is taken of the motion of the recoiling atom. He has shown that no sensible error has been introduced in this way even for atoms of such low atomic weight as carbon. Mr. Darwin has also worked out the scattering to be expected if the law of force is not that of the inverse square, and has shown that it is not in accord with experiment either with regard to the variation of scattering with velocity. The general evidence certainly indicates that the law of force between the α particle and the nucleus is that of the inverse square.

It is of interest to note that C.T.R. Wilson*, by photographing the trails of the α particle, later showed that the α particle occasionally suffering a sudden deflection does occasionally occur as a result of an encounter with a single atom.

* Wilson, C. T. R., (1912), *Proc. Roy. Soc. A.*, lxxxvii. p. 277.

On the theory outlined, the large deflections of the α particle are supposed to be due to its passage close to the nucleus where the field is very intense and to be not appreciably affected by its passage through the external distribution of electrons. This assumption seems to be legitimate when we remember that the mass and energy of the α particle are very large compared with that of an electron even moving with a velocity comparable with that of light. Simple considerations show that the deflections which an α particle would experience even in passing through the complex electronic distribution of a heavy atom like gold, must be small compared with the large deflections actually observed. In fact, the passage of swift α particles through matter affords the most definite and straightforward method of throwing light on the gross structure of the atom, for the α particle is able to penetrate the atom without serious disturbance from the electronic distribution, and this is only affected by the intense field associated with the nucleus of the atom.

This independence of the large angle scattering on the external distribution of electrons is only true for charged particles whose kinetic energy is very large. It is not to be expected that it will hold for particles moving at very much lower speeds and with much less energy--such, for example, as ordinary cathode particles or the recoil atoms from active matter. In such cases it is probable that the external electronic distribution plays a far more prominent part in governing the scattering than in the case under consideration.

## Scattering of β particles

It is to be anticipated on the nucleus theory that swift β particles should suffer deflections through large angles in their passage close to the nucleus. There seems to be no doubt that such large deflections are actually produced, and I showed in my previous paper that the results of scattering of β particles found by Crowther* could be generally explained on the nucleus theory of atomic structure.

* Crowther, (1910), *Proc. Roy. Soc. A.*, lxxxiv. p. 226.

It should be borne in mind, however, that there are several important points of distinction between the effects to be expected for an α particle and a β particle. Since the force between the nucleus and β particle is attractive, the β particle increases rapidly in speed in approaching the nucleus. On the ordinary electrodynamics, this entails a loss of energy by radiation, and also an increase of the apparent mass of the electron. Darwin** has worked out mathematically the result of these effects on the orbit of the electron, and has shown that, under certain conditions, the β particle does not escape from the atom but describes a spiral orbit ultimately falling into the nucleus.

** Darwin, C. G. (1913) XXV. On some orbits of an electron. Phil. Mag., 1, 25, 201-10.

This result is of great interest, for it may offer an explanation of the disappearance of swift β particles in their passage through matter. In addition, it must be borne in mind that the swiftest β particle expelled from radium C possess only about one-third of the energy of the corresponding α particle, while the average energy of the β particle is less than one-sixth of that of the α particle. It is thus to be anticipated that the large angle scattering of a β particle by the nucleus will take place in regions where the α particle will only suffer a small deflection -- regions for which the application of the simple theory may not have been accurately tested. For these reasons, *it is of great importance to determine the laws of large angle scattering of β particles of different speeds in passing throug*h *matter*, as it should throw light on a number of important points connected with atomic structure. Experiments are at present in progress in the laboratory to examine the scattering of such β particles in detail.

It is obvious that a β particle in passing close to an electron will occasionally suffer a large deflection. The problem is mathematically similar to that for a close encounter of an α particle with a helium atom of the same mass, which is discussed by Mr. Darwin in the accompanying paper. Such large deflections due to electronic encounter, however, should be relatively small in number compared with those due to the nucleus of a heavy atom.

**Scattering in Hydrogen.**

Special interest attaches to the effects to be expected when α particles pass through light gases like hydrogen and helium. In a previous paper by Mr. Nuttal and the author*, it has been shown that the scattering of α particles in hydrogen and helium is in good agreement with the view that the hydrogen nucleus has one positive charge, while the α particle, or helium, has two.

* Rutherford & Nuttall, (1913), *Phil Mag.*, xxvi. p. 702.

Mr. Darwin has worked out in detail the simple scattering to be anticipated when α particles pass through hydrogen and helium. It is only necessary here to refer to the fact that on the nucleus theory a small number of hydrogen atoms should acquire, as the result of close encounters with α particles, velocities about 1.6 times that of the velocity of the α particle itself. On account of the fact that the hydrogen atom carries one positive charge while the α particle carries two, it can be calculated that some of the hydrogen atoms should have a range in hydrogen of nearly four times that of the α particle which sets them in motion.

Mr. Marsden has kindly made experiments for me to test whether the presence of such hydrogen atoms can be detected. A detailed account of his experiments will appear later, but it suffices to mention here that undoubted evidence has been obtained by him that some of the hydrogen atoms are set in such swift motion that they are able to produce a visible scintillation on a zinc sulphide screen and are able to travel through hydrogen a distance three or four times greater than the colliding α particle. The general method employed was to place a thin α-ray tube containing about 100 millicuries of purified emanation in a tube filled with hydrogen. The scintillations due to the α particle from the tube disappeared in air after traversing a distance of about 5 cm. When the air was displaced by hydrogen the great majority of the scintillations disappeared at about 20 cm. from the source, which corresponds to the range of the α particle in hydrogen. A small number of scintillations observed is of the order of magnitude to be anticipated on the theory of single scattering, supposing that the nucleus in hydrogen and helium has such small dimensions, and that they behave like point charges for distances up to $10^{-13}$ cm.

There appears to be no doubt that the scintillations observed beyond 20 cm. are due to charged hydrogen atoms which are set in swift motion by a close encounter with an α particle. Experiments are at present in progress by Mr. Marsden to determine the number of hydrogen atoms set in motion, and the variation of the number of hydrogen atoms set in motion, and the variation of the number with the scattering angle.

It does not appear possible to explain the appearance of such swift hydrogen atoms unless it be supposed that the forces of repulsion between the α particle and the hydrogen atom are exceedingly intense. Such intense forces can only arise if the positive nuclei have exceedingly small dimensions, so that a close approach between them is possible.

**Dimensions and Constitution of the Nucleus.**

In my previous paper I showed that the nucleus must have exceedingly small dimensions, and calculated that in the case of gold its radius was not greater than $3 \times 10^{-12}$ cm. In order to account for the velocity given to hydrogen atoms by the collision with α particles, it can be simply calculated (see Darwin) that the centers of nuclei of helium and hydrogen must approach within a distance of $1.7 \times 10^{-13}$ cm of each other. Supposing for simplicity the nuclei to have dimensions and to be spherical in shape, it is clear that the sum of the radii of the hydrogen and helium nuclei is not greater than $1.7 \times 10^{-13}$ cm. This is an exceedingly small quantity, even *smaller* than the ordinarily accepted value of the diameter of the electron, viz. $2 \times 10^{-13}$ cm. It is obvious that the method we have considered gives a maximum estimate of the dimensions of the nuclei, and it is not improbable that the hydrogen nucleus itself may have still smaller dimensions. This raises the question whether the hydrogen nucleus is so small that its mass may be accounted for in the same way as the mass of the negative electron.

It is well known from the experiments of J.J. Thomson and others, that no positively charged carrier has been observed of mass less than that of the hydrogen atom. The exceedingly small dimensions found for the hydrogen nucleus ***add weight to the suggestion that the hydrogen nucleus is the positive electron***, and that its mass is entirely electromagnetic in origin. According to the electromagnetic theory, the electrical mass of a charged body, supposed spherical, is $(2/3) \, e^2/a$ where $e$ is the charge and $a$ the radius. The hydrogen nucleus consequently must have a radius about $1/1830$ of the electron if its mass is to be explained in this way. There is no experimental evidence at present contrary to such an assumption.

The helium nucleus has a mass nearly four times that of hydrogen. If one supposes that the positive electron, i.e. the hydrogen atom, is a unit of which all atoms are composed, it is to be anticipated that the helium atom contains four positive electrons and two negative.

It is well known that a helium atom is expelled in many cases in the transformation of radioactive matter, but no evidence has so far been obtained of the expulsion of a hydrogen atom. In conjunction with Mr. Robinson, I have examined whether any other charged atoms are expelled from radioactive matter except helium atoms, and the recoil atoms which accompany the expulsion of a particles. The examination showed that if such particles are expelled, their number is certainly less than 1 in 10,000 of the number of helium atoms. It thus follows that the helium nucleus is a very stable configuration which survives the intense disturbances resulting in its expulsion with high velocity from the radioactive atom, and is one of the units, of which possibly the great majority of the atoms are composed. The radioactive evidence indicates that the atomic weight of successive products decreases by four units consequent on the expulsion of an α particle, and it has often been pointed out that the atomic weights of many of the permanent atoms differ by about four units.

It will be seen later that *the resultant positive charge on the nucleus determines the main physical and chemical properties of the atom*. The *mass of the atom* is, however, dependent on the number and arrangement of the positive and negative electrons constituting the atom. Since the experimental evidence indicates that the nucleus has very small dimensions, the constituent positive and negative electrons must be very close together. As Lorentz has pointed out, the electrical mass of a system of charged particles, if close together, will depend not only on the number of these particles, but on the way their fields interact. For the dimensions of the positive and negative electrons considered, the packing must be very close in order to produce an appreciable alteration in the mass due to this cause. This may, for example, be the explanation of the fact that the helium atom has not quite four times the mass of the hydrogen atom. Until, however, the nucleus theory has been more definitely tested, *it would appear premature to discuss the possible structure of the nucleus itself*. The general theory would indicate that the nucleus of a heavy atom is an exceedingly complicated system, although its dimensions are very minute.

An important question arises whether the atomic nuclei, which all carry a positive charge, contain negative electrons. This question has been discussed by Bohr*, who concluded from the radioactive evidence that the high speed β particles have their origin in the nucleus.

* Bohr, (1913), *Phil. Mag.*, xxvi. pp. 476, 857.

The general radioactive evidence certainly supports such a conclusion. It is well known that the radioactive transformations which are accompanied by the expulsion of high speed β particles are, like the α ray changes, unaffected by wide ranges of temperature or by physical and chemical conditions. On the nucleus theory, there can be no doubt that the a particle has its origin in the nucleus and gains a great part, if not all, or its energy of motion in escaping from the atom. It seems reasonable, therefore, to suppose that a β ray transformation also originates from the expulsion of a negative electron from the nucleus. It is well known that the energy expelled in the form of β and γ rays during the transformation of radium C is about the one-quarter of the energy of the expelled α particle.

It does not seem easy to explain this large emission of energy by supposing it to have its origin in the electronic distribution. It seems more likely that a very high-speed electron is liberated from the nucleus, and in its escape from the atom sets the electronic distribution in violent vibration, given rise to intense γ rays and also to secondary β particles. The general evidence certainly indicates that many of the high-speed electrons form radioactive matter are liberated from the electronic distribution in consequence of the disturbance due to the primary electron escaping from the nucleus.

**Charge on the Nucleus**

We have seen that from an examination of the scattering of *a* particles by matter, that the positive charge on the nucleus is approximately equal to 1/2 A*e*, when A is the atomic weight and *e* the unit charge. This is equivalent to the statement that the number of electrons in the external distribution is about half the atomic weight in terms of hydrogen. It is of interest to note that this is the value deduced by Barkla* from entirely different evidence viz. the scattering of X rays in their passage through matter.

* Barkla, (1911), *Phil. Mag.*, xxi. p. 648.

This is founded on the theory of scattering given by Sir J. J. Thomson, which supposes that each electron in an atom scatters as an independent unit. It seems improbable that the electrons within the nucleus would contribute to this scattering, for they are packed together with positive nuclei and must be held in equilibrium by forces of a different order of magnitude from those which bind the external electrons.

It is obvious from the consideration of the cases of hydrogen and helium, where hydrogen has one electron and helium two, that the number of electrons cannot be exactly half the atomic weight in all cases. This has led to an interesting suggestion by van den Broek** that the number of units of charge on the nucleus, and consequently the number of external electrons, may be equal to the number of the elements when arranged in order of increasing atomic weight.

** van den Broek, (1913), *Phys. Zeit.*, xiv. p. 32.

On this view, the nucleus charges of hydrogen, helium, and carbon are 1, 2, 6 respectively, and so on for the other elements, provided there is no gap due to a missing element. This view has been taken by Bohr in his theory of the constitution of simple atoms and molecules.

Recently strong evidence of two distinct kinds has been brought in support of such a contention. Soddy* has pointed out that the recent generalization of the relation between the chemical properties of the elements and the radiations can be interpreted by supposing that the atom loses two positive charges by the expulsion of an α particle, and one negative by the expulsion of a high-speed electron.

* Soddy, (1913), *Jahr. d. Rad.*, x. p.188.

From a consideration of the series of products of the three main radioactive branches of uranium, thorium, and actinium, it follows that some of the radioactive elements may be arranged so that the nucleus charge decreases by one unit as we pass from one element to another. It would thus appear that van den Broek's suggestion probably holds for some if not all of the heavy radioactive elements. Recently Moseley** has supplied very valuable evidence that this rule also holds for a number of the lighter elements.

** Moseley, (1913), *Phil. Mag.*, xxvi. p.1024.

By examination of the wave-length of the characteristic X rays emitted by twelve elements varying in atomic weight between calcium (40) and zinc (65.4), he has shown that the variation of wave-length can be simply explained by supposing that the charge on the nucleus increases from element to element by exactly one unit. This holds true for cobalt and nickel, although it has long been known that they occupy an anomalous relative position in the periodic classification of the elements according to atomic weights.

There appears to be no reason why this new and powerful method of analysis, depending on an examination of the frequency of the characteristic X ray spectra of the elements, should not be extended to a large number of elements, so that further definite data on the point may be expected in the near future.

*It is clear on the nucleus theory that the physical and chemical properties of the ordinary elements are for the most part dependent entirely on the charge of the nucleus*, for the latter determines *the number and distribution of the external electrons on which the chemical and physical properties must mainly depend*. As Bohr has pointed out, the properties of gravitation and radioactivity, which are entirely uninfluenced by chemical or physical agencies, must be ascribed mainly if not entirely to the nucleus, while the ordinary physical and chemical properties are determined by the number and distribution of the external electrons. On this view, the nucleus charge is a fundamental constant of the atom, while the atomic mass of an atom may be a complicated function of the arrangement of the units which make up the nucleus.

It should be borne in mind that there is no inherent impossibility on the nucleus theory that atoms may differ considerably in atomic weight and yet have the same nucleus charge. This is most simply illustrated by radioactive evidence. In the following table the atomic weight and nucleus charge are given for a few of the successive elements arising from the transformation of uranium. The actual nucleus charge of uranium is known, but for simplicity it is assumed to be 100.

| Successive Elements ve | Ur1 --> | UrX1 --> | UrX2 --> | Ur2 --> | Io --> | Ra |
|---|---|---|---|---|---|---|
| Atomic weights | 238.5 | 234.5 | 234.5 | 234.5 | 230.5 | 226.5 |
| Charge on nucleus | 100 | 98 | 99 | 100 | 98 | 96 |

Following the recent theories, it is supposed that the emission of a $a$ particle lowers the nucleus charge by two units, while the emission of a $\beta$ particle raises it by one unit. It is seen that $Ur_1$ and $Ur_2$ have the same nucleus charge although they differ in atomic weight by four units.

If the nucleus is supposed to be composed of a mixture of hydrogen nuclei with one charge and of helium nuclei with two charges, it is *a priori* conceivable that a number of atoms may exist with the same nucleus charge but of different atomic masses. The radioactive evidence certainly supports such a view, but probably only a few of such possible atoms would be stable enough to survive for a measurable time.

Bohr* has drawn attention to the difficulties of constructing atoms on the "nucleus" theory, and has shown that the stable positions of the external electrons cannot be deducted from the classical mechanics.

* Bohr, (1913), *Phil. Mag.*, xxvi. pp. 476, 857.

By the introduction of a conception connected with Planck's quantum, he has shown that on certain assumptions it is possible to construct simple atoms and molecules out of positive and negative nuclei, e.g. the hydrogen atom and molecule and the helium atom, which behave in many respects like the actual atoms or molecules. While there may be much difference of opinion as to the validity and of the underlying physical meaning of the assumptions made by Bohr, there can be no doubt that the theories of Bohr are of great interest and importance to all physicists as the first definite attempt to construct simple atoms and molecules and to explain their spectra.

University of Manchester,

February 1914.

# PART IV    The neutron.

# Rutherford, E. (June, 1919). Collision of α particles with light atoms. I-IV.

*Philosophical Magazine.* 37: 537-87.

142                *The Journal of the Röntgen Society.* OCTOBER, 1919.

COLLISION OF α-PARTICLES WITH LIGHT ATOMS. I.-IV. E. *Rutherford.* (Phil. Mag. 37, pp. 537-587, June, 1919.)—I. *Hydrogen.* II. *Velocity of the Hydrogen Atom.* III. *Nitrogen and Oxygen Atoms.* IV. *An Anomalous Effect in Nitrogen.*

These four papers are of an extremely important nature, as they throw considerable light on the internal structure of atoms. The method of atom-analysis employed, in which α-particles are in intimate collision with the nuclei of gaseous atoms, is unique, supplying, as it does, atomic data which could not be obtained by other known methods.

I. *Hydrogen.* The production of high-speed hydrogen atoms due to close collisions between α-particles and atoms of hydrogen has been studied, using the α-particles from RaC as a homogeneous source of radiation. In such close collisions, where the nuclei approach within a distance of about $3 \times 10^{-13}$ cm., the number and distribution of H atoms are entirely different from those calculated on the assumption that the nuclei are to be regarded as point charges repelling each other according to the law of inverse squares. The H atoms produced by swift α-particles of range 7 cm. are shot forward mainly in the direction of the α-particles and are nearly uniform in velocity. The distribution with velocity of H atoms becomes more and more heterogeneous with decrease of velocity of the α-particles. For α-particles of range less than 4 cm. of air, the distribution and absorption of H atoms are in fair accord with the simple theory although the observed numbers are greater than those calculated on the theory.

The number of swift H atoms produced by α-particles of range 7 cm. is thirty times greater than the theoretical number. The number falls off rapidly for ranges of α-particles between 3 and 2 cm. On an average $10^5$ α-particles give rise to one swift hydrogen atom in traversing 1 cm. of hydrogen. It has been calculated that all α-particles of range 7 cm. projected within a perpendicular distance $p = 2 \cdot 4 \times 10^{-13}$ cm. of the centre of the hydrogen nucleus give rise to swift H atoms. The corresponding apsidal distance is about $3 \cdot 5 \times 10^{-13}$ cm. As observed by Marsden {Abs. 1421 (1915)}, hydrogen atoms are emitted by the radio-active source. The number observed is small, and it is difficult to decide whether these H atoms arise from the radio-active transformation or from occluded hydrogen in the source. An important theoretical discussion follows the statement of the above results. For this the original paper should be consulted.

II. *Velocity of the Hydrogen Atoms.* Measurements of the magnetic and electrostatic deflection of H atoms produced by the α-particles from RaC are described. The deflection in a magnetic field alone was first determined, then in a combined magnetic and electrostatic field. From the magnetic deflections it was found $m u_0 /e = 3 \cdot 15 \times 10^5$, and from the electric deflection $u_0 = 3 \cdot 12 \times 10^9$; consequently $e/m = 10^4$ e.m. units. The value of $e/m$ for the hydrogen atom in the electrolysis of water is 9570. The agreement is sufficiently close to show that the long-range scintillations produced by α-particles in hydrogen are due to hydrogen atoms carrying a unit positive charge. The agreement between the calculated and observed velocities of the H atom shows that within the margin of experimental error, the conservation of momentum and energy hold for close collisions between the atomic nuclei and that there is no sensible loss of energy due to radiation.

The relative brightness of H atoms and α-particles scintillation is discussed from the point of view of energy loss along their paths due to ionization.

In the course of counting H scintillations, it was often noted that a number of the scintillations appeared as *instantaneous* "doubles;" *i.e.*, two points of light of about equal brightness appeared in the field of view at the same instant. This question has been systematically examined by E. Marsden, but experimental conditions were such that the effects observed were too small and uncertain to allow of drawing definite conclusions.

III. *Nitrogen and Oxygen Atoms.* From simple theoretical considerations it may be proved that all atoms of atomic weight up to oxygen should be detected beyond the range of the α-particle. Supposing that α-particles of range 7 cm. are used, the max. ranges to be expected for unit charge are for He, 28·0 ; Li, 19·6 ; Be, 15·4 ; B, 12·4 ; C, 11·2 ; N, 9·3 ; O, 7·8 cm. The case of helium has already been examined, when it was found that this atom carried a double charge, like the α-particle. It is intended later to make a systematic examination of the elements Li, Be, and B—the results so far obtained being merely preliminary. The elements nitrogen and oxygen have received very careful attention, and some startling results have been revealed. It was found that the scintillations in pure oxygen and $CO_2$ were of about the same brightness for corresponding ranges, and had nearly the same equivalent ranges in air as those due presumably to N atoms from the air. This is surprising, as it is to be expected that the O atom would have considerably less range than the N atom ; the calculated ranges are 7·8 and 9·3 cm. respectively. There seems to be no doubt that the effects produced by the collision of α-particles with N and O atoms are very similar to those observed in hydrogen. These observations only receive an explanation by assuming that the N and O atoms, like the H atoms, are thrown forward mainly in the direction of the α-particles.

IV. *An Anomalous Effect in Nitrogen.* In making absorption experiments with H particles in pure oxygen it was found that the number of scintillations diminished to the amount to be expected from the stopping power of the column of gas. A surprising effect was noticed, however, when dried air was introduced. Instead of diminishing, the number of scintillations was increased, and for an absorption corresponding to about 19 cm. of air the number was about twice that observed when the air was exhausted. A systematic series of observations was undertaken to account for the origin of these scintillations which were eventually proved to originate in the nitrogen. *Both as regards range and brightness of scintillations the long-range atoms from nitrogen closely resemble H atoms, and in all probability are H atoms.* This extremely important conclusion

# Rutherford, E. (October, 1919). Collision of α particles with light atoms. LIV. Collision of α particles with light atoms. IV. An anomalous effect in nitrogen. *

*Philosophical Magazine*. 37: 581-7; https://zenodo.org/records/1430800.

* Communicated by the Author.

It has been shown in paper I. that a metal source, coated with a deposit of radium C, always gives rise to a number of scintillations on a zinc sulphide screen far beyond the range of the α particles. The swift atoms causing these scintillations carry a positive charge and are deflected by a magnetic field, and have about the same range and energy as the swift H atoms produced by the passage of a particles through hydrogen. These "natural" scintillations are believed to be due mainly to swift H atoms from the radioactlve source, but it is difficult to decide whether they are expelled from the radioactive source itself or are due to the action of α particles on occluded hydrogen.

The apparatus employed to study these "natural" scintillations is the same as that described in paper I. The intense source of radium C was placed inside a metal box about 3 cm from the end, and an opening in the end of the box was covered with a silver plate of stopping power equal to about 6 cm of air. The zinc sulphide screen was mounted outside, about 1 mm distant from the silver plate, to admit of the introduction of absorbing foils between them. The whole apparatus was placed in a strong magnetic field to deflect the β rays. The variation in the number of these "natural" scintillations with absorption in terms of cms of air is shown in fig. 1, curve A. In this case, the air in the box was exhausted and absorbing foils of aluminum were used. When dried oxygen or carbon dioxide was admitted into the vessel, the number of scintillations diminished to about the amount to be expected from the stopping power of the column of gas.

A surprising effect was noticed, however, when dried air was introduced. Instead of diminishing, the number of scintillations was increased, and for an absorption corresponding to about 19 cm of air the number was about twice that observed when the air was exhausted. It was clear from this experiment that the α particles in their passage through air gave rise to long-range scintillations which appeared to the eye to be about equal in brightness to H scintillations. A systematic series of observations was undertaken to account for the origin of these scintillations. In the first place we have seen that the passage of a particles through nitrogen and oxygen gives rise to numerous bright scintillations which have a range of about 9 cm in air. These scintillations have about the range to be expected if they are due to swift N or O atoms, carrying unit charge, produced by collision with α particles. All experiments have consequently been made with an absorption greater than 9 cm of air, so that these atoms are completely stopped before reaching the zinc sulphide screen.

It was found that these long-range scintillations could not be due to the presence of water vapor in the air; for the number was only slightly reduced by thoroughly drying the air. This is to be expected, since on the average the number of the additional scintillations due to air was equivalent to the number of H atoms produced by the mixture of hydrogen at 6 cm. pressure with oxygen. Since on the average the vapor pressure of water in air was not more than 1 cm., the effects of complete drying would not reduce the number by more than one sixth. Even when oxygen and carbon dioxide saturated with water vapor at 20°C were introduced in place of dry air, the number of scintillations was much less than with dry air.

It is well known that the amount of hydrogen or gases containing hydrogen is normally very small in atmospheric air. No difference was observed whether the air was taken directly from the room or from outside the laboratory or was stored for some days over water.

There was the possibility that the effect in air might be due to liberation of H atoms from the dust nuclei in the air. No appreciable difference, however, was observed when the dried air was filtered through long plugs of cotton-wool, or by storage over water for some days to remove dust nuclei.

Since the anomalous effect was observed in air, but not in oxygen, or carbon dioxide, it must be due either to nitrogen or to one of the other gases present in atmospheric air. The latter possibility was excluded by comparing the effects produced in air and in chemically prepared nitrogen. The nitrogen was obtained by the well-known method of adding ammonium chloride to sodium nitrite, and stored over water. It was carefully dried before admission to the apparatus. With pure nitrogen, the number of long-range scintillations under similar conditions was greater than in air. As a result of careful experiments, the ratio was found to be 1.25, the value to be expected if the scintillations are due to nitrogen.

The results so far obtained show that the long-range scintillations obtained from air must be ascribed to nitrogen, but it is important, in addition, to show that they are due to collision of a particles with atoms of nitrogen through the volume of the gas. In the first place, it was found that the number of the scintillations varied with the pressure of the air in the way to be expected if they resulted from collision of α particles along the column of gas. In addition, when an absorbing screen of gold or aluminum was placed close to the source, the range of the scintillations was found to be reduced by the amount to be expected if the range of the expelled atom was proportional to the range of the colliding α particles. These results show that the scintillations arise from the volume of the gas and are not due to some surface effect in the radioactive source.

In fig. 1 curve A the results of a typical experiment are given showing the variation in the number of natural scintillations with the amount of absorbing matter in their path measured in terms of centimeters of air for α particles. In these experiments carbon dioxide was introduced at a pressure calculated to give the same absorption of the α rays as ordinary air. In curve B the corresponding curve is given when air at N.T.P. is introduced in place of carbon dioxide. The difference curve C shows the corresponding variation of the number of scintillations arising from the nitrogen in the air. It was generally observed that the ratio of the nitrogen effect to the natural effect was somewhat greater for 19 cm than for 12 cm absorption.

In order to estimate the magnitude of the effect, the space between the source and screen was filled with carbon dioxide at diminished pressure and a known pressure of hydrogen was added. The pressure of the carbon dioxide and of hydrogen were adjusted so that the total absorption of α particles in the mixed gas should be equal to that of the air. In this way it was found that the curve of absorption of H atoms produced under these conditions was somewhat steeper than curve C of fig. 1. As a consequence, the amount of hydrogen mixed with carbon dioxide required to produce a number of scintillations equal to that of air, increased with the increase of absorption. For example, the effect in air was equal to about 4 cm of hydrogen at 12 cm absorption, and about 8 cm at 19 cm absorption. For a mean value of the absorption, the effect was equal to about 6 cm of hydrogen. This increased absorption of H atoms under similar conditions indicated either that (1) the swift atoms from air had a somewhat greater range than the H atoms, or (2) that the atoms from air were projected more in the line of flight of the α particles.

While the maximum range of the scintillations from air using radium C as a source of α rays appeared to be about the same, viz. 28 cm as for H atoms produced from hydrogen, it was difficult to fix the end of the range with certainty on account of the smallness of the number and the weakness of the scintillations. Some special experiments were made to test whether, under favorable conditions, any scintillations due to nitrogen could be observed beyond 28 cm of air absorption. For this purpose, a strong source (about 60 mg Ra activity) was brought within 2.5 cm of the zinc sulphide screen, the space between containing dry air. On still further reducing the distance, the screen became too bright to detect very feeble scintillations. No certain evidence of scintillations was found beyond a range of 28 cm. *It would therefore appear that (2) above* [that the atoms from air were projected more in the line of flight of the α particles] *is the more probable explanation*.

In a previous paper (III) we have seen that the number of swift atoms of nitrogen or oxygen produced per unit path by collision with α particles is about the same as the corresponding number of H atoms in hydrogen. Since the number of long-range scintillations in air is equivalent to that produced under similar conditions in a column of hydrogen at 6 cm pressure, we may consequently conclude that only one long-range atom is produced for every 12 close collisions giving rise to a swift nitrogen atom of maximum range 9 cm.

It is of interest to give data showing the number of long-range scintillations produced in nitrogen at atmospheric pressure under definite conditions. For a column of nitrogen 3.3 cm long, and for a total absorption of 19 cm of air from the source, the number due to nitrogen per milligram of activity is .6 per minute on a screen of 3.14 sq. mm area.

Both as regards range and brightness of scintillations, *the long-range atoms from nitrogen closely resemble H atoms, and in all probability are hydrogen atoms*. In order, however, to settle this important point definitely, it is necessary to determine the deflection of these atoms in a magnetic field. Some preliminary experiments have been made by a method similar to tbat employed in measuring the velocity of the H atom (see paper II). The main difficulty is to obtain a sufficiently large deflection of the stream of atoms and yet have a sufficient number of scintillations per minute for counting. The α rays from a strong source passed through dry air between two parallel horizontal plates 3 cm long and 1.6 mm apart, and the number of scintillations on the screen placed near the end of the plates was observed for different strengths of the magnetic field. Under these conditions, when the scintillations arise from the whole length of the column of air between the plates, the strongest magnetic field available reduced the number of scintillations by only 30 per cent. When the air was replaced by a mixture of carbon dioxide and hydrogen of the same stopping power for α rays, about an equal reduction was noted. As far as the experiment goes, this is an indication that the scintillations are due to H atoms; but the actual number of scintillations and the amount of reduction was too small to place much reliance on the result. In order to settle this question definitely, it will probably prove necessary to employ a solid nitrogen compound, free from hydrogen, as a source, and to use much stronger sources of α rays. In such experiments, it will be of importance to discriminate between the deflections due to H atoms and possible atoms of atomic weight 2. From the calculations given in paper III, it is seen that a collision of an α particle with a free atom of mass 2 should give rise to an atom of range about 32 cm. in air, and of initial energy about .89 of that of the H atom produced under similar conditions. The deflection of the pencil of these rays in a magnetic field should be about .6 of that shown by a corresponding pencil of H atoms.

**Discussion of results.**

*From the results so far obtained it is difficult to avoid the conclusion that the long-range atoms arising from collision of α particles with nitrogen are not nitrogen atoms but probably atoms of hydrogen, or atoms of mass 2.* If this be the case, *we must conclude that the nitrogen atom is disintegrated under the intense forces developed in a close collision with a swift a particle, and that the hydrogen atom which is liberated formed a constituent part of the nitrogen nucleus.* We have drawn attention in paper III. to the rather surprising observation that the range of the nitrogen atoms in air is about the same as the oxygen atoms, although we should expect a difference of about 19 per cent. If in collisions which give rise to swift nitrogen atoms, the hydrogen is at the same time disrupted, such a difference might be accounted for, for the energy is then shared between two systems.

It is of interest to note, that while the majority of the light atoms, as is well known, have atomic weights represented by 4n or 4n + 3 where n is a whole number, nitrogen is the only atom which is expressed by 4n+2. We should anticipate from radioactive data that the nitrogen nucleus consists of three helium nuclei each of atomic mass 4 and either two hydrogen nuclei or one of mass 2. If the H nuclei were outriders of the main system of mass 12, the number of close collisions with the bound H nuclei would be less than if the latter were free, for the α particle in a collision comes under the combined field of the H nucleus and of the central mass. Under such conditions, it is to be expected that the α particle would only occasionally approach close enough to the H nucleus to give it the maximum velocity, although in many cases it may give it sufficient energy to break its bond with the central mass. Such a point of view would explain why the number of swift H atoms from nitrogen is less than the corresponding number in free hydrogen and less also than the number of swift nitrogen atoms. The general results indicate that the H nuclei, which are released, are distant about twice the diameter of the electron ($7 \times 10^{-13}$ cm) from the center of the main atom. Without a knowledge of the laws of force at such small distances, it is difficult to estimate the energy required to free the H nucleus or to calculate the maximum velocity that can be given to the escaping H atom. It is not to be expected, a priori, that the velocity or range of the H atom released from the nitrogen atom should be identical with that due to a collision in free hydrogen.

Taking into account the great energy of motion of the α particle expelled from radium C, the close collision of such an α particle with a light atom seems to be the most likely agency to promote the disruption of the latter; for the forces on the nuclei arising from such collisions appear to be greater than can be produced by any other agency at present available. Considering the enormous intensity of the forces brought into play, it is not so much a matter of surprise that the nitrogen atom should suffer disintegration as that the α particle itself escapes disruption into its constituents. The results as a whole suggest that, if α particles--or similar projectiles--of still greater energy were available for experiment, we might expect to break down the nucleus structure of many of the lighter atoms.

I desire to express my thanks to Mr. William Kay for his invaluable assistance in counting scintillations.

University of Manchester,
April, 1919.

# Rutherford, E. (July, 1920). Nuclear Constitution of Atoms.

*Roy. Soc. Proc. A*, 97, 686, 374–400; https://doi.org/10.1098/rspa.1920.0040; http://royalsocietypublishing.org/rspa/article-pdf/97/686/374/32944/rspa.1920.0040.pd

Bakerian lecture at the Royal Society.

Lecture delivered June 3, 1920.

**Introduction.** —The conception of the nuclear constitution of atoms arose initially from attempts to account for the scattering of α-particles through large angles in traversing thin sheets of matter. *

* Geiger and Marsden, (1909). *Roy. Soc. Proc. A*, 82, p. 495.

Taking into account the large mass and velocity of the α-particles, these large deflections were very remarkable, and indicated that very intense electric or magnetic fields exist within the atom. To account for these results, it was found necessary to assume** that ***the atom consists of a charged massive nucleus of dimensions very small, compared with the ordinarily accepted magnitude of the diameter of the atom***.

** Rutherford, (1911), *Phil. Mag.*, 21, p. 669; (1914), 27, p. 488.

***This positively charged nucleus contains most of the mass of the atom, and is surrounded at a distance by a distribution of negative electrons equal in number to the resultant positive charge on the nucleus***. Under these conditions, a very intense electric field exists close to the nucleus, and the large deflection of the α-particle in an encounter with a single atom happens when the particle passes close to the nucleus. Assuming that the electric forces between the α-particle and the nucleus varied according to an inverse square law in the region close to the nucleus, the writer worked out the relations connecting the number of α-particles scattered through any angle with the charge on the nucleus and the energy of the α-particle. Under the central field of force, the α-particle describes a hyperbolic orbit round the nucleus, and the magnitude of the deflection depends on the closeness of approach to the nucleus. From the data of scattering of α-particles then available, it was deduced that the resultant charge on the nucleus was about ½ Ae, where A is the atomic weight and e the fundamental unit of charge. Geiger and Marsden***; made an elaborate series of experiments to test the correctness of the theory, and confirmed the main conclusions.

*** Geiger and Marsden, (1913), *Phil. Mag.*, 25, p. 604.

They found the nucleus charge was about ½ Ae, but, from the nature of the experiments, it was difficult to fix the actual value within about 20 per cent. C. G. Darwin****

**** Darwin, (1914), *Phil, Mag.*, 27, p. 499.

worked out completely the deflection of the α-particle and of the nucleus, taking into account the mass of the latter, and showed that the scattering experiments of Geiger and Marsden could not be reconciled with any law of central force, except the inverse square. The nuclear constitution of the atom was thus very strongly supported by the experiments on scattering of α-rays. Since the atom is electrically neutral, the number of external electrons surrounding the nucleus must be equal to the number of units of resultant charge on the nucleus. It should be noted that, from the consideration of the scattering of X-rays by light elements, Barkla* had shown, in 1911, that the number of electrons was equal to about half the atomic weight.

* Barkla, (1911). *Phil. Mag.*, 21, p. 648.

This was deduced from **the theory of scattering of Sir J. J. Thomson**, in which it was assumed that each of the external electrons in an atom acted as an independent scattering unit. Two entirely different methods had thus given similar results with regard to the number of external electrons in the atom, but **the scattering of α-rays had shown in addition that the positive charge must be concentrated on a massive nucleus of small dimensions**. It was suggested by Van den Broek** that the scattering of α-particles by the atoms was **not inconsistent with the possibility that the charge on the nucleus was equal to the atomic number of the atom**, i.e., to the number of the atom when arranged in order of increasing atomic weight.

** Van den Broek, (1913). *Phys. Zeit.*, 14, p. 32.

The importance of the atomic number in fixing the properties of an atom was shown by the remarkable work of Moseley *** on the X-ray spectra of the elements.

*** Moseley, (1913), *Phil. Mag.*, 26, p. 1024; (1914), 27, p. 703.

He showed that **the frequency of vibration of corresponding lines in the X-ray spectra of the elements depended on the square of a number which varied by unity in successive elements**. This relation received an interpretation by supposing that the nuclear charge varied by unity in passing from atom to atom, and was given numerically by the atomic number. I can only emphasize in passing the great importance of Moseley's work, not only in fixing the number of possible **elements**, and the position of undetermined **elements**, but in showing that the properties of an atom were defined by a number which varied by unity in successive atoms. This gives a new method of regarding the *periodic classification of the elements*, for the atomic number, or its equivalent the nuclear charge, is of more fundamental importance than its atomic weight. In Moseley's work, *the frequency of vibration of the atom was not exactly proportional to N, where N is the atomic number, but to $(N-a)^2$*, where $a$ was a constant which had different values, depending on whether the K or L series of characteristic radiations were measured. It was supposed that this constant depended on the number and position of the electrons close to the nucleus.

**Charge on the Nucleus.** —The question whether the atomic number of an element is the actual measure of its nuclear charge is a matter of such fundamental importance that all methods of attack should be followed up. Several researches are in progress in the Cavendish Laboratory to test the accuracy of this relation. The two most direct methods depend on the scattering of swift α- and β-rays. The former is under investigation, using new methods, by Mr. Chadwick, and the latter by Dr. Crowther. The results so far obtained by Mr. Chadwick strongly support the *identity of the atomic number with the nuclear charge* within the possible accuracy of experiment, viz., about 1 per cent.

It thus seems clear that we are on firm ground in supposing that the nuclear charge is numerically given by the atomic number of the element. Incidentally, these results, combined with the work of Moseley, indicate that the law of the inverse square holds with considerable accuracy in the region surrounding the nucleus. It will be of great interest to determine the extent of this region, for it will give us definite information as to the distance of the inner electrons from the nucleus. A comparison of the scattering of slow and swift β-rays should yield important information on this point. The agreement of experiment with theory for the scattering of α-rays between 5° and 150° shows that the law of inverse square holds accurately in the case of a heavy element like gold for distances between about 36 x $10^{-12}$ cm and 3 x $10^{-12}$ cm from the center of the nucleus. *We may consequently conclude that few, if any, electrons are present in this region.*

An α-particle in a direct collision with a gold atom of nuclear charge 79 will be turned back in its path at a distance of 3 x $10^{-12}$ cm, indicating that *the nucleus may be regarded as a point charge* even for such a short distance. Until swifter α-particles are available for experiment, we are unable in the case of heavy elements to push further the question of dimensions of heavy atoms. We shall see later, however, that the outlook is more promising in the case of lighter atoms, where the α-particle can approach closer to the nucleus.

It is hardly necessary to emphasize the great importance of the nuclear charge in fixing the physical and chemical properties of an element, for obviously *the number and the arrangements of the external electrons on which the great majority of the physical and chemical properties depend, is conditioned by the resultant charge on the nucleus*. It is to be anticipated theoretically, and is. confirmed by experiment, that the actual mass of the nucleus exercises only a second order effect on the arrangement of the external electrons and their rates of vibration.

It is thus quite possible to imagine the existence of *elements* of almost identical physical and chemical properties, but which differ from one another in mass, for, provided the resultant nuclear charge is the same, a number of possible stable modes of combination of the different units which make up a complex nucleus may be possible. *The dependence of the properties of an atom on its nuclear charge and not on its mass thus offers a rational explanation of the existence of isotopes in which the chemical and physical properties may be almost indistinguishable, but the mass of the isotopes may vary within certain limits*. This important question will be considered in more detail later in the paper in the light of evidence as to the nature of the units which make up the *nucleus*.

The general problem of the structure of the atom thus naturally divides itself into two parts: — 1. *Constitution of the nucleus itself*. 2. *The arrangement and modes of vibration of the external electrons*.

I do not propose to-day to enter into (2), for it is a very large subject in which there is room for much difference of opinion. This side of the problem was first attacked by Bohr and Nicholson, and substantial advances have been made. Recently, Sommerfeld and others have applied Bohr's general method with great success in explaining the fine structure of the spectral lines and the complex modes of vibration of simple atoms involved in the Stark effect. Recently, *Langmuir* and others have attacked the problem of the arrangement of the external electrons from the chemical standpoint, and have emphasized the importance of assuming a more or less cubical arrangement of the electrons in the atom. No doubt each of these theories has a definite sphere of usefulness, but our knowledge is as yet too scanty to bridge over the apparent differences between them.

I propose to-day to discuss in some detail experiments that have been made with a view of throwing light on *the constitution and stability of the nuclei* of some of the simpler atoms. From a study of radio-activity we know that the nuclei of the radio-active elements consist in part of helium nuclei of charge 2e. *We also have strong reason for believing that the nuclei of atoms contain electrons* as well as positively charged bodies, and that the positive charge on the nucleus represents the excess positive charge. It is of interest to note the very different role played by the electrons in the outer and inner atom. In the former case, *the electrons arrange themselves at a distance from the nucleus, controlled no doubt mainly by the charge on the nucleus and the interaction of their own fields*. In the case of the nucleus, the electron forms a very close and powerful combination with the positively charged units and, as far as we know, there is a region just outside the nucleus where no electron is in stable equilibrium. While no doubt each of the external electrons acts as a point charge in considering the forces between it and the nucleus, this cannot be the case for the electron in the nucleus itself. It is to be anticipated that under the intense forces in the latter, the electrons are much deformed and the forces may be of a very different character from those to be expected from an undeformed electron, as in the outer atom. It may be for this reason that the electron can play such a different part in the two cases and yet form stable systems.

*It has been generally assumed, on the nucleus theory, that electric forces and charges play a predominant part in determining the structure of the inner and outer atom.* The considerable success of this theory in explaining fundamental phenomena is an indication of the general correctness of this point of view. At the same time if the electrons and parts composing the nucleus are in motion, *magnetic fields* must arise which will have to be taken into account in any complete theory of the atom. In this sense the magnetic fields are to be regarded as a secondary rather than a primary factor, even though such fields may be shown to have an important bearing on the conditions of equilibrium of the atom.

**Dimensions of Nuclei.**

We have seen that in the case of atoms of large nuclear charge the swiftest α-particle is unable to penetrate to the actual structure of the nucleus so that it is possible to give only a maximum estimate of its dimensions. In the case of light atoms, *however, when the nucleus charge is small, there is so close an approach during a direct collision with an α-particle that we are able to estimate its dimensions and form some idea of the forces in operation.* This is best shown in the case of a direct collision between an α-particle and an atom of hydrogen. In such a case, the H atom is set in such swift motion that it travels four times as far as the colliding α-particle and can be detected by the scintillation produced on a zinc sulphide screen. *

* Marsden, (1914). *Phil. Mag.*, 27, p. 824.

The writer** has shown that these scintillations are due to hydrogen atoms carrying unit positive charge recoiling with the velocity to be expected from the simple collision theory, viz., 1.6 times the velocity of the α-particle.

** Rutherford, (1919), *Phil. Mag.*, 37, I and II, pp. 538-71.

*The relation between the number and velocity of these H atoms is entirely different from that to be expected if the α-particle and H atom are regarded as point charges for the distances under consideration*. The result of the collision with swift α-particles is to produce H atoms which have a narrow range of velocity, and which travel nearly in the direction of the impinging particles. It was deduced that the law of inverse squares no longer holds when the nuclei approach to within a distance of 3 x $10^{-13}$ cm of each other. *This is an indication that the nuclei have dimensions of this order of magnitude* and that the forces between the nuclei vary very rapidly in magnitude and in direction for a distance of approach comparable with the diameter of the electron as ordinarily calculated. It was pointed out that in such close encounters there were enormous forces between the nuclei, and probably the structure of the nuclei was much deformed during the collision. *The fact that the helium nucleus, which may be supposed to consist of four H atoms and two electrons, appeared to survive the collision is an indication that it must be a highly stable structure*. Similar results\*\*\* were observed in the collision between α-particles and atoms of nitrogen and oxygen for the recoil atoms appeared to be shot forward mainly in the direction of the α-particles and the region where special forces come into play is of the same order of magnitude as in the case of the collision of an α-particle with hydrogen.

\*\*\* Rutherford, (1919), *Phil. Mag.*, 37, III, p. 571. [See above.]

No doubt the space occupied by a nucleus and the distance at which the forces become abnormal increase with the complexity of the nucleus structure. We should expect the H nucleus to be the simplest of all and, if it be the positive electron, it may have exceedingly small dimensions compared with the negative electron. In the collisions between α-particles and H atoms, the α-particle is to be regarded as the more complex structure.

The diameter of the nuclei of the light atoms except hydrogen are probably of the order of magnitude 5 x $10^{-13}$ cm and in a close collision the nuclei come nearly in contact and may possibly penetrate each other's structure. Under such conditions, only very stable nuclei would be expected to survive the collision and it is thus of great interest to examine whether evidence can be obtained of their disintegration.

**Long Range Particles from Nitrogen.**

In previous papers, *loc. cit.*, I have given an account of the effects produced by close collisions of swift α-particles with light atoms of matter with the view of determining whether the nuclear structure of some of the lighter atoms could be disintegrated by the intense forces brought into play in such close collisions. Evidence was given that the passage of α-particles through dry nitrogen gives rise to swift particles which closely resembled in brilliancy of the scintillations and distance of penetration hydrogen atoms set in motion by close collision with α-particles. It was shown that these swift atoms which appeared only in dry nitrogen and not in oxygen or carbon dioxide could not be ascribed to the presence of water vapor or other hydrogen material, but must arise from the collision of α-particles with nitrogen atoms. The number of such scintillations due to nitrogen was small, viz., about 1 in 12 of the corresponding number in hydrogen, but was two to three times the number of natural scintillations from the source. The number observed in nitrogen was on an average equal to the number of scintillations when hydrogen at about 6 cm pressure was added to oxygen or carbon dioxide at normal pressure.

While the general evidence indicated that these long-range atoms from nitrogen were charged atoms of hydrogen, the preliminary experiments to test the mass of the particles by bending them in a strong magnetic field yielded no definite results.

From the data given in my previous paper (*loc. cit.*) several theories could be advanced to account for these particles. The calculated range of a singly charged atom set in motion by a close collision with an α-particle of range R cm in air was shown to be for

Mass 1 .......... Range 3.91 R
Mass 2 .......... Range 4.6 R
Mass 3 .......... Range 5.96 R
Mass 4 .......... Range 4.0 R

On account of the small number and weakness of the scintillations under the experimental conditions, the range of the swift atoms from nitrogen could not be determined with sufficient certainty to decide definitely between any of these possibilities. The likelihood that the particles were the original α-particles which had lost one of their two charges, i.e., atoms of charge 1 and mass 4, was suggested by me to several correspondents, but there appeared to be no obvious reason why nitrogen, of all the elements examined, should be the only one in which the passage of a swift a-particle led to the capture of a single electron. If, however, a sufficient number of scintillations could be obtained under the experimental conditions, there should be no inherent difficulty in deciding between the various possibilities by examining the deflection of the swift atoms by a magnetic field. The amount of deflection of charged atoms in a magnetic field perpendicular to the direction of flight is proportional to $e/mu$. Assuming that the particles were liberated by a direct collision with an α-particle, the relative values of this quantity for different recoiling masses are easily calculated. Taking values MY/E for the α-particle as unity, the corresponding values of $mu/e$ for atoms of charge 1 and mass 1, 2, 3, and 4 are 1.25, 0.75, 0.58, and 0.50 respectively. Consequently, the H atoms should be more bent than the α-particles which produced them while the atoms of mass 2 or 3, or 4 would be more difficult to deflect than the parent α-particle.

On my arrival in Cambridge, this problem was attacked in several ways. By the choice of objectives of wide aperture, the scintillations were increased in brilliancy and counting thus made easier. A number of experiments were also made to obtain more powerful sources of radiation with the radium at my command, but finally it was found best, for reasons which need not be discussed here, to obtain the active source of radiation of radium C in the manner described in my previous paper. After a number of observations with solid nitrogen compounds, described later, a simple method was finally devised to estimate the mass of the particle by the use of nitrogen in the gaseous state. The use of the gas itself for this purpose had several advantages over the use of solid nitrogen compounds, for not only was the number of scintillations greater, but the absence of hydrogen or other hydrogen compounds could be ensured.

The arrangement finally adopted is shown in fig. 1. The essential point Fig. 1. lay in the use of wide slits, between which the α-particles passed. Experiment showed that the ratio of the number of scintillations on the screen arising from the gas to the number of natural scintillations from the source, increased rapidly with increased depth of the slits. For plates 1 mm apart, this ratio was less than unity, but for slits 8 mm apart the ratio had a value 2 to 3. Such a variation is to be anticipated on theoretical grounds if the majority of the particles are liberated at an angle with the direction of the incident α-particles.

The horizontal slits A, B were 6.0 cm long, 1.5 cm wide, and 8 mm deep, with the source, C oi the active deposit of radium placed at one end and the zinc sulphide screen near the other. The carrier of the source and slits were placed in a rectangular brass box, through which a current of dry air or other gas was continuously passed to avoid the danger of radio-active contamination. The box was placed between the poles of a large electro magnet, so that the uniform field was parallel to the plane of the plates and perpendicular to their length. A distance piece, D, of length 1.2 cm, was added between the source and end of the slits, in order to increase the amount of deflection of the radiation issuing from the slits. The zinc sulphide screen, S, was placed on a glass plate covering the end of the box. The distance between the source and the screen was 7.4 cm. The recoil atoms from oxygen or nitrogen of range 9 cm could be stopped by inserting an aluminum screen of stopping power about 2 cm of air placed at the end of the slits.

With such deep slits it was impossible to bend the wide beam of radiation to the sides, but the amount of deflection of the radiation issuing near the bottom of the slit was measured. For this purpose, it was essential to observe the scintillations at a fixed point of the screen near M. The method of fixing the position of the counting microscope was as follows: The source, C, was placed in position, and the air exhausted to a pressure of a few centimeters. Without the field, the bottom edge of the beam was fixed by the straight line PM cutting

174

the screen at M. The microscope was adjusted so that the boundary line of scintillations appeared above the horizontal cross wire in the microscope, marking the center of the field.

On exciting the magnet to bend the rays upward (called the + field), the path of the limiting α-particles is marked by the curve PLKN cutting the screen at 1ST, so that the, boundary of the scintillations appears to be displaced downwards in the field of view. On reversing the field (called the − field), the path of the limiting α-particle PQRT cuts the screen at T, and the band of scintillations appears to be bent upwards. The strength of the magnetic field was adjusted so that, with a negative field, the scintillations were observed all over the screen, while, with a positive field, they were mainly confined below the cross wire. The appearance in the field of view of the microscope for the two fields is illustrated in fig. 2, where the dots represent approximately the density of distribution of the scintillations. The horizontal boundaries of the field of view were given by a rectangular opening in a plate fixed in the position of the cross wires. A horizontal wire, which bisected the field of view, was visible under the conditions of counting, and allowed the relative numbers of scintillations in the two halves of the field to be counted if required. Since the number of scintillations in the actual experiments with nitrogen was much too small to mark directly the boundary of the scintillations, in order to estimate the bending of the rays, it was necessary to determine the ratio of the number of scintillations with the + and − field.

The position of the microscope and the strength of the magnetic field were in most experiments so adjusted that this ratio was about one-third. Preliminary observations showed that this ratio was sensitive to changes of the field and it thus afforded a suitable method for estimating the relative bending of any radiations under examination.

After the position of the microscope was fixed, air was let in, and a continuous flow of dry air maintained through the apparatus. The absorbing-screen was introduced at E to stop the atoms from N and 0 of range 9 cm. The number of scintillations was then systematically counted for the two directions of the field, and a correction, if required, made for any slight radio-active contamination of the screen. The deflection due to the unknown radiation was directly compared with that produced by a known radiation of α-rays. For this purpose, after removal of the source and absorbing screen, a similar plate, coated with a weak distribution of the active deposit of thorium, was substituted for the radium source. The α-particles from thorium C of range 8.6 cm produced bright scintillations in the screen after traversing the 7.4 cm of air in their path. The ratio of the number of scintillations with + and − fields was determined as before.

An example of such comparison is given below. For a current of 4.0 amp. through the electromagnet, the ratio for particles from nitrogen was found to be 0.33. The corresponding ratio for α-particles from thorium C was 0.44 for a current of 4 amp. and

0.31 for a current of 5 amp. It is thus seen that on the average, the particles from nitrogen are more bent in a given field than the α-particles from thorium C. In order, however, to make a quantitative comparison, it is necessary to take into account the reduction in velocity of the radiations in passing through the air. The value $mu/e$ or the α-ray of range 8.6 cm from thorium C is known to be $4.28 \times 10^5$. Since the rays pass through 7.4 cm of air in a uniform field before striking the screen, it can be calculated that the actual deflection corresponds to α-rays in a vacuum for which $mu/e = 3.7 \times 10^5$, about. Taking the deflection of the α-particles for a current of 4.8 amp. to be the same as for the nitrogen particles for a field of 4 amp. —ratio of fields 1.17—it is seen that the average deflection of the nitrogen particles under the experimental conditions corresponds to a radiation in a vacuum for which the value of $mu/e = 3.1 \times 10^5$.

Bearing in mind that the particles under examination are produced throughout the volume of the gas between the slits, and that their distribution is unknown, and also that the particles are shot forward on an average at an angle with the incident α-particles, the experimental data are quite insufficient to calculate the average value of $mu/e$ to be expected under the experimental conditions for any assumed mass of projected particles. It seems probable that the majority of the particles which produce scintillations are generated in the first few centimeters of the air next the source. The actual deflection of a given particle by the magnetic field will depend on the distance of its point of origin from the source. These factors will obviously tend to make the average deflection of the particles to appear less than if they were all expelled with constant velocity from the source itself. Assuming that the correction for reduction of velocity of the long-range particles in traversing the gas is 10 per cent., the average value of $mu/e$ is about $3.4 \times 10^5$. Since the value of MY/E for the α-particle from radium C is $3.98 \times 10^5$, it is seen that under the experimental conditions the average value of $mu/e$ for the nitrogen particles is less than that of the α-particles which produce them.

Erom the data given earlier in the paper, this should only be true if the particles are comparable in mass with an atom of hydrogen, for singly charged particles of mass 2, 3, or 4 should suffer less deflection than the α-particles. For example, if we assume that the particles were helium atoms carrying one charge, we should expect them to be deflected to about one-half of the extent of the α-particle. The experimental results thus afford strong presumptive evidence that the particles liberated from nitrogen are atoms of hydrogen.

A far more decisive test, however, can be made by comparing the deflection of the nitrogen particles with that of H atoms under similar conditions. For this purpose, a mixture of about one volume of hydrogen to two of carbon dioxide was stored in a gas-holder and circulated in place of air through the testing apparatus. The proportions of the two gases were so adjusted that the stopping power of the mixture for α-rays was equal to that of air. Under these conditions, the H atoms, like the nitrogen particles, are produced throughout the volume of the gas, and probably the relative distribution of H atoms along the path of the α-rays is not very different from that of the nitrogen particles under examination. If the nitrogen particles are H atoms, we should expect the average deflection to be nearly the same as for the H atoms liberated from the hydrogen mixture. A number of careful experiments showed that the ratio of the number of scintillations in + and − fields of equal value was so nearly identical in the two cases that the experiments were unable to distinguish between them. Since the two experiments were carried out under as nearly as possible identical conditions, the equality of the ratio shows that the long-range particles liberated from nitrogen are atoms of hydrogen. The possibility that the particles may be of mass 2, 3, or 4 is definitely excluded.

In a previous paper I have given evidence that the long-range particles observed in dry air and pure nitrogen must arise from the nitrogen atoms themselves. It is thus clear that some of the nitrogen atoms are disintegrated by their collision with swift α-particles and that swift atoms of positively charged hydrogen are expelled. It is to be inferred that the charged atom of hydrogen is one of the components of which the nucleus of nitrogen is built up. While it has long been known that helium is a product of the spontaneous transformation of some of the radio-active elements, the possibility of disintegrating the structure of stable atoms by artificial methods has been a matter of uncertainty. This is the first time that evidence has been obtained that hydrogen is one of the components of the nitrogen nucleus. It should be borne in mind that the amount of disintegration effected in nitrogen by the particles is excessively small, for probably on an average only one α-particle in about 300,000 is able to get near enough to the nitrogen nucleus to liberate the atom of hydrogen with sufficient energy to be detected by the scintillation method. Even if the whole α-radiation from 1 gramme of radium were absorbed in nitrogen gas, the volume of hydrogen set free would be only about 1/300000 of the volume of helium due to the collected α-particles, viz., about $5 \times 10^{-4}$ cub. mm per year. It may be possible that the collision of an α-particle is effective in liberating the hydrogen from the nucleus without necessarily giving it sufficient velocity to be detected by scintillations. If this should prove to be the case, the amount of disintegration may be much greater than the value given above.
*Pages 385-400:*


**Experiments with Solid Nitrogen Compounds.**
…


**Short Range Atoms from Oxygen and Nitrogen.**
…


**Energy Considerations.**

**Properties of the new Atom.**

...

**Constitution of Nuclei and Isotopes.**

...

**Structure of Carbon, Oxygen, and Nitrogen Nuclei.**

...

It is intended to continue experiments, to test whether any evidence can be obtained of the disintegration of other light atoms besides nitrogen and oxygen. The problem is more difficult in the case of elements which cannot be conveniently obtained in the gaseous state, since it is not an easy matter to ensure the absence of hydrogen or to prepare uniform thin films of such substances. For these reasons, and the strain involved in counting scintillations under difficult conditions, further progress is not likely to be rapid.

I am indebted to my assistant, G. A. E. Crowe, for the preparation of the radio-active sources and his help in counting; also, to Mr. J. Chadwick and Dr. Ishida for assistance in counting scintillations in some of the later experiments.

## Bohr, N. (1922). The structure of the atom

Nobel Lecture, December 11, 1922;
https://www.nobelprize.org/prizes/physics/1922/bohr/lecture/

The Nobel Prize in Physics 1922 was awarded to Niels Henrik David Bohr "for his services in the investigation of the structure of *atoms* and of the radiation emanating from them". The Nobel Prize in Physics 1922. NobelPrize.org; https://www.nobelprize.org/prizes/physics/1922/ summary/. The discovery of the electron and radioactivity in the late 19th century led to different models being proposed for the atom's structure. In 1913, Niels Bohr proposed a theory for the hydrogen atom, based on quantum theory that some physical quantities only take discrete values. Electrons move around a nucleus, but only in prescribed orbits, and If electrons jump to a lower-energy orbit, the difference is sent out as radiation. Bohr's model explained why atoms only emit light of fixed wavelengths, and later incorporated the theories on light quanta. Niels Bohr – Facts. NobelPrize.org. https://www.nobelprize.org/prizes/physics/1922/bohr/facts/.

Ladies and Gentlemen.

Today, as a consequence of the great honor the Swedish Academy of Sciences has done me in awarding me this year's Nobel Prize for Physics for my work on the structure of the atom, it is my duty to give an account of the results of this work and I think that I shall be acting in accordance with the traditions of the Nobel Foundation if I give this report in the form of a survey of the development which has taken place in the last few years within the field of physics to which this work belongs.

### The general picture of the atom

The present state of atomic theory is characterized by the fact that we not only believe the existence of atoms to be proved beyond a doubt, but also, we even believe that we have an intimate knowledge of the constituents of the individual atoms. I cannot on this occasion give a survey of the scientific developments that have led to this result; I will only recall the discovery of the electron towards the close of the last century, which furnished the direct verification and led to a conclusive formulation of the conception of the atomic nature of electricity which had evolved since the discovery by Fara day of the fundamental laws of electrolysis and Berzelius's electrochemical theory, and had its greatest triumph in the electrolytic dissociation theory of Arrhenius. This discovery of the electron and elucidation of its properties was the result of the work of a large number of investigators, among whom Lenard and J. J. Thomson may be particularly mentioned. The latter especially has made very important contributions to our subject by his ingenious attempts to develop ideas about atomic constitution on the basis of the electron theory. The present state of our knowledge of the elements of atomic structure was reached, however, by the discovery of the atomic nucleus, which we owe to Rutherford, whose work on the radioactive substances discovered towards the close of the last century has much enriched physical and chemical science.

According to our present conceptions, an atom of an element is built up of a nucleus that has a positive electrical charge and is the seat of by far the greatest part of the atomic mass, together with a number of electrons, all having the same negative charge and mass, which move at distances from the nucleus that are very great compared to the dimensions of the nucleus or of the electrons themselves. In this picture we at once see a striking resemblance to a planetary system, such as we have in our own solar system. Just as the simplicity of the laws that govern the motions of the solar system is intimately connected with the circumstance that the dimensions of the moving bodies are small in relation to the orbits, so the corresponding relations in atomic structure provide us with an explanation of an essential feature of natural phenomena in so far as these depend on the properties of the elements. It makes clear at once that these properties can be divided into two sharply distinguished classes.

To the first class belong most of the ordinary physical and chemical properties of substances, such as their state of aggregation, color, and chemical reactivity. These properties depend on the motion of the electron system and the way in which this motion changes under the influence of different external actions. On account of the large mass of the nucleus relative to that of the electrons and its smallness in comparison to the electron orbits, the electronic motion will depend only to a very small extent on the nuclear mass, and will be determined to a close approximation solely by the total electrical charge of the nucleus. Especially the inner structure of the nucleus and the way in which the charges and masses are distributed among its separate particles will have a vanishingly small influence on the motion of the electron system surrounding the nucleus. On the other hand, the structure of the nucleus will be responsible for the second class of properties that are shown in the radioactivity of substances. In the radioactive processes we meet with an explosion of the nucleus, whereby positive or negative particles, the so-called α- and β-particles, are expelled with very great velocities.

Our conceptions of atomic structure afford us, therefore, an immediate explanation of the complete lack of interdependence between the two classes of properties, which is most strikingly shown in the existence of substances which have to an extraordinarily close approximation the same ordinary physical and chemical properties, even though the atomic weights are not the same, and the radioactive properties are completely different. Such sub stances, of the existence of which the first evidence was found in the work of Soddy and other investigators on the chemical properties of the radioactive elements, are called isotopes, with reference to the classification of the elements according to ordinary physical and chemical properties. It is not necessary for me to state here how it has been shown in recent years that iso topes are found not only among the radioactive elements, but also among ordinary stable elements; in fact, a large number of the latter that were previously supposed simple have been shown by Aston's well-known investigations to consist of a mixture of isotopes with different atomic weights.

The question of the inner structure of the nucleus is still but little under stood, although a method of attack is afforded by Rutherford's experiments on the disintegration of atomic nuclei by bombardment with α-particles. Indeed, these experiments may be said to open up a new epoch in natural philosophy in that for the first time the artificial transformation of one element into another has been accomplished. In what follows; however, we shall confine ourselves to a consideration of the ordinary physical and chemical properties of the elements and the attempts which have been made to explain them on the basis of the concepts just outlined.

It is well known that the elements can be arranged as regards their ordinary physical and chemical properties in a natural system which displays most suggestively the peculiar relationships between the different elements. It was recognized for the first time by Mendeleev and Lothar Meyer that when the elements are arranged in an order which is practically that of their atomic weights, their chemical and physical properties show a pronounced periodicity. A diagrammatic representation of this so-called *Periodic Table* is given in Fig. 1, where, however, the elements are not arranged in the ordinary way but in a somewhat modified form of a table first given by Julius Thomsen, who has also made important contributions to science in this domain. In the figure the elements are denoted by their usual chemical symbols, and the different vertical columns indicate the so-called periods. The elements in successive columns which possess homologous chemical and physical properties are connected with lines. The meaning of the square brackets around certain series of elements in the later periods, the properties of which exhibit typical deviations from the simple periodicity in the first periods, will be discussed later.

In the development of the theory of atomic structure the characteristic features of the natural system have found a surprisingly simple interpretation. Thus, we are led to assume that the ordinal number of an *element* in the *Periodic Table*, the so-called *atomic number*, is just equal to the number of *electrons* which move about the *nucleus* in the *neutral atom*. In an imperfect form, this law was first stated by Van den Broek; it was, however, foreshadowed by J. J. Thomson's investigations of the number of electrons in the atom, as well as by Rutherford's measurements of the charge on the atomic nucleus. As we shall see, convincing support for this law has since been obtained in various ways, especially by Moseley's famous investigations of the X-ray spectra of the elements. We may perhaps also point out, how the simple connection between atomic number and nuclear charge offers an explanation of the laws governing the changes in chemical properties of the elements after expulsion of or β-particles, which found a simple formulation in the so-called ***radioactive displacement law***.

**Atomic stability and electrodynamic theory**

As soon as we try to trace a more intimate connection between the properties of the elements and atomic structure, we encounter profound difficulties, in that essential differences between an atom and a planetary system show themselves here in spite of the analogy we have mentioned.

The motions of the bodies in a planetary system, even though they obey the general law of gravitation, will not be completely determined by this law alone, but will depend largely on the previous history of the system. Thus, the length of the year is not determined by the masses of the sun and the earth alone, but depends also on the conditions that existed during the formation of the solar system, of which we have very little knowledge. Should a sufficiently large foreign body someday traverse our solar system, we might among other effects expect that from that day the length of the year would be different from its present value.

It is quite otherwise in the case of atoms. The definite and unchangeable properties of the ***elements*** demand that the state of an atom cannot undergo permanent changes due to external actions. As soon as the atom is left to itself again, its constituent particles must arrange their motions in a manner which is completely determined by the electric charges and masses of the particles. We have the most convincing evidence of this in spectra, that is, in the properties of the radiation emitted from substances in certain circumstances, which can be studied with such great precision. It is well known that the wavelengths of the spectral lines of a substance, which can in many cases be measured with an accuracy of more than one part in a million, are, in the same external circumstances, always exactly the same within the limit of error of the measurements, and quite independent of the previous treatment of this substance. It is just to this circumstance that we owe the great importance of spectral analysis, which has been such an invaluable aid to the chemist in the search for new elements, and has also shown us that even on the most distant bodies of the universe there occur elements with exactly the same properties as on the earth.

On the basis of our picture of the constitution of the atom it is thus impossible, so long as we restrict ourselves to the ordinary mechanical laws, to account for the characteristic atomic stability which is required for an explanation of the properties of the ***elements***.

The situation is by no means improved if we also take into consideration the well-known electrodynamic laws which Maxwell succeeded in formulating on the basis of the great discoveries of Oersted and Faraday in the first half of the last century. Maxwell's theory has not only shown itself able to account for the already known electric and magnetic phenomena in all their details, but has also celebrated its greatest triumph in the prediction of the electromagnetic waves which were discovered by Hertz, and are now so extensively used in wireless telegraphy.

For a time, it seemed as though this theory would also be able to furnish a basis for an explanation of the details of the properties of the elements, after it had been developed, chiefly by Lorentz and Larmor, into a form consistent with the atomistic conception of electricity. I need only remind you of the great interest that was aroused when Lorentz, shortly after the discovery by Zeeman of the characteristic changes that spectral lines undergo when the emitting substance is brought into a magnetic field, could give a natural and simple explanation of the main features of the phenomenon. Lorentz assumed that the radiation which we observe in a spectral line is sent out from an electron executing simple harmonic vibrations about a position of equilibrium, in precisely the same manner as the electromagnetic waves in radiotelegraphy are sent out by the electric oscillations in the antenna. He also pointed out how the alteration observed by Zeeman in the spectral lines corresponded exactly to the alteration in the motion of the vibrating electron which one would expect to be produced by the magnetic field.

It was, however, impossible on this basis to give a closer explanation of the spectra of the elements, or even of the general type of the laws holding with great exactness for the wavelengths of lines in these spectra, which had been established by Balmer, Rydberg, and Ritz. After we obtained details as to the constitution of the atom, this difficulty became still more manifest; in fact, so long as we confine ourselves to the classical electrodynamic theory we cannot even understand why we obtain spectra consisting of sharp lines at all. This theory can even be said to be incompatible with the assumption of the existence of atoms possessing the structure we have described, in that the motions of the electrons would claim a continuous radiation of energy from the atom, which would cease only when the electrons had fallen into the nucleus.

**The origin of the quantum theory**

It has, however, been possible to avoid the various difficulties of the electro dynamic theory by introducing concepts borrowed from the so-called quantum theory, which marks a complete departure from the ideas that have hitherto been used for the explanation of natural phenomena. This theory was originated by Planck, in the year 1900, in his investigations on the law of heat radiation, which, because of its independence of the individual properties of substances, lent itself peculiarly well to a test of the applicability of the laws of classical physics to atomic processes.

Planck considered the equilibrium of radiation between a number of systems with the same properties as those on which Lorentz had based his theory of the Zeeman effect, but he could now show not only that classical physics could not account for the phenomena of heat radiation, but also that a complete agreement with the experimental law could be obtained if - in pronounced contradiction to classical theory - it were assumed that the energy of the vibrating electrons could not change continuously, but only in such a way that the energy of the system always remained equal to a whole number of so-called energy-quanta. The magnitude of this quantum was found to be proportional to the frequency of oscillation of the particle, which, in accordance with classical concepts, was supposed to be also the frequency of the emitted radiation. The proportionality factor had to be regarded as a new universal constant, since termed Planck's constant, similar to the velocity of light, and the charge and mass of the electron.

Planck's surprising result stood at first completely isolated in natural science, but with Einstein's significant contributions to this subject a few years after, a great variety of applications was found. In the first place, Einstein pointed out that the condition limiting the amount of vibrational energy of the particles could be tested by investigation of the specific heat of crystalline bodies, since in the case of these we have to do with similar vibrations, not of a single electron, but of whole atoms about positions of equilibrium in the crystal lattice. Einstein was able to show that the experiment confirmed Planck's theory, and through the work of later investigators this agreement has proved quite complete. Furthermore, Einstein emphasized another consequence of Planck's results, namely, that radiant energy could only be emitted or absorbed by the oscillating particle in so-called "quanta of radiation", the magnitude of each of which was equal to Planck's constant multiplied by the frequency.

In his attempts to give an interpretation of this result, Einstein was led to the formulation of the so-called "hypothesis of light-quanta" according to which the radiant energy, in contradiction to Maxwell's electromagnetic theory of light, would not be propagated as electromagnetic waves, but rather as concrete light atoms, each with an energy equal to that of a quantum of radiation. This concept led Einstein to his well-known theory of the photoelectric effect. This phenomenon, which had been entirely unexplainable on the classical theory, was thereby placed in a quite different light, and the predictions of Einstein's theory have received such exact experimental confirmation in recent years, that perhaps the most exact determination of Planck's constant is afforded by measurements on the photoelectric effect. In spite of its heuristic value, however, the hypothesis of light-quanta, which is quite irreconcilable with so-called interference phenomena, is not able to throw light on the nature of radiation. I need only recall that these interference phenomena constitute our only means of investigating the properties of radiation and therefore of assigning any closer meaning to the frequency which in Einstein's theory fixes the magnitude of the light-quantum.

In the following years many efforts were made to apply the concepts of the quantum theory to the question of atomic structure, and the principal emphasis was sometimes placed on one and sometimes on the other of the consequences deduced by Einstein from Planck's result. As the best known of the attempts in this direction, from which, however, no definite results were obtained, I may mention the work of Stark, Sommerfeld, Hasenöhrl, Haas, and Nicholson.

From this period also dates an investigation by Bjerrum on infrared absorption bands, which, although it had no direct bearing on atomic structure, proved significant for the development of the quantum theory. He directed attention to the fact that the rotation of the molecules in a gas might be investigated by means of the changes in certain absorption lines with temperature. At the same time, he emphasized the fact that the effect should not consist of a continuous widening of the lines such as might be expected from classical theory, which imposed no restrictions on the molecular rotations, but in accordance with the quantum theory he predicted that the lines should be split up into a number of components, corresponding to a sequence of distinct possibilities of rotation. This prediction was confirmed a few years later by Eva von Bahr, and the phenomenon may still be regarded as one of the most striking pieces of evidence of the reality of the quantum theory, even though from our present point of view the original explanation has under gone a modification in essential details.

**The quantum theory of atomic constitution**

The question of further development of the quantum theory was in the meantime placed in a new light by Rutherford's discovery of the ***atomic nucleus*** (1911). As we have already seen, this discovery made it quite clear that by classical conceptions alone it was quite impossible to understand the most essential properties of atoms. One was therefore led to seek for a formulation of the principles of the quantum theory that could immediately account for the stability in atomic structure and the properties of the radiation sent out from atoms, of which the observed properties of substances bear witness. Such a formulation was proposed (1913) by the present lecturer in the form of two postulates, which may be stated as follows:

(1). Among the conceivably possible states of motion in an atomic system there exist a number of so-called ***stationary states*** which, in spite of the fact that the motion of the particles in these states obeys the laws of classical mechanics to a considerable extent, possess a peculiar, mechanically unexplainable stability, of such a sort that every permanent change in the motion of the system must consist in a complete transition from one stationary state to another. (2). While in contradiction to the classical electromagnetic theory no radiation takes place from the atom in the stationary states themselves, a process of transition between two *stationary states* can be accompanied by the emission of electromagnetic radiation, which will have the same properties as that which would be sent out according to the classical theory from an electrified particle executing a harmonic vibration with constant frequency. This frequency v has, however, no simple relation to the motion of the particles of the atom, but is given by the relation

$$hv = E' - E''$$

where h is Planck's constant, and E' and E'' are the values of the energy of the atom in the two stationary states that form the initial and final state of the radiation process. Conversely, irradiation of the atom with electromagnetic waves of this frequency can lead to an absorption process, whereby the atom is transformed back from the latter stationary state to the former.

While the first postulate has in view the general stability of the atom, the second postulate has chiefly in view the existence of spectra with sharp lines. Furthermore, the quantum-theory condition entering in the last postulate affords a starting-point for the interpretation of the laws of series spectra. The most general of these laws, the *combination principle* enunciated by Ritz, states that the frequency v for each of the lines in the spectrum of an element can be represented by the formula

$$v = T'' - T'$$

where T'' and T' are two so-called spectral terms belonging to a manifold of such terms characteristic of the substance in question.

According to our postulates, this law finds an immediate interpretation in the assumption that the spectrum is emitted by transitions between a number of *stationary states* in which the numerical value of the energy of the atom is equal to the value of the spectral term multiplied by Planck's constant. This explanation of the *combination principle* is seen to differ fundamentally from the usual ideas of electrodynamics, as soon as we consider that there is no simple relation between the motion of the atom and the radiation sent out. The departure of our considerations from the ordinary ideas of natural philosophy becomes particularly evident, however, when we observe that the occurrence of two spectral lines, corresponding to combinations of the same spectral term with two other different terms, implies that the nature of the radiation sent out from the atom is not determined only by the motion of the atom at the-beginning of the radiation process, but also depends on the state to which the atom is transferred by the process.

At first glance one might, therefore, think that it would scarcely be possible to bring our formal explanation of the combination principle into direct relation with our views regarding the constitution of the atom, which, in deed, are based on experimental evidence interpreted on classical mechanics and electrodynamics. A closer investigation, however, should make it clear that a definite relation may be obtained between the spectra of the elements and the structure of their atoms on the basis of the postulates.

**The hydrogen spectrum**

The simplest spectrum we know is that of hydrogen. The frequencies of its lines may be represented with great accuracy by means of Balmer's formula:

$$v = K \left( 1/n''^2 - 1/n'^2 \right),$$

where K is a constant and n' and n" are two integers. In the spectrum we accordingly meet a single series of spectral terms of the form $K/n^2$, which decrease regularly with increasing term number n. In accordance with the postulates, we shall therefore assume that each of the hydrogen lines is emitted by a transition between two states belonging to a series of stationary states of the hydrogen atom in which the numerical value of the atom's energy is equal to $hK/n^2$.

Following our picture of atomic structure, a hydrogen atom consists of a positive nucleus and an electron which - so far as ordinary mechanical conceptions are applicable - will with great approximation describe a periodic elliptical orbit with the nucleus at one focus. The major axis of the orbit is inversely proportional to the work necessary completely to remove the electron from the nucleus, and, in accordance with the above, this work in the stationary states is just equal to $hK/n^2$. We thus arrive at a manifold of stationary states for which the major axis of the electron orbit takes on a series of discrete values proportional to the squares of the whole numbers.

The accompanying Fig. 2 shows these relations diagrammatically. For the sake of simplicity, the electron orbits in the stationary states are represented by circles, although in reality the theory places no restriction on the eccentricity of the orbit, but only determines the length of the major axis. The arrows represent the transition processes that correspond to the red and green hydrogen lines, $H_\alpha$ and $H_\beta$ the frequency of which is given by means of the Balmer formula when we put n" = 2 and n' = 3 and 4 respectively. The transition processes are also represented which correspond to the first three lines of the series of ultraviolet lines found by Lyman in 1914, of which the frequencies are given by the formula when n is put equal to 1, as well as to the first line of the infrared series discovered some years previously by Paschen, which are given by the formula if n" is put equal to 3.

This explanation of the origin of the hydrogen spectrum leads us quite naturally to interpret this spectrum as the manifestation of a process *whereby the electron is bound to the nucleus*. While the largest spectral term with term number corresponds to the final stage in the binding process, the small spectral terms that have larger values of the term number correspond to stationary states which represent the initial states of the binding process, where the electron orbits still have large dimensions, and where the work required to remove an electron from the nucleus is still small. The final stage in the binding process we may designate as the normal state of the atom, and it is distinguished from the other stationary states by the property that, in accordance with the postulates, the state of the atom can only be changed by the addition of energy whereby the electron is transferred to an orbit of larger dimensions corresponding to an earlier stage of the binding process.

The size of the electron orbit in the normal state calculated on the basis of the above interpretation of the spectrum agrees roughly with the value for the dimensions of the atoms of the elements that have been calculated by the kinetic theory of matter from the properties of gases. Since, however, as an immediate consequence of the stability of the stationary states that is claimed by the postulates, we must suppose that the interaction between two atoms during a collision cannot be completely described with the aid of the laws of classical mechanics, such a comparison as this cannot be carried further on the basis of such considerations as those just outlined.

A more intimate connection between the spectra and the atomic model has been revealed, however, by an investigation of the motion in those stationary states where the term number is large, and where the dimensions of the electron orbit and the frequency of revolution in it vary relatively little when we go from one stationary state to the next following. It was possible to show that the frequency of the radiation sent out during the transition between two stationary states, the difference of the term numbers of which is small in comparison to these numbers themselves, tended to coincide in frequency with one of the harmonic components into which the electron motion could be resolved, and accordingly also with the frequency of one of the wave trams in the radiation which would be emitted according to the laws of ordinary electrodynamics. The condition that such a coincidence should occur in this region where the stationary states differ but little from one another proves to be that the constant in the Balmer formula can be expressed by means of the relation

$$K = 2\pi^2 e^4 m/h^3,$$

where e and m are respectively the charge and mass of the electron, while h is Planck's constant. This relation has been shown to hold to within the considerable accuracy with which, especially through the beautiful investigations of Millikan, the quantities e, m, and h are known.

This result shows that there exists a connection between the hydrogen spectrum and the model for the hydrogen atom which, on the whole, is as close as we might hope considering the departure of the postulates from the classical mechanical and electrodynamic laws. At the same time, it affords some indication of how we may perceive in the quantum theory, in spite of the fundamental character of this departure, a natural generalization of the fundamental concepts of the classical electrodynamic theory. To this most important question we shall return later, but first we will discuss how the interpretation of the hydrogen spectrum on the basis of the postulates has proved suitable in several ways, for elucidating the relation between the properties of the different elements.

**Relationships between the elements**

The discussion above can be applied immediately to the process whereby an electron is bound to a nucleus with any given charge. The calculations show that, in the stationary state corresponding to a given value of the number n, the size of the orbit will be inversely proportional to the nuclear charge, while the work necessary to remove an electron will be directly proportional to the square of the nuclear charge. The spectrum that is emit ted during the binding of an electron by a nucleus with charge N times that of the hydrogen nucleus can therefore be represented by the formula:

$$v = N^2 K \left(1/n''^2 - 1/n'^2\right),$$

If in this formula we put $N = 2$, we get a spectrum which contains a set of lines in the visible region which was observed many years ago in the spectrum of certain stars. Rydberg assigned these lines to hydrogen because of the close analogy with the series of lines represented by the Balmer formula. It was never possible to produce these lines in pure hydrogen, but just before the theory for the hydrogen spectrum was put forward, Fowler succeeded in observing the series in question by sending a strong discharge through a mixture of hydrogen and helium. This investigator also assumed that the lines were hydrogen lines, because there existed no experimental evidence from which it might be inferred that two different substances could show properties resembling each other so much as the spectrum in question and that of hydrogen. After the theory was put forward, it became clear, however, that the observed lines must belong to a spectrum of helium, but that they were not like the ordinary helium spectrum emitted from the neutral atom. They came from an ionized helium atom which consists of a single electron moving about a nucleus with double charge. In this way there was brought to light a new feature of the relationship between the elements, which corresponds exactly with our present ideas of atomic structure, according to which the physical and chemical properties of an element depend in the first instance only on the electric charge of the atomic nucleus.

Soon after this question was settled the existence of a similar general relationship between the properties of the elements was brought to light by Moseley's well-known investigations on the characteristic X-ray spectra of the elements, which was made possible by Laue's discovery of the interference of X-rays in crystals and the investigations of W. H. and W. L. Bragg on this subject. It appeared, in fact, that the X-ray spectra of the different elements possessed a much simpler structure and a much greater mutual resemblance than their optical spectra. In particular, it appeared that the spectra changed from element to element in a manner that corresponded closely to the formula given above for the spectrum emitted during the binding of an electron to a nucleus, provided N was put equal to the atomic number of the element concerned. This formula was even capable of expressing, with an approximation that could not be without significance, the frequencies of the strongest X-ray lines, if small whole numbers were substituted for n' and n".

This discovery was of great importance in several respects. In the first place, the relationship between the X-ray spectra of different elements proved so simple that it became possible to fix without ambiguity the atomic number for all known substances, and in this way to predict with certainty the atomic number of all such hitherto unknown elements for which there is a place in the natural system. Fig. 3 shows how the square root of the frequency for two characteristic X-ray lines depends on the atomic number. These lines belong to the group of so-called K-lines, which are the most penetrating of the characteristic rays. With very close approximation the points lie on straight lines, and the fact that they do so is conditioned not only by our taking account of known elements, but also by our leaving an open place between molybdenum (42) and ruthenium (44), just as in Mendeleev's original scheme of the natural system of the elements. Further, the laws of X-ray spectra provide a confirmation of the general theoretical conceptions, both with regard to the constitution of the atom and the ideas that have served as a basis for the interpretation of spectra. Thus, the similarity between X-ray spectra and the spectra emitted during the binding of a single electron to a nucleus may be simply interpreted from the fact that the transitions between stationary states with which we are concerned in X-ray spectra are accompanied by changes in the motion of an electron in the inner part of the atom, where the influence of the attraction of the nucleus is very great compared with the repulsive forces of the other electrons.

The relations between other properties of the elements are of a much more complicated character, which originates in the fact that we have to do with processes concerning the motion of the electrons in the outer part of the atom, where the forces that the electrons exert on one another are of the same order of magnitude as the attraction towards the nucleus, and where, therefore, the details of the interaction of the electrons play an important part. A characteristic example of such a case is afforded by the spatial extension of the atoms of the elements. Lothar Meyer himself directed attention to the characteristic periodic change exhibited by the ratio of the atomic weight to the density, the so-called atomic volume, of the elements in the natural system. An idea of these facts is given by Fig. 4, in which the atomic volume is represented as a function of the atomic number. A greater difference between this and the previous figure could scarcely be imagined. While the X-ray spectra vary uniformly with the atomic number, the atomic volumes show a characteristic periodic change which corresponds exactly to the change in the chemical properties of the elements.

Ordinary optical spectra behave in an analogous way. In spite of the dissimilarity between these spectra, Rydberg succeeded in tracing a certain general relationship between the hydrogen spectrum and other spectra. Even though the spectral lines of the elements with higher atomic number appear as combinations of a more complicated manifold of spectral terms which is not so simply co-ordinated with a series of whole numbers, still the spectral terms can be arranged in series each of which shows a strong similarity to the series of terms in the hydrogen spectrum. This similarity appears in the fact that the terms in each series can, as Rydberg pointed out, be very accurately represented by the formula $K/(n + \alpha)^2$, where K is the same constant that occurs in the hydrogen spectrum, often called the Rydberg constant, while n is the term number, and $\alpha$ a constant which is different for the different series.

This relationship with the hydrogen spectrum leads us immediately to regard these spectra as the last step of a process whereby the neutral atom is built up by the capture and binding of electrons to the nucleus, one by one. In fact, it is clear that the last electron captured, so long as it is in that stage of the binding process in which its orbit is still large compared to the orbits of the previously bound electrons, will be subjected to a force from the nucleus and these electrons, that differs but little from the force with which the electron in the hydrogen atom is attracted towards the nucleus while it is moving in an orbit of corresponding dimensions.

The spectra so far considered, for which Rydberg's laws hold, are excited by means of electric discharge under ordinary conditions and are often called *arc spectra*. The *elements* emit also another type of spectrum, the so-called *spark spectra*, when they are subjected to an *extremely powerful discharge*. Hitherto it was impossible to disentangle the spark spectra in the same way as the arc spectra. Shortly after the above view on the origin of arc spectra was brought forward, however, Fowler found (1914) that an empirical expression for the spark spectrum lines could be established which corresponds exactly to Rydberg's laws with the single difference that the constant K is replaced by a constant four times as large. Since, as we have seen, the constant that appears in the spectrum sent out during the binding of an electron to a helium nucleus is exactly equal to K, it becomes evident that *spark spectra are due to the ionized atom*, and that their emission corresponds to the last step but one in the formation of the *neutral atom* by the successive capture and binding of electrons.

## Absorption and excitation of spectral lines

The interpretation of the origin of the spectra was also able to explain the characteristic laws that govern absorption spectra. As Kirchhoff and Bunsen had already shown, there is a close relation between the selective absorption of substances for radiation and their emission spectra, and it is on this that the application of spectrum analysis to the heavenly bodies essentially rests. Yet on the basis of the classical electromagnetic theory, it is impossible to understand why substances in the form of vapor show absorption for certain lines in their emission spectrum and not for others.

On the basis of the postulates given above we are, however, led to assume that the absorption of radiation corresponding to a spectral line emitted by a transition from one stationary state of the atom to a state of less energy is brought about by the return of the atom from the last-named state to the first. We thus understand immediately that in ordinary circumstances a gas or vapor can only show selective absorption for spectral lines that are produced by a transition from a state corresponding to an earlier stage in the binding process to the normal state. Only at higher temperatures or under the influence of electric discharges whereby an appreciable number of atoms are being constantly disrupted from the normal state, can we expect absorption for other lines in the emission spectrum in agreement with the experiments.

A most direct confirmation for the general interpretation of spectra on the basis of the postulates has also been obtained by investigations on the excitation of spectral lines and ionization of atoms by means of impact of free electrons with given velocities. A decided advance in this direction was marked by the well-known investigations of Franck and Hertz (1914). It appeared from their results that by means of electron impacts it was impossible to impart to an atom an arbitrary amount of energy, but only such amounts as corresponded to a transfer of the atom from its normal state to another stationary state of the existence of which the spectra assure us, and the energy of which can be inferred from the magnitude of the spectral term.

Further, striking evidence was afforded of the independence that, according to the postulates, must be attributed to the processes which give rise to the emission of the different spectral lines of an element. Thus, it could be shown directly that atoms that were transferred in this manner to a stationary state of greater energy were able to return to the normal state with emission of radiation corresponding to a single spectral line.

Continued investigations on electron impacts, in which a large number of physicists have shared, have also produced a detailed confirmation of the theory concerning the excitation of series spectra. Especially it has been possible to show that for the ionization of an atom by electron impact an amount of energy is necessary that is exactly equal to the work required, according to the theory, to remove the last electron captured from the atom. This work can be determined directly as the product of Planck's constant and the spectral term corresponding to the normal state, which, as mentioned above, is equal to the limiting value of the frequencies of the spectral series connected with selective absorption.

**The quantum theory of multiply-periodic systems**

While it was thus possible by means of the fundamental postulates of the quantum theory to account directly for certain general features of the properties of the elements, a closer development of the ideas of the quantum theory was necessary in order to account for these properties in further detail. In the course of the last few years a more general theoretical basis has been attained through the development of formal methods that permit the fixation of the stationary states for electron motions of a more general type than those we have hitherto considered. For a simply periodic motion such as we meet in the pure harmonic oscillator, and at least to a first approximation, in the motion of an electron about a positive nucleus, the manifold of stationary states can be simply coordinated to a series of whole numbers. For motions of the more general class mentioned above, the so-called *multiply periodic motions*, however, the stationary states compose a more complex manifold, in which, according to these formal methods, each state is characterized by several whole numbers, the so-called "*quantum numbers*".

In the development of the theory a large number of physicists have taken part, and the introduction of several quantum numbers can be traced back to the work of Planck himself. But the definite step which gave the impetus to further work was made by Sommerfeld (1915) in his explanation of the *fine structure* shown by the hydrogen lines when the spectrum is observed with a spectroscope of high resolving power. The occurrence of this fine structure must be ascribed to the circumstance that we have to deal, even in hydrogen, with a motion which is not exactly simply periodic. In fact, *as a consequence of the change in the electron's mass with velocity that is claimed by the theory of relativity*, the electron orbit will undergo a very slow precession in the orbital plane. The motion will therefore be doubly periodic, and besides a number characterizing the term in the Balmer formula, which we shall call the ***principal quantum number*** because it determines in the main the energy of the atom, the fixation of the stationary states demands another quantum number which we shall call the ***subordinate quantum number***.

A survey of the motion in the stationary states thus fixed is given in the diagram (Fig. 5), which reproduces the relative size and form of the electron orbits. Each orbit is designated by a symbol where n is the principal quantum number and k the subordinate quantum number. All orbits with the same principal quantum number have, to a first approximation, the same major axis, while orbits with the same value of k have the same parameter, i.e. the same value for the shortest chord through the focus. Since the energy values for different states with the same value of n but different values of k differ a little from each other, we get for each hydrogen line corresponding to definite values of n' and n" in the Balmer formula a number of different transition processes, for which the frequencies of the emitted radiation as calculated by the second postulate are not exactly the same. As Sommerfeld was able to show, the components this gives for each hydrogen line agree with the observations on the fine structure of hydrogen lines to within the limits of experimental error. In the figure the arrows designate the processes that give rise to the components of the red and green lines in the hydrogen spectrum, the frequencies of which are obtained by putting n" = 2 and n' = 3 or 4 respectively in the Balmer formula.

In considering the figure, it must not be forgotten that the description of the orbit is there incomplete, in so much as with the scale used the slow precession does not show at all. In fact, this precession is so slow that even for the orbits that rotate most rapidly the electron performs about 40,000 revolutions before the perihelion has gone round once. Nevertheless, it is this precession alone that is responsible for the multiplicity of the stationary states characterized by the subordinate quantum number. If, for example, the hydrogen atom is subjected to a small disturbing force which perturbs the regular precession, the electron orbit in the stationary states will have a form altogether different from that given in the figure. This implies that the fine structure will change its character completely, but the hydrogen spectrum will continue to consist of lines that are given to a close approximation by the Balmer formula, due to the fact that the approximately periodic character of the motion will be retained. Only when the disturbing forces become so large that even during a single revolution of the electron the orbit is appreciably disturbed, will the spectrum undergo essential changes. The statement often advanced that the introduction of two quantum numbers should be a necessary condition for the explanation of the Balmer formula must therefore be considered as a misconception of the theory.

Sommerfeld's theory has proved itself able to account not only for the fine structure of the hydrogen lines, but also for that of the lines in the helium spark spectrum. Owing to the greater velocity of the electron, the intervals between the components into which a line is split up are here much greater and can be measured with much greater accuracy. The theory was also able to account for certain features in the fine structure of X-ray spectra, where we meet frequency differences that may even reach a value more than a million times as great as those of the frequency differences for the components of the hydrogen lines.

Shortly after this result had been attained, Schwarzschild and Epstein (1916) simultaneously succeeded, by means of similar considerations, in accounting for the characteristic changes that the hydrogen lines undergo in an electric field, which had been discovered by Stark in the year 1914. Next, an explanation of the essential features of the Zeeman effect for the hydrogen lines was worked out at the same time by Sommerfeld and Debye (1917). In this instance the application of the postulates involved the consequence that only certain orientations of the atom relative to the magnetic field were allowable, and this characteristic consequence of the quantum theory has quite recently received a most direct confirmation in the beautiful researches of Stern and Gerlach on the deflection of swiftly moving silver atoms in a nonhomogeneous magnetic field.

**The correspondence principle**

While this development of the theory of spectra was based on the working out of formal methods for the fixation of stationary states, the present lecturer succeeded shortly afterwards in throwing light on the theory from a new viewpoint, by pursuing further the characteristic connection between the quantum theory and classical electrodynamics already traced out in the hydrogen spectrum. In connection with the important work of Ehrenfest and Einstein these efforts led to the formulation of the so-called *correspondence principle*, according to which the occurrence of transitions between the stationary states accompanied by emission of radiation is traced back to the harmonic components into which the motion of the atom may be resolved and which, according to the classical theory, determine the properties of the radiation to which the motion of the particles gives rise.

According to the *correspondence principle*, it is assumed that every transition process between two stationary states can be coordinated with a corresponding harmonic vibration component in such a way that the probability of the occurrence of the transition is dependent on the amplitude of the vibration. The state of polarization of the radiation emitted during the transition depends on the further characteristics of the vibration, in a manner analogous to that in which on the classical theory the intensity and state of polarization in the wave system emitted by the atom as a consequence of the presence of this vibration component would be determined respectively by the amplitude and further characteristics of the vibration.

With the aid of the ***correspondence principle***, it has been possible to confirm and to extend the above-mentioned results. Thus, it was possible to develop a complete quantum theory explanation of the ***Zeeman effect*** for the hydrogen lines, which, in spite of the essentially different character of the assumptions that underlie the two theories, is very similar throughout to Lorentz's original explanation based on the classical theory. In the case of the ***Stark effect***, where, on the other hand, the classical theory was completely at a loss, the quantum theory explanation could be so extended with the help of the correspondence principle as to account for the polarization of the different components into which the lines are split, and also for the characteristic intensity distribution exhibited by the components. This last question has been more closely investigated by Kramers, and the accompanying figure will give some impression of how completely it is possible to account for the phenomenon under consideration.

Fig. 6 reproduces one of Stark's well-known photographs of the splitting up of the hydrogen lines. The picture displays very well the varied nature of the phenomenon, and shows in how peculiar a fashion the intensity varies from component to component. The components below are polarized perpendicular to the field, while those above are polarized parallel to the field.

Fig. 7 gives a diagrammatic representation of the experimental and theoretical results for the line the frequency of which is given by the Balmer formula with n" = 2 and n' = 5. The vertical lines denote the components into which the line is split up, of which the picture on the right gives the components which are polarized parallel to the field and that on the left those that are polarized perpendicular to it. The experimental results are represented in the upper half of the diagram, the distances from the dotted line representing the measured displacements of the components, and the lengths of the lines being proportional to the relative intensity as estimated by Stark from the blackening of the photographic plate. In the lower half is given for comparison a representation of the theoretical results from a drawing in Kramers' paper.

The symbol (n'$_{s'}$− n"$_{s''}$) attached to the lines gives the transitions between the stationary states of the atom in the electric field by which the components are emitted. Besides the principal quantum integer n, the stationary states are further characterized by a subordinate quantum integer s, which can be negative as well as positive and ***has a meaning quite different from that of the quantum number k occurring in the relativity theory of the fine structure of the hydrogen lines***, which fixed the form of the electron orbit in the undisturbed atom. Under the influence of the electric field both the form of the orbit and its position undergo large changes, but certain properties of the orbit remain unchanged, and the subordinate quantum number s is connected with these. In Fig. 7 the position of the components corresponds to the frequencies calculated for the different transitions, and the lengths of the lines are proportional to the probabilities as calculated on the basis of the correspondence principle, by which also the polarization of the radiation is determined. It is seen that the theory reproduces completely the main feature of the experimental results, and in the light of the correspondence principle we can say that the Stark effect reflects down to the smallest details the action of the electric field on the orbit of the electron in the hydrogen atom, even though in this case the reflection is so distorted that, in contrast with the case of the Zeeman effect, it would scarcely be possible directly to recognize the motion on the basis of the classical ideas of the origin of electromagnetic radiation.

Results of interest were also obtained for the spectra of elements of higher atomic number, the explanation of which in the meantime had made important progress through the work of Sommerfeld, who introduced several quantum numbers for the description of the electron orbits. Indeed, it was possible, with the aid of the correspondence principle, to account completely for the characteristic rules which govern the seemingly capricious occurrence of combination lines, and it is not too much to say that the quantum theory has not only provided a simple interpretation of the combination principle, but has further contributed materially to the clearing up of the mystery that has long rested over the application of this principle.

The same viewpoints have also proved fruitful in the investigation of the so-called **band spectra**. These do not originate, as do series spectra, from individual atoms, but from **molecules**; and the fact that these spectra are so rich in lines is due to the complexity of the motion entailed by the vibrations of the atomic nuclei relative to each other and the rotations of the molecule as a whole. The first to apply the postulates to this problem was Schwarzschild, but the important work of Heurhnger especially has thrown much light on the origin and structure of band spectra. The considerations employed here can be traced back directly to those discussed at the beginning of this lecture in connection with Bjerrum's theory of the influence of molecular rotation on the infrared absorption lines of gases. It is true we no longer think that the rotation is reflected in the spectra in the way claimed by classical electrodynamics, but rather that the line components are due to transitions between stationary states which differ as regards rotational motion. That the phenomenon retains its essential feature, however, is a typical consequence of the correspondence principle.

**The natural system of the elements**

The ideas of the origin of spectra outlined in the preceding have furnished the basis for a theory of the structure of the atoms of the elements which has shown itself suitable for a general interpretation of the main features of the properties of the *elements*, as exhibited in the natural system. This theory is based primarily on considerations of the manner in which the atom can be imagined to be built up by the capture and binding of electrons to the nucleus, one by one. As we have seen, the optical spectra of elements provide us with evidence on the progress of the last steps in this building-up process.

An insight into the kind of information that the closer investigation of the spectra has provided in this respect may be obtained from Fig. 8, which gives a diagrammatic representation of the orbital motion in the stationary states corresponding to the emission of the *arc-spectrum of potassium*. The curves show the form of the orbits described in the stationary states by the last electron captured in the potassium atom, and they can be considered as stages in the process whereby the 19th electron is bound after the 18 previous electrons have already been bound in their normal orbits. In order not to complicate the figure, no attempt has been made to draw any of the orbits of these inner electrons, but the region in which they move is enclosed by a dotted circle. In an atom with several electrons the orbits will, in general, have a complicated character. Because of the symmetrical nature of the field of force about the nucleus, however, *the motion of each single electron can be approximately described as a plane periodic motion on which is superimposed a uniform rotation in the plane of the orbit*. The orbit of each electron will therefore be to a first approximation doubly periodic, and will be fixed by two quantum numbers, as are the stationary states in a hydrogen atom when the relativity precession is taken into account.

In Fig. 8, as in Fig. 5, the electron orbits are marked with the symbol $n_k$, where n is the principal quantum number and k the subordinate quantum number. While for the initial states of the binding process, where the quantum numbers are large, the orbit of the last electron captured lies completely outside of those of the previously bound electrons, this is not the case for the last stages. Thus, in the potassium atom, the electron orbits with subordinate quantum numbers 2 and 1 will, as indicated in the figure, penetrate partly into the inner region. Because of this circumstance, the orbits will deviate very greatly from a simple Kepler motion, since they will consist of a series of successive outer loops that have the same size and form, but each of which is turned through an appreciable angle relative to the preceding one. Of these outer loops only one is shown in the figure. Each of them coincides very nearly with a piece of a Kepler ellipse, and they are connected, as indicated, by a series of inner loops of a complicated character in which the electron approaches the nucleus closely. This holds especially for the orbit with subordinate quantum number 1, which, as a closer investigation shows, will approach nearer to the nucleus than any of the previously bound electrons.

On account of this penetration into the inner region, the strength with which an electron in such an orbit is bound to the atom will - in spite of the fact that for the most part it moves in a field of force of the same character as that surrounding the hydrogen nucleus - be much greater than for an electron in a hydrogen atom that moves in an orbit with the same principal quantum number, the maximum distance of the electron from the nucleus at the same time being considerably less than in such a hydrogen orbit. As we shall see, ***this feature of the binding process in atoms with many electrons is of essential importance in order to understand the characteristic periodic way in which many properties of the elements as displayed in the natural system vary with the atomic number***.

In the accompanying table (Fig. 9) is given a summary of the results concerning the structure of the atoms of the elements to which the author has been led by a consideration of successive capture and binding of electrons to the atomic nucleus. The figures before the different elements are the atomic numbers, which give the total number of electrons in the neutral atom. The figures in the different columns give the number of electrons in orbits corresponding to the values of the principal and subordinate quantum numbers standing at the top. In accordance with ordinary usage, we will, for the sake of brevity, designate an orbit with principal quantum number n as an n quantum orbit. The first electron bound in each atom moves in an orbit that corresponds to the normal state of the hydrogen atom with quantum symbol 1. In the hydrogen atom there is of course only one electron; but we must assume that in the atoms of other elements the next electron also will be bound in such a r-quantum orbit of type 1. As the table shows, the following electrons are bound in 2-quantum orbits. To begin with, the binding will result in a $2_1$ orbit, but later electrons will be bound in $2_2$ orbits, until, after binding the first 10 electrons in the atom, we reach a closed configuration of the r-quantum orbits in which we assume there are four orbits of each type. This configuration is met for the first time in the neutral neon atom, which forms the conclusion of the second period in the system of the elements. When we proceed in this system, the following electrons are bound in 3-quantum orbits, until, after the conclusion of the third period of the system, we encounter for the first time, in elements of the fourth period, electrons in 4-quantum orbits, and so on.

This picture of atomic structure contains many features that were brought forward by the work of earlier investigators. Thus, the attempt to interpret the relations between the elements in the natural system by the assumption of a division of the electrons into groups goes as far back as the work of J. J. Thomson in 1904. Later, this viewpoint was developed chiefly by Kossel (1916), who, moreover, has connected such a grouping with the laws that investigations of X-ray spectra have brought to light.

Also G. R. Lewis and *I. Langmuir* have sought to account for the relations between the properties of the elements on the basis of a grouping inside the atom. These investigators, however, assumed that the electrons do not move about the nucleus, but occupy positions of equilibrium. In this way, though, no closer relation can be reached between the properties of the elements and the experimental results concerning the constituents of the atoms. Statical positions of equilibrium for the electrons are in fact not possible in cases in which the, forces between the electrons and the nucleus even approximately obey the laws that hold for the attractions and repulsions between electrical charges.

The possibility of an interpretation of the properties of the elements on the basis of these latter laws is quite characteristic for the picture of atomic structure developed by means of the quantum theory. As regards this picture, the idea of connecting the grouping with a classification of electron orbits according to increasing quantum numbers was suggested by Moseley's discovery of the laws of X-ray spectra, and by Sommerfeld's work on the fine structure of these spectra. This has been principally emphasized by Vegard, who some years ago in connection with investigations of X-ray spectra proposed a grouping of electrons in the atoms of the elements, which in many ways shows a likeness to that which is given in the above table.

*A satisfactory basis for the further development of this picture of atomic structure has, however, only recently been created by the study of the binding processes of the electrons in the atom*, of which we have experimental evidence in optical spectra, and the characteristic features of which have been elucidated principally by the correspondence principle. It is here an essential circumstance that the restriction on the course of the binding process, which is expressed by the presence of electron orbits with higher quantum numbers in the normal state of the atom, can be naturally connected with the general condition for the occurrence of transitions between stationary states, formulated in that principle.

*Another essential feature of the theory is the influence, on the strength of binding and the dimensions of the orbits, of the penetration of the later bound electrons into the region of the earlier bound ones*, of which we have seen an example in the discussion of the origin of the potassium spectrum. Indeed, this circumstance may be regarded as the essential cause of the pronounced periodicity in the properties of the elements, in that it implies that the atomic dimensions and chemical properties of homologous substances in the different periods, as, for example, the alkali-metals, show a much greater similarity than that which might be expected from a direct comparison of the orbit of the last electron hound with an orbit of the same quantum number in the hydrogen atom.

*The increase of the principal quantum number which we meet when we proceed in the series of the elements, affords also an immediate explanation of the characteristic deviations from simple periodicity which are exhibited by the natural system* and are expressed in Fig. 1 by the bracketing of certain series of elements in the later periods. The first time such a deviation is met with is in the 4th period, and the reason for it can be simply illustrated by means of our figure of the orbits of the last electron bound in the atom of potassium, which is the first element in this period. Indeed, in potassium we encounter for the first time in the sequence of the elements a case in which the principal quantum number of the orbit of the last electron bound is, in the normal state of the atom, larger than in one of the earlier stages of the binding process. The normal state corresponds here to a $4_1$ orbit, which, because of the penetration into the inner region, corresponds to a much stronger binding of the electron than a 4-quantum orbit in the hydrogen atom. The binding in question is indeed even stronger than for a 2-quantum orbit in the hydrogen atom, and is therefore more than twice as strong as in the circular $3_3$ orbit which is situated completely outside the inner region, and for which the strength of the binding differs but little from that for a 3-quantum orbit in hydrogen.

This will not continue to be true, however, when we consider the binding of the 19th electron in substances of higher atomic number, because of the much smaller relative difference between the field of force outside and inside the region of the first eighteen electrons bound. As is shown by the investigation of the spark spectrum of calcium, the binding of the 19th electron orbit is here but little stronger than in $3_3$ orbits, and as soon as we in the $4_1$ reach scandium, we must assume that the $3_3$ orbit will represent the orbit of the 19th electron in the normal state, since this type of orbit will correspond to a stronger binding than a $4_1$ orbit. While the group of electrons in 2-quantum orbits has been entirely completed at the end of the 2nd period, the development that the group of 3-quantum orbits undergoes in the course of the 3rd period can therefore only be described as a provisional completion, and, as shown in the table, this electron group will, in the bracketed elements of the 4th period, undergo a stage of further development in which electrons are added to it in 3-quantum orbits.

This development brings in new features, in that the development of the electron group with 4-quantum orbits comes to a standstill, so to speak, until the 3-quantum group has reached its final closed form. Although we are not yet in a position to account in all details for the steps in the gradual development of the 3-quantum electron group, still we can say that with the help of the quantum theory we see at once why it is in the 4th period of the system of the elements that there occur for the first time successive elements with properties that resemble each other as much as the properties of the iron group; indeed, *we can even understand why these elements show their well-known paramagnetic properties*. Without further reference to the quantum theory, Eadenburg had on a previous occasion already suggested the idea of relating the chemical and magnetic properties of these elements with the development of an inner electron group in the atom.

I will not enter into many more details, but only mention that the peculiarities we meet with in the 5th period are explained in much the same way as those in the 4th period. Thus, the properties of the bracketed elements in the 5th period as it appears in the table, depend on a stage in the development of the 4-quantum electron group that is initiated by the entrance in the normal state of electrons in $4_3$ orbits. In the 6th period, however, we meet new features. In this period, we encounter not only a stage of the development of the electron groups with 5- and 6-quantum orbits, but also the final completion of the development of the 4-quantum electron group, which is initiated by the entrance for the first time of electron orbits of the $4_4$ type in the normal state of the atom. This development finds its characteristic expression in the occurrence of the peculiar family of elements in the 6th period, known as the *rare-earths*. These show, as we know, a still greater mutual similarity in their chemical properties than the elements of the iron family. This must be ascribed to the fact that we have here to do with the development of an electron group that lies deeper in the atom. It is of interest to note that the theory can also naturally account for the fact that these elements, which resemble each other in so many ways, still show great differences in their magnetic properties.

The idea that the occurrence of the rare-earths depends on the development of an inner electron group has been put forward from different sides. Thus, it is found in the work of Vegard, and at the same time as my own work, it was proposed by Bury in connection with considerations of the systematic relation between the chemical properties and the grouping of the electrons inside the atom from the point of view of Langmuir's static atomic model. While until now it has not been possible, however, to give any theoretical basis for such a development of an inner group, we see that our extension of the quantum theory provides us with an unforced explanation. Indeed, it is scarcely an exaggeration to say that if the existence of the rare earths had not been established by direct chemical investigation, the occurrence of a family of elements of this character within the 6th period of the natural system of the elements might have been theoretically predicted.

When we proceed to the 7th period of the system, we meet for the first time with 7-quantum orbits, and we shall expect to find within this period features that are essentially similar to those in the 6th period, in that besides the first stage in the development of the 7-quantum orbits, we must expect to encounter further stages in the development of the group with 6- or 5 quantum orbits. However, it has not been possible directly to confirm this expectation, because only a few elements are known in the beginning of the 7th period. The latter circumstance may be supposed to be intimately connected with the ***instability of atomic*** *nuclei with large charges*, which is expressed in the prevalent radioactivity among elements with high atomic number.

## X-ray spectra and atomic constitution

In the discussion of the conceptions of atomic structure we have hitherto placed the emphasis on the formation of the atom by successive capture of electrons. Our picture would, however, be incomplete without some reference to the confirmation of the theory afforded by the study of X-ray spectra. Since the interruption of Moseley's fundamental researches by his untimely death, the study of these spectra has been continued in a most admirable way by Prof. Siegbahn in Lund. On the basis of the large amount of experimental evidence adduced by him and his collaborators, it has been possible recently to give a classification of X-ray spectra that allows an immediate interpretation on the quantum theory. In the first place it has been possible, just as in the case of the optical spectra, to represent the frequency of each of the X-ray lines as the difference between two out of a manifold of spectral terms characteristic of the element in question. Next, a direct connection with the atomic theory is obtained by the assumption that each of these spectral terms multiplied by Planck's constant is equal to the work which must be done on the atom to remove one of its inner electrons. In fact, the removal of one of the inner electrons from the completed atom may, in accordance with the above considerations on the formation of atoms by capture of electrons, give rise to transition processes by which the place of the electron removed is taken by an electron belonging to one of the more loosely bound electron groups of the atom, with the result that after the transition an electron will be lacking in this latter group.

The X-ray lines may thus be considered as giving evidence of stages in a process by which the atom undergoes a reorganization after a disturbance in its interior. According to our views on the stability of the electronic configuration such a disturbance must consist in the removal of electrons from the atom, or at any rate in their transference from normal orbits to orbits of higher quantum numbers than those belonging to completed groups; a circumstance which is clearly illustrated in the characteristic difference between selective absorption in the X-ray region, and that exhibited in the optical region.

The classification of the X-ray spectra, to the achievement of which the above-mentioned work of Sommerfeld and Kossel has contributed materially, has recently made it possible, by means of a closer examination of the manner in which the terms occurring in the X-ray spectra vary with the atomic number, to obtain a very direct test of a number of the theoretical conclusions as regards the structure of the atom. In Fig. 9 the abscissæ are the atomic numbers and the ordinates are proportional to the square roots of the spectral terms, while the symbols K, L, M, N, O, for the individual terms refer to the characteristic discontinuities in the selective absorption of the elements for X-rays; these were originally found by Barkla before the discovery of the interference of X-rays in crystals had provided a means for the closer investigation of X-ray spectra. Although the curves generally run very uniformly, they exhibit a number of deviations from uniformity which have been especially brought to light by the recent investigation of Coster, who has for some years worked in Siegbahn's laboratory.

These deviations, the existence of which was not discovered until after the publication of the theory of atomic structure discussed above, correspond exactly to what one might expect from this theory. At the foot of the figure the vertical lines indicate where, according to the theory, we should first expect, in the normal state of the atom, the occurrence of $n_k$ orbits of the type designated. We see how it has been possible to connect the occurrence of every spectral term with the presence of an electron moving in an orbit of a definite type, to the removal of which this term is supposed to correspond. That in general there corresponds more than one curve to each type of orbit $n_k$ is due to a complication in the spectra which would lead us too far afield to enter into here, and may be attributed to the deviation from the previously described simple type of motion of the electron arising from the interaction of the different electrons within the same group.

The intervals in the system of the elements, in which a further development of an inner electron group takes place because of the entrance into the normal atom of electron orbits of a certain type, are designated in the figure by the horizontal lines, which are drawn between the vertical lines to which the quantum symbols are affixed. It is clear that such a development of an inner group is everywhere reflected in the curves. Particularly the course of the N- and O-curves may be regarded as a direct indication of that stage in the development of the electron groups with 4-quantum orbits of which the occurrence of the rare-earths bears witness. Although the apparent complete absence of a reflection in the X-ray spectra of the complicated relationships exhibited by most other properties of the elements was the typical and important feature of Moseley's discovery, we can recognize, nevertheless, in the light of the progress of the last years, an intimate connection between the X-ray spectra and the general relationships between the elements within the natural system.

Before concluding this lecture, I should like to mention one further point in which X-ray investigations have been of importance for the test of the theory. This concerns the properties of the hitherto unknown element with atomic number 72. On this question opinion has been divided in respect to the conclusions that could be drawn from the relationships within the Periodic Table, and in many representations of the table a place is left open for this element in the rare-earth family. In Julius Thomsen's representation of the natural system, however, this hypothetical element was given a position homologous to titanium and zirconium in much the same way as in our representation in Fig. 1. Such a relationship must be considered as a necessary consequence of the theory of atomic structure developed above, and is expressed in the table (Fig. 9) by the fact that the electron configurations for titanium and zirconium show the same sort of resemblances and differences as the electron configurations for zirconium and the element with atomic number 72. A corresponding view was proposed by Bury on the basis of his above-mentioned systematic considerations of the connection between the grouping of the electrons in the atom and the properties of the elements.

Recently, however, a communication was published by Dauvillier announcing the observation of some weak lines in the X-ray spectrum of a preparation containing rare-earths. These were ascribed to an element with atomic number 72 assumed to be identical with an element of the rare-earth family, the existence of which in the preparation used had been presumed by Urbain many years ago. This conclusion would, however, if it could be maintained, place extraordinarily great, if not unsurmountable, difficulties in the way of the theory, since it would claim a change in the strength of the binding of the electrons with the atomic number which seems incompatible with the conditions of the quantum theory. In these circumstances Dr. Coster and Prof. Hevesy, who are both for the time working in Copenhagen, took up a short time ago the problem of testing a preparation of zircon-bearing minerals by X-ray spectroscopic analysis. These investigators have been able to establish the existence in the minerals investigated of appreciable quantities of an element with atomic number 72, the chemical properties of which show a great similarity to those of zirconium and a decided difference from those of the rare-earths. *

> * For the result of the continued work of Coster and Hevesy with the new element, for which they have proposed the name hafnium, the reader may be referred to their letters in *Nature* of January 20, February 10 and 24, and April 7.

I hope that I have succeeded in giving a summary of some of the most important results that have been attained in recent years in the field of atomic theory, and I should like, in concluding, to add a few general remarks concerning the viewpoint from which these results may be judged, and particularly concerning the question of how far, with these results, it is possible to speak of an explanation, in the ordinary sense of the word. By a theoretical explanation of natural phenomena, we understand in general a classification of the observations of a certain domain with the help of analogies pertaining to other domains of observation, where one presumably has to do with simpler phenomena. The most that one can demand of a theory is that this classification can be pushed so far that it can contribute to the development of the field of observation by the prediction of new phenomena.

When we consider the atomic theory, we are, however, in the peculiar position that there can be no question of an explanation in this last sense, since here we have to do with phenomena which from the very nature of the case are simpler than in any other field of observation, where the phenomena are always conditioned by the combined action of a large number of atoms. We are therefore obliged to be modest in our demands and content ourselves with concepts which are formal in the sense that they do not provide a visual picture of the sort one is accustomed to require of the explanations with which natural philosophy deals. Bearing this in mind I have sought to convey the impression that the results, on the other hand, fulfill, at least in some degree, the expectations that are entertained of any theory; in fact, I have attempted to show how the development of atomic theory has contributed to the classification of extensive fields of observation, and by its predictions has pointed out the way to the completion of this classification. It is scarcely necessary, however, to emphasize that the theory is yet in a very preliminary stage, and many fundamental questions still await solution.

**Uhlenbeck, G. E. & Goudsmit, S. (November, 1925). Ersetzung der Hypothese vom unmechanischen Zwang durch eine Forderung bezuglich des inneren Verhaltens jedes einzelnen Elektrons. (Replacement of the hypothesis of unmechanical coercion by a requirement regarding the internal behavior of each individual electron.)**

[*Naturw.*, 13, 47, 953-4; https://doi.org/10.1007/BF01558878; (translation by T. G. Underwood).]

October 17, 1925.

Instituut voor Theoretische Natuurkunde, Leiden.

The idea of a quantized spinning of the electron was put forward for the first time by Compton in August 1921, who pointed out the possible bearing of this idea on the origin of the natural unit of magnetism, without being aware of Compton's suggestion Uhlenbeck and Goudsmit notes doublets in the alkali spectra that did not conform to current models of the atom, proposes possibility of applying the model of spinning electron to interpret a number of features of the quantum theory of the *anomalous Zeeman effect*, applies classical formula for spherical rotating electron with finite radius and surface charge.

[Uhlenbeck, G. E. & Goudsmit, S. (February 20, 1926). *Spinning Electrons and the Structure of Spectra. Nature*, 117, 264-5; https://doi.org/10.1038/117264a0; "*Abstract.* So far as we know, the idea of a quantized spinning of the electron was put forward for the first time by A. K. Compton [(August, 1921). *The magnetic electron. Journ. Frankl. Inst.*, 192, 145-55], who pointed out the possible bearing of this idea on the origin of the natural unit of magnetism. Without being aware of Compton's suggestion, we have directed attention in a recent note (*Naturw.*, November 20, 1925) to the possibility of applying the spinning electron to interpret a number of features of the quantum theory of the Zeeman effect, which were brought to light by the work especially of van Lohuizen, Sommerfeld, Landé and Pauli, and also of the analysis of complex spectra in general. In this letter we shall try to show how our hypothesis enables us to overcome certain fundamental difficulties which have hitherto hindered the interpretation of the results arrived at by those authors."]

[*The discovery of the electron spin*, lecture by Samuel Goudsmit on the golden jubilee of the Dutch Physical Society in April 1971: https://www.lorentz. leidenuniv.nl/history/spin/goudsmit.html: "… when I went to Leiden, I ended up with Ehrenfest. Ehrenfest's classes were small and one had a very good interaction with one's professor. And Ehrenfest was always worried when we interrupted our classes when we had to go somewhere. Once I had to accompany my father to Germany, because of his business, and then Ehrenfest said: "Do you again have to interrupt your classes?" But my father could not travel alone. Then he asked: "Where are you going?" When I told him, he said: "Nearby is a university and there is a spectroscopist, Paschen. You are interested in spectroscopy (I had become interested in it through my high-school teacher Lohuizen), go and have a look". … I went to visit Paschen, who did not treat me as a student but as a colleague. And he showed me the experimental set up which he had for the study of the spectral line of ionized helium, which entirely confirmed Sommerfeld's relativistic electron orbits. I did not understand a bit of it. But, I think, I managed to hide my lack of understanding and after my return to Leiden I have nicely studied all this. One of the things which stuck to me is that in Paschen's experiments on the helium line, its fine structure and the relativistic explanation, there was a forbidden component which was obviously present. The following summer I was sent for a stay to Paschen, and Paschen and Back have taught me the techniques of spectroscopy. And when I talked to the theoreticians about that forbidden component ......... but you know how theoreticians are ...... they then say: "Poor experiments". That forbidden line already was an important milestone. …

I was interested in spectral lines and the first thing I did .... I found a formula for the doublets in the spectra, claiming that it was exactly the same formula as used by Sommerfeld for the X-ray doublets. And I told this to Ehrenfest. At that stage it was all wrong but Ehrenfest never discouraged anyone and said: "That's nice, we'll publish it". And there was a short little piece in "Naturwissenschaften" and a very lengthy article in "Archives Néerlandaises des Sciences exactes et naturelles", which was published in Holland in French to be sure that nobody would read it. Of course, as a young student I was very proud of it.

… Two and a half years later exactly the same work was done, the very same formula, by Millikan in America, and Koster gave a seminar about it in Leiden. Of course, he did not know that I had already done so. At the end of the seminar, I said: "I have spoken about the very same, here, two and a half years ago".

… "I had simply guessed it while Millikan, when he obtained the formula, had new experimental material which demonstrated its correctness. One did not understand that the formula was correct, but the new experimental data made it clear that he was the one who had the right formula. He had reasons for it, I had simply guessed, I could not even convince Ehrenfest, and it was published in French ...". (George Uhlenbeck was Ehrenfest's assistant, assigned to work with his graduate student, Goudsmit.)]

———————————————

§ 1. As is well known, the structure and magnetic behavior of the spectra can be described in detail with the help of Landé's vector model R, K, J and m[1].

[1] See Back, E. & Landé, A. (1925). *Zeemaneffekt und Multiplettstruktur der Spektrallinien.* (Zeeman effect and Multiplet structure of the spectral lines.) Berlin: Verlag von Julius Springer).

Here, R denotes the *momentum* moment of the atomic remnant ~ i.e. of the atom without the luminous electron - K the *momentum* moment of the luminous electrons, J their resultant and m the projection of J on the direction of an external magnetic field, all expressed in the branch quantum units:

(a) that for the rest of the atom the behavior of the magnetic moment to the mechanical is twice as large as you would expect classically.

(b) that in the formulae, where $R^2$, $K^2$, $J^2$ occurs, you can do this by using these expressions $R^2 - \frac{1}{4}$, $K^2 - \frac{1}{4}$, $J^2 - \frac{1}{4}$. [The Heisenberg Averaging[2])].

[2] Heisenberg, W. (1924). *Über eine Änderung der formalen Regeln der Quantentheorie in einem Problem anomaler Zeeman-Effekte.* (On an alteration to the formal rules of quantum theory in a problem of anomalous Zeeman effect.) *Zeit. Phys.*, 26, 291-307.

This model has shown itself to be very robust and has, among other things, fought to unravel the most complicated spectra.

§ 2. However, one starts to encounter difficulties as soon as one tries to connect the Landé's vector model to our ideas our ideas about the formation of the atom from electrons. E.g.:

a) Pauli[3] has already shown that in the case of the alkali atoms, the atomic radical must be magnetically ineffective, otherwise the influence of *relativity* correction would cause a dependency of the Zeeman effect on the nuclear charge, which is not perceived in these spectra.

[3] Pauli Jr., W. (1925). *Über den Einfluss der Geschwindigkeitsabhängigkeit der Elektronenmasse auf den Zeemaneffekt.* (On the influence of the velocity dependence of the electron mass on the Zeeman effect.) *Zeit. Phys.*, 31, 373.

b) In Lande's model, one must not identify the momentum moment of the atomic radical with that of the positive ions, as one would expect it according to the definition of the atomic radical. [Landé-Heisenberg branching theorem[4] — unmechanical coercion].

[4] See Back, E. & Landé, A. (1925). *Zeemaneffekt und Multiplettstruktur der Spektrallinien. Loc. cit.*, pages 55ff.

c) For some spectra recently analyzed with the help of Lande's scheme (e.g. vanadium, titanium), the K of the basic term did not correspond at all with the values expected from the Bohr-Stone periodic system.

§ 3. The above-mentioned difficulties point all in the same direction, namely that the meaning of which is attributed to Lande's vectors is probably not correct. Pauli[5] has already embarked on a new path, which is particularly difficult.

[5] Pauli Jr., W. (1925). *Über den Zusammenhang des Abschlusses der Elektronengruppen im Atom mit der Komplexstruktur der Spektren.* (On the relationship between the completion of the electron groups in the atom and the complex structure of the spectra.) *Zeit. Phys.*, 31, 765.

From this he concluded that in the case of alkali spectra, all quantum numbers must be written to the luminous electron alone. According to Pauli, each electron in the magnetic field then gets 4 independent quantum numbers. With the help of Bohr's construction principle and a few general sentences, he was then able to achieve the same results as Landé in a simple way[6].

[6] Compare: Goudsmit, S. (December, 1925). *Über die Komplexstruktur der Spektren. Zeit. Phys.*, 32, 1, 794-98; https://doi.org/10.1007/BF01331715; Heisenberg, W. (1925). *Quantentheorie der multiplen Struktur und des abnormalen Zeeman-Effekts.* (Quantum theory of multiple structure and the abnormal Zeeman effect.) *Zeit. Phys.*, 32, 841-60; Hund, F. (1925). *Zur Deutung verwickelter Spektren, insbesondere der Elemente Scandium bis Nickel.* (On the interpretation of entangled spectra, in particular the elements scandium to nickel.) *Zeit. Phys.*, 33, 345-71; http://dx.doi.org/ 10.1007/BF01328319.

The difficulties mentioned in § 2 disappear completely in the Pauli procedure. The connection to the Bohr-Stoner periodic system is achieved, and new aspects are still opened[7].

[7] See those in 5) below.

§ 4. In both cases, however, the appearance of the so-called *relativistic doublet* in the rontgen and alkali spectra remains an enigma. To explain this fact, one has recently come to the assumption of a classically indescribable ambiguity in the quantum theoretical properties of the electron[1].

[1] Heisenberg, W. (1925). *Quantentheorie der multiplen Struktur und des abnormalen Zeeman-Effekts.* (Quantum theory of multiple structure and the abnormal Zeeman effect.) *Zeit. Phys.*, 32, 841-60; *loc. cit.*

§ 5 There seems to us to be another way open. Pauli does not bind himself to a model idea. The 4 quantum numbers assigned to each electron have lost their original Landé meaning. It is now obvious to give to each electron with its 4 quantum numbers 4 degrees of freedom. One can then give the quantum numbers, for example, the following meaning: n and k remain as before the main and azimuthal quantum number of the electron in its orbit. *R, however, will be assigned its own rotation of the electron*[2].

[2] Note that the quantum numbers of the electron occurring here must be taken from the alkali spectra. R therefore has only the value 1 for each electron (in Landé standardization).

The other quantum numbers retain their old meaning. Through our imagination, the conceptions of Landé and Pauli with all their advantages have formally merged with each other[3].

[3] For example, the meaning of the Heisenberg's Scheme III is now becoming more understandable, in which one has to assemble both the R and the K of the electrons for an entire atom.

The electron must now take over the still misunderstood property (referred to in § I under a), which Landé attributed to the atomic remnant.

The closer quantitative implementation of this idea will probably depend heavily on the choice of the electron model. In order to come into line with the facts, the following demands must therefore be made of this model:

a) The ratio of the magnetic moment of the electron to the mechanical one must be twice as large for the self-rotation as for the orbital motion[4].

[4] For example, for a spherical rotating electron with surface charge can be used to the Abraham formulas [Abraham, M. (1903). *Prinzipien der Dynamik des Elektrons*. (Principles of electron dynamics.) *Ann. Phys.*, 315, 105-79] read:

Rotational energy $1/9 \ e^2 a/c^2 \ \dot{\varphi}^2$      ($a$ = electron radius),

also: $p_\varphi = 2/9 \ e^2 a/c^2 \ \dot{\varphi}$
Magnetic moment: $\Phi = 1/3 \ ea^2/c \ \dot{\varphi}$

Mass: $m = 2/3 \ e^2/c^2 a$

Also: $\Phi/p_\varphi = 3/2 \ ac/e = 2 \times e/2mc$ in fact, twice as much as in the orbital motion.

Note, however, that when quantizing this rotational motion, the peripheral speed of the electron is far from the speed of light.

b) The different orientations from the R to the orbital plane (or K) of the electron must be able to provide the explanation of *relativity-doublets*, perhaps in connection with a Heisenberg -Wentzel averaging rule[5].

[5] Heisenberg, W. *loc. cit.* Wentzel, G. (1925). *Ann. Phys.*, 76, 803.

## Schrödinger, E. R. J. A. (August 12, 1887 – January 4, 1961)

Schrödinger was a Nobel Prize-winning Austrian-Irish physicist who developed a number of fundamental results in quantum theory: the Schrödinger equation provides a way to calculate the wave function of a system and how it changes dynamically in time.

In addition, he wrote many works on various aspects of physics: statistical mechanics and thermodynamics, physics of dielectrics, color theory, electrodynamics, general relativity, and cosmology, and he made several attempts to construct a unified field theory. In his book *What Is Life?* Schrödinger addressed the problems of genetics, looking at the phenomenon of life from the point of view of physics. He paid great attention to the philosophical aspects of science, ancient, and oriental philosophical concepts, ethics, and religion. He also wrote on philosophy and theoretical biology. In popular culture, he is most known for his "Schrödinger's cat" thought experiment.

Schrödinger was born in Erdberg, Vienna, Austria, on August 12, 1887, to Rudolf Schrödinger (cerecloth producer, botanist) and Georgine Emilia Brenda Schrödinger (née Bauer) (daughter of Alexander Bauer, professor of chemistry, TU Wien). He was their only child. His mother was of half Austrian and half English descent; his father was Catholic and his mother was Lutheran. He was also able to learn English outside school, as his maternal grandmother was British.

Between 1906 and 1910 (the year he earned his doctorate) Schrödinger studied at the University of Vienna under the physicists Franz S. Exner and Friedrich Hasenöhrl. He received his doctorate at Vienna under Hasenöhrl. He also conducted experimental work with Karl Wilhelm Friedrich "Fritz" Kohlrausch. In 1911, Schrödinger became an assistant to Exner. In 1914 Schrödinger achieved habilitation (venia legendi).

Between 1914 and 1918 he participated in war work as a commissioned officer in the Austrian fortress artillery (Gorizia, Duino, Sistiana, Prosecco, Vienna).

On 6 April 1920, Schrödinger married Annemarie (Anny) Bertel. Schrödinger suffered from tuberculosis and several times in the 1920s stayed at a sanatorium in Arosa. It was there that he formulated his wave equation.

In 1920 he became the assistant to Max Wien, in Jena, and in September 1920 he attained the position of ao. Prof. (ausserordentlicher Professor) in Stuttgart, roughly equivalent to reader (UK) or associate professor (US).

In 1921, he became o. Prof. (ordentlicher Professor, i.e. full professor), in Breslau (now Wrocław, Poland). In 1921, he moved to the University of Zürich. In the first years of his career Schrödinger became acquainted with the ideas of the old quantum theory, developed in the works of Max Planck, Albert Einstein, Niels Bohr, Arnold Sommerfeld, and others. This knowledge helped him work on some problems in theoretical physics, but the Austrian scientist at the time was not yet ready to part with the traditional methods of classical physics.

The first publications of Schrödinger about atomic theory and the theory of spectra began to emerge only from the beginning of the 1920s, after his personal acquaintance with Sommerfeld and Wolfgang Pauli and his move to Germany. In January 1921, Schrödinger finished his first article on this subject, about the framework of the Bohr-Sommerfeld effect of the interaction of electrons on some features of the spectra of the alkali metals. *Of particular interest to him was the introduction of relativistic considerations in quantum theory.*

In autumn 1922 he analyzed the electron orbits in an atom from a geometric point of view, using methods developed by the mathematician Hermann Weyl (1885–1955). This work, in which it was shown that quantum orbits are associated with certain geometric properties, was an important step in predicting some of the features of wave mechanics. Earlier in the same year he created the Schrödinger equation of the *relativistic* Doppler effect for spectral lines, based on the hypothesis of light quanta and considerations of energy and momentum. He liked the idea of his teacher Exner on the statistical nature of the conservation laws, so he enthusiastically embraced the articles of Bohr, Kramers, and Slater, which suggested the possibility of violation of these laws in individual atomic processes (for example, in the process of emission of radiation). Although the experiments of Hans Geiger and Walther Bothe soon cast doubt on this, the idea of *energy as a statistical concept* was a lifelong attraction for Schrödinger and he discussed it in some reports and publications.

In March 1926, Schrödinger published his first paper on *wave mechanics* and presented what is now known as the *Schrödinger equation*. [Schrodinger, E. (March, 1926). *Quantisierung als Eigenwertproblem. (Erste Mitteilung)* (Quantization as an eigenvalue problem. (First communication).) *Ann. Physik*, 384, 4, 79, 261-376; https://doi.org/ 10.1002/andp.19263840404.] In this paper, he gave a "derivation" of *the wave equation for time-independent systems and showed that it gave the correct energy eigenvalues for a hydrogen-like atom*. This paper has been universally celebrated as one of the most important achievements of the twentieth century and created a revolution in most areas of quantum mechanics and indeed of all physics and chemistry.

A second paper was submitted just four weeks later that solved the quantum harmonic oscillator, rigid rotor, and diatomic molecule problems and gave a new derivation of the Schrödinger equation. [Schrodinger, E. (1926). *Quantisierung als Eigenwertproblem (Zweite Mitteilung).* (Quantization as an eigenvalue problem. (Second communication).) *Ann. Physik*, 4, 79, 489-527.]

A third paper, published in May, showed the equivalence of his approach to that of Heisenberg and gave the treatment of the Stark effect. [Schrodinger, E. (1926). *Quantisierung als Eigenwertproblem (Dritte Mitteilung: Störungstheorie, mit Anwendung auf den Starkeffekt der Balmerlinien)*. (Quantization as an eigenvalue problem. (Third communication: Perturbation theory, with application to the strong effect of Balmer lines)).) *Ann. Physik,* 4, 80, 437-90.]

A fourth paper in this series showed how to treat problems in which the system changes with time, as in scattering problems. [Schrodinger, E. (1926). *Quantisierung als Eigenwertproblem (Vierte Mitteilung)*. (Quantization as an eigenvalue problem. (Fourth communication)).) *Ann. Physik,* 4, 81, 109-39.]

In this paper he introduced a complex solution to the wave equation in order to prevent the occurrence of fourth and sixth order differential equations. (***This was arguably the moment when quantum mechanics switched from real to complex numbers*.**) When he introduced complex numbers in order to lower the order of the differential equations, something magical happened, and all of wave mechanics was at his feet. (He eventually reduced the order to one.)

These papers were his central achievement and were at once recognized as having great significance by the physics community. An account of the four papers in English was published in December of that year. [Schrodinger, E. (December, 1926). *A Wave Theory of the Mechanics of Atoms and Molecules. Phys. Rev.,* 28, 1049-70.]

Schrödinger was not entirely comfortable with the implications of quantum theory referring to his theory as "wave mechanics." He wrote about the probability interpretation of quantum mechanics, saying: "I don't like it, and I'm sorry I ever had anything to do with it."

In 1927, he succeeded Max Planck at the Friedrich Wilhelm University in Berlin. In 1933, Schrödinger decided to leave Germany because he disliked the Nazis' antisemitism. He became a Fellow of Magdalen College at the University of Oxford. Soon after he arrived, he received the Nobel Prize for the formulation of the Schrödinger equation, which he shared with Dirac.

His position at Oxford did not work out well; his unconventional domestic arrangements, sharing living quarters with two women, were not met with acceptance. In 1934, Schrödinger lectured at Princeton University; he was offered a permanent position there, but did not accept it. Again, his wish to set up house with his wife and his mistress may have created a problem. He had the prospect of a position at the University of Edinburgh but visa delays occurred, and in the end, he took up a position at the University of Graz in Austria in 1936. In the midst of these tenure issues in 1935, after extensive correspondence with Albert Einstein, he proposed what is now called the Schrödinger's cat thought experiment.

In 1938, after the Anschluss, Schrödinger had problems in Graz because of his flight from Germany in 1933 and his known opposition to Nazism. He issued a statement recanting this opposition (he later regretted doing so and explained the reason to Einstein). However, this did not fully appease the new dispensation and the University of Graz dismissed him from his post for political unreliability. He suffered harassment and was instructed not to leave the country. He and his wife, however, fled to Italy. From there, he went to visiting positions in Oxford and Ghent University.

In the same year he received a personal invitation from Ireland's Taoiseach, Éamon de Valera – a mathematician himself – to reside in Ireland and agree to help establish an Institute for Advanced Studies in Dublin. When he migrated to Ireland in 1938, he obtained visas for himself, his wife and also another woman, Mrs. Hilde March. March was the wife of an Austrian colleague with whom Schrödinger had fathered a daughter in 1934. Schrödinger wrote personally to de Valera to obtain the visa for Mrs. March. In October 1939 the ménage à trois duly took up residence in Dublin. He moved to Kincora Road, Clontarf, Dublin and lived modestly. Schrödinger fathered two further daughters by two different women during his time in Ireland.

He became the Director of the School for Theoretical Physics in 1940 and remained there for 17 years. He became a naturalized Irish citizen in 1948, but also retained his Austrian citizenship. He wrote around 50 further publications on various topics, including his explorations of unified field theory.

In 1944, he wrote *What Is Life?*, which contains a discussion of negentropy and the concept of a complex molecule with the genetic code for living organisms. According to James D. Watson's memoir, *DNA, the Secret of Life*, Schrödinger's book gave Watson the inspiration to research the gene, which led to the discovery of the DNA double helix structure in 1953. Similarly, Francis Crick, in his autobiographical book *What Mad Pursuit*, described how he was influenced by Schrödinger's speculations about how genetic information might be stored in molecules.

Following his work on quantum mechanics, Schrödinger devoted considerable effort to working on a unified field theory that would unite gravity, electromagnetism, and nuclear forces within the basic framework of General Relativity, doing the work with an extended correspondence with Albert Einstein. In 1947, he announced a result, "Affine Field Theory," in a talk at the Royal Irish Academy, but the announcement was criticized by Einstein as "preliminary" and failed to lead to the desired unified theory. Following the failure of his attempt at unification, Schrödinger gave up his work on unification and turned to other topics

In 1956, he returned to Vienna to take up his appointment as Chair of Physics at the University of Vienna. At an important lecture during the World Energy Conference, he refused to speak on nuclear energy because of his skepticism about it and gave a philosophical lecture instead. During this period Schrödinger turned from mainstream quantum mechanics' definition of wave–particle duality and promoted the wave idea alone, causing much controversy.

On 4 January 1961, Schrödinger died of tuberculosis, aged 73, in Vienna.

## Schrodinger, E. (December, 1926). A Wave Theory of the Mechanics of Atoms and Molecules.

[*Phys. Rev.*, 28, 6, 1049-70; https://doi.org/10.1103/PhysRev.28.1049; (first published as a series of papers in German from March, 1926).]

September 3, 1926.

Zurich, Physikalisches Institut der Universitiit.

*Non-relativistic* development of de Broglie's *relativistic* wave mechanics in which *phase-waves* associated with motion of material points, in particular with motion of an electron or proton, assumes material points are wave-systems, *wave-equation* $\Delta\psi + 8\pi^2 m(E - V)\psi/h^2 = 0$, *laws of motion* and *quantum conditions* deduced simultaneously from Hamiltonian principle, *wave function* converts atom into system of fluctuating charges spread out continuously in space, generates electric moment that changes in time, discrepancy between frequency of motion and frequency of emission disappears, frequency of emission coincides with differences of frequency of motion, superposition of frequencies, definite localization of electric charge in space and time associated with the wave-system, solutions of *wave equation* for simplified hydrogen atom or one body problem correspond to Bohr's stationary energy levels of the elliptic orbits, the selected values called "*eigenvalues*" and the solutions that belong to them "*eigenfunctions*", the charge of the electron is spread out through space but the *wave-phenomenon* is restricted to a small sphere of a few Angstroms diameter constituting the atom, also possible to calculate *amplitudes* of harmonic components of the *electric moment* for any direction in space, in the case of the *Stark effect* (perturbation of the hydrogen-atom caused by an external homogeneous electric field) parallel to the electric field or perpendicular to the field, shows that squares of these *amplitudes* are proportional to the *intensities* of the several line components polarized in either direction, *wave mechanics has been developed without reference to relativity modifications of classical mechanics or to action of a magnetic field on the atom, not been possible to extend the relativistic theory to a system of more than one electron, relativistic theory of hydrogen atom in grave contradiction with experiment*, how to take into account *electron spin* is yet unknown.

*This paper gives an account of the author's work on a new form of quantum theory.*

§1. The Hamiltonian analogy between mechanics and optics. §2. The analogy is to be extended to include real "physical" or "wave" mechanics instead of mere geometrical mechanics. §3. The significance of wave-length; macro-mechanical and micro-mechanical problems. §4. The wave-equation and its application to the hydrogen atom. §5. The intrinsic reason for the appearance of discrete characteristic frequencies. §6. Other problems; intensity of emitted light. §7. The wave-equation derived from a Hamiltonian variation-principle; generalization to an arbitrary conservative system. §8. The wave-function physically means and determines a continuous distribution of electricity in space, the fluctuations of which determine the radiation by the laws of ordinary electrodynamics. §9. Non-conservative systems. Theory of dispersion and scattering and of the "transitions" between the "stationary states." §10. The question of *relativity* and the action of a magnetic field. Incompleteness of that part of the theory.

1. The theory which is reported in the following pages is based on the very interesting and fundamental researches of L. de Broglie[1] on what he called "*phase-waves*" ("ondes de phase") and ***thought to be associated with the motion of material points, especially with the motion of an electron or proton***. The point of view taken here, which was first published in a series of German papers,[2] is rather that ***material points consist of, or are nothing but, wave-systems***.

[1] de Broglie, L. (February, 1925). *Recherches sur la théorie des quanta.* Thesis, Paris, 1924. *Ann. de Physique*, 10, 3, 22 [; de Broglie describes a ***relativistic*** theory of ***wave mechanics*** for a moving particle, applies Einstein's ***equivalence of mass and energy*** and ***relativistic change of mass when moving relative to the observer*** to an electron to obtain ***total energy***, sets ***energy*** of electron in rest frame equal to quantum of energy with a frequency given by Planck's ***quantum relationship***, calculates ***frequency of moving electron*** measured by fixed observer by applying ***clock retardation***, differs from frequency calculated from ***quantum relation***, resolves by showing that the phases of the moving electron and its associated *wave* remain the same, represents wave as ***phase wave*** with velocity greater than the velocity of light, applies to the periodic motion of an electron in a BOHR atom, stability conditions of a BOHR orbit seen as identical to ***resonance condition*** of the associated ***phase wave***, applies to the mutual interaction of electrons and protons in the hydrogen atom, does not address transitions from one stable orbit to another, requires a modified version of electrodynamics].

[2] Schrodinger, E. (March, 1926). *Quantisierung als Eigenwertproblem. (Erste Mitteilung)* (Quantization as an eigenvalue problem. (First communication).) *Ann. Physik*, 384, 4, 79, 361-376; https://doi.org/ 10.1002/andp.19263840404; Schrodinger, E. (1926). *Quantisierung als Eigenwertproblem (Zweite Mitteilung).* (Quantization as an eigenvalue problem. (Second communication).) *Ann. Physik*, 384, 4, 79, 489-527; Schrodinger, E. (1926). *Quantisierung als Eigenwertproblem (Dritte Mitteilung: Störungstheorie, mit Anwendung auf den Starkeffekt der Balmerlinien).* (Quantization as an eigenvalue problem. (Third communication: Perturbation theory, with application to the strong effect of Balmer lines).) *Ann. Physik,* 384, 4, 80, 437-90; Schrodinger, E. (1926). *Quantisierung als Eigenwertproblem (Vierte Mitteilung).* (Quantization as an eigenvalue problem. (Fourth communication).) *Ann. Physik,* 384, 4, 81, 109-39; Schrodinger, E. (1926). *Naturw.*, 14, 664.

This extreme conception may be wrong; indeed, it does not offer as yet the slightest explanation of *why only such wave-systems seem to be realized in nature as correspond to mass-points of definite mass and charge*. On the other hand, the opposite point of view, which neglects altogether the waves discovered by L. de Broglie and treats only the motion of material points, has led to such grave difficulties in the theory of atomic mechanics - and this after century-long development and refinement - that it seems not only not dangerous but even desirable, for a time at least, to lay an exaggerated stress on its counterpart. In doing this we must of course realize that a thorough correlation of all features of physical phenomena can probably be afforded only by a harmonic union of these two extremes.

The chief advantages of the present *wave-theory* are the following.

a. The *laws of motion* and the *quantum conditions* are deduced simultaneously from one simple Hamiltonian principle.

b. The discrepancy hitherto existing in quantum theory between the *frequency of motion* and the *frequency of emission* disappears in so far as *the latter frequencies coincide with the differences of the former*. *A definite localization of the electric charge in space and time can be associated with the wave-system* and this with the aid of ordinary electrodynamics accounts for the *frequencies, intensities* and *polarizations* of the emitted light and makes superfluous all sorts of *correspondence and selection principles*.

c. It seems possible by the new theory to pursue in all detail the so-called "*transitions*", which up to date have been wholly mysterious.

d. There are several instances of disagreement between the new theory and the older one as to the particular values of the *energy* or *frequency* levels. In these cases, it is the new theory that is better supported by experiment.

To explain the main lines of thought, I will take as an example of a mechanical system a material point, mass *m,* moving in a conservative field of force V(x, y, z). All the following treatment may very easily be extended to the motion of the "image-point," picturing the motion of a wholly arbitrary conservative system in its "configuration-space" (q-space, not pq-space). We shall effect this generalization in a somewhat different manner in Section 7. Using the usual notations, the *kinetic energy* T is

$$T = \tfrac{1}{2}m(\dot{x}^2 + \dot{y}^2 + \dot{z}^2) = (1/2m)(p_x^2 + p_y^2 + p_z^2). \qquad (1)$$

The well-known *Hamiltonian function of action* **W**,

$$W = \int_{t_0}^{t} (T - V)dt \qquad (2)$$

taken as a function of the upper limit *t* and of the final values of the coordinates *x, y, z* satisfies the Hamiltonian partial differential equation,

$$\partial W/\partial t + (1/2m)\left[(\partial W/\partial x)^2 + (\partial W/\partial y)^2 + (\partial W/\partial z)/^2\right] + V(x,y,z,) = 0. \quad (3)$$

To solve this equation, we put as usual

$$W = -Et + S(x, y, z) \qquad (4)$$

E being an integration constant, viz., the total energy, and S a function of $x$, $y$, $z$ only. Eq. (3) may then be written

$$|grad\ W| = [2m(E - V)]^{1/2}. \qquad (5)$$

In this form it lends itself to a very simple geometrical interpretation. Assume $t$ constant for the moment. Any function W of space alone can be described by giving geometrically the system of surfaces on which W is constant and by writing down on each one of these surfaces the constant value, say $W_0$, which the function W takes on it. On the other hand, we can easily construct a solution of Eq. (5) starting from an arbitrary surface and an arbitrarily chosen value $W_0$, which we ascribe to it. For after having chosen starting surface and starting value and after-still arbitrarily-having designated one of its two sides or "shores" as the positive one, we simply have to extend the normal at every point of the chosen surface to the length, say

$$dn = dW_0/[2m(E - V)]^{1/2}.$$

The totality of points arrived at in this way will fill a surface to which we obviously have to ascribe the value $W_0 + d\,W_0$. The continuation of this procedure will supply us the whole system of surfaces and values of constants belonging to them, i.e. the whole distribution in space of the function W, at first for $t$ constant.

Now let the time vary, Eq. (4) shows that the system of surfaces will not vary, but that the values of the constants will travel along the normals from surface to surface with a certain speed $u$, given by

$$u = E/[2m(E - V)]^{1/2}. \qquad (6)$$

The *velocity u* is a function of the energy-constant E and besides, since it contains $V(x, y, z)$ is a function of the **coordinates**.

Instead of thinking of the surfaces as fixed in space and letting the values of the constant wander from surface to surface, we may equally well think of a certain numerical value of W as attached to a certain individual surface and let the surfaces wander in such a way that each of them continually takes the place and exact form of the following one. Then the quantity $u$, given by Eq. (6)

$$[u = E/[2m(E - V)]^{1/2}. \qquad (6)]$$

will denote the normal velocity of any surface at any one of its points. *Adopting this view, we arrive at a picture which exactly coincides with the propagation of a stationary wave-*

*system in an optically non-homogeneous (but isotropic) medium, W being proportional to the phase and u, being the phase velocity*. (The index of refraction would have to be taken proportional to $u^{-1}$.) The above-mentioned construction of normals d$n$ is obviously equivalent to Huygens' principle. The orthogonal curves of our system of W- surfaces form a system of rays in our optical picture; they are possible orbits of the material point in the mechanical problem. Indeed, it is well known that

$$px = m\dot{x} = \partial W/\partial x \qquad (7)$$

(with two analogous equations for y and z). It may be useful, to remark, that the phase-velocity $u$ is not the velocity of the material point [$v$]. The latter is, by (7) and (5)

$$[|\text{grad } W| = [2m(E - V)]^{1/2}. \qquad (5)]$$

$$v = (\dot{x}^2 + \dot{y}^2 + \dot{z}^2)^{1/2} = [2(E - V)/m]^{1/2}. \qquad (8)$$

Comparing (6) and (8) we see, that $u$ and $v$ vary even inversely to each other. The well-known mechanical principle due to and named after Hamilton can very easily be shown to correspond to the equally well known optical principle of Fermat.

2. Nothing of what has hitherto been said is in any way new. All this was very much better known to Hamilton himself than it is in our day to a good many physicists. Indeed, the theory of the propagation of light in a non-homogeneous medium, which Hamilton had developed about ten years earlier, became, by the striking analogy which occurred to him, the starting-point for his famous theories in pure mechanics. Notwithstanding the great popularity reached by the latter, the way which had led to them was nearly forgotten.[3]

[3] See Klein, F. (1891). *Jahresber. d. Deutsch. Math. Ver. 1*; (1901). *Zeits. f. Math. u. Phys.*, 46; (Ges. Abh; II, 601, 603): Whittaker, E. T. *Analytical Dynamics*, Chap. 11. Sommerfeld, *A. Atombau*. German ed., p. 803. The analogy has been rediscovered in relativistic mechanics in the paper of L. de Broglie, quoted above.

Stress must now be laid on the fact, that though in our above-stated reasoning such conceptions as "wave-surfaces," "Huygens' principle," "Fermat's principle" come into play, nevertheless the whole established analogy deals rather with geometrical optics than with real physical or wave optics. Indeed, *the chief and fundamental mechanical conception is that of the path or orbit of the material particle*, and it corresponds to the conception of rays in the optical analogy. Now the conception of rays is thoroughly well defined only in pure abstract geometrical optics. It loses nearly all significance in real physical optics as soon as the dimensions of the beam or of material obstacles in its path become comparable with the wave length. And even when this is not the case, the notion of rays is, in physical optics, merely an approximate one. It is wholly incapable of being applied to the fine structure of real optical phenomena, i.e. to the phenomena of diffraction. Even in extending geometrical optics somewhat by adding the notion of Huygens' principle (in the simple form, used above) one is not able to account for the most simple phenomena of diffraction without adding some further very strange rules concerning the circumstances under which Huygens' envelope-surface is or is not physically significant. (I mean the construction of "Fresnel's zones.") These rules would be wholly incomprehensible to one versed in geometrical optics alone. Furthermore, it may be observed that the notions which are fundamental to real physical optics, i.e. the wave-function itself (W is merely the *phase*), the equation of wave-propagation, the wave length and frequency of, the waves, do not enter at all into the above stated analogy. The *phase-velocity* $u$ does enter but we have seen that it is not very intimately connected with the mechanical velocity $\upsilon$.

At first sight it does not seem at all tempting, to work out in detail the Hamiltonian analogy as in real wave optics. By giving the wave-length a proper well-defined meaning, the well-defined meaning of rays is lost at least in some cases, and by this the analogy would seem to be weakened or even to be wholly destroyed for those cases in which the dimensions of the mechanical orbits or their radii of curvature be come comparable with the wave-length. To save the analogy it would seem necessary to attribute an exceedingly small value to the wave-length, small in comparison with all dimensions that may ever become of any interest in the mechanical problem. But then again, the working out of a wave picture would seem superfluous, for geometrical optics is the real limiting case of wave optics for vanishing wave-length.[4]

[4] Sommerfeld, A. & Runge, I. (1911). *Anwendung der Vektorrechnung auf die Grundlagen der Geometrischen Optik.* (Application of vector calculus to the fundamentals of geometric optics.) *Ann. Phys.*, 35, 277-298, p. 290.

Now compare with these considerations the very striking fact, of which we have today irrefutable knowledge, that ordinary mechanics is really not applicable to mechanical systems of very small, viz. of atomic dimensions. Taking into account this fact, which impresses its stamp upon all modern physical reasoning, is one not greatly tempted to investigate whether the non-applicability of ordinary mechanics to micro-mechanical problems is perhaps of exactly the same kind as the non-applicability of geometrical optics to the phenomena of diffraction or interference and may, perhaps, be overcome in an exactly similar way? The conception is: *the Hamiltonian analogy has really to be worked out towards wave optics and a definite size is to be attributed to the wave-length in every special case.* This quantity has a real meaning for the mechanical problem, viz. that ordinary mechanics with its conception of a moving point and its linear path (or more generally of an "image-point" moving in the coordinate space) is only approximately applicable so long as they supply a path, which is (and whose radii of curvature are) large in comparison with the wave-length. If this is not the case, it is a phenomenon of wave-propagation that has to be studied. In the simple case of one material point moving in an external field of force the wave-phenomenon may be thought of as taking place in the ordinary three-dimensional space; in the case of a more general mechanical system, it will primarily be located in the coordinate space ($q$-space, not $pq$-space) and will have to be projected somehow into ordinary space. *At any rate the equations of ordinary mechanics will be of no more use for the study of these micro-mechanical wave-phenomena than the rules of geometrical optics are for the study of diffraction phenomena. Well known methods of wave-theory, somewhat generalized, lend themselves readily.* The conceptions, roughly sketched in the preceding are fully justified by the success which has attended their development.

3.  Let us return to the system of W-surfaces, dealt with in Section 1 and let us associate with them the idea of stationary sinusoidal waves whose *phase* is given by the quantity W, Eq. (4)

$$[W = - Et + S(x, y, z). \tag{4}]$$

The *wave-function*, say $\psi$, will be of the form

$$\psi = A(x, y, z) \sin(W/K)$$
$$= A(x, y, z) \sin [- Et/K + S(x, y, z)/K], \tag{9}$$

A being an "*amplitude*" function.   The constant K must be introduced and must have the physical dimension of *action* (energy x time), since the argument of a sine must always be a pure number. Now, since the *frequency* of the wave (9) is obviously

$$\nu = E/2\pi K \tag{10}$$

one cannot resist the temptation of supposing K to be a *universal constant*, independent of E and independent of the nature of the mechanical system, because if this be done and $K$ be given the value $h/2\pi$, then the frequency $\nu$ will be given by

$$\nu = E/h, \tag{11}$$

*h* being Planck's constant. Thus, *the well-known universal relation between energy and frequency is arrived at in a rather simple and unforced way*.

In ordinary mechanics the absolute value of the energy has no definite meaning, only energy-differences have.  This difficulty can be met and a zero-level of energy can be defined in an entirely satisfactory way by using *relativistic* mechanics and the conception of equivalence of mass and energy.  But it is unnecessary to dwell on this subject here. While the *frequency v* of our waves by Eq. (10) or (11) is indeed dependent on the zero-level of energy, *their wave-length is not*. And after what has been said above, *it is the wave-length that is of greatest interest.* The comparison of this quantity with the dimensions of the path or orbit of the material particle, calculated according to ordinary mechanics, will tell us whether the latter calculation is or is not of physical significance, whether the methods of ordinary mechanics are approximately applicable to the special problem or not.

The *wave-length* $\lambda$ by (11) and (6)

$$[\nu = E/h, \tag{11}$$

is

$$u = E/[2m(E − V)]^{1/2} \tag{6]}$$

$$\lambda = u/v = h/[2m(E − V)]^{1/2} \tag{12}$$

Here E − V is the kinetic energy ½ m$v^2$ which indeed is independent of the zero-level of the total energy. Inserting its value, we have

$$\lambda = h/mv. \tag{13}$$

To test the question whether an electron, moving in a Keplerian orbit of atomic dimensions may, following our hypotheses, still be dealt with by ordinary mechanics, let $a$ be a length of atomic dimensions and compare $\lambda$ with $a$.

$$\lambda/a = h/mva. \tag{14}$$

The denominator on the right is certainly of the order of magnitude of the **moment of momentum** of the electron, and the latter is well known to be of the order of magnitude of Planck's constant for a Keplerian orbit of atomic dimensions. So $\lambda/a$ becomes of the order of unity and, following our conceptions, ordinary mechanics will be no more applicable to such an orbit than geometrical optics is to the diffraction of light by a disk of diameter equal to the wave-length. Were a physicist to try to understand the latter phenomenon by the conception of rays, with which he is acquainted from macroscopic geometrical optics, he would meet with most serious difficulties and apparent contradictions. The "rays" (stream lines of the flow of energy) would no longer be rectilinear and would influence one another in a most curious way, in full contradiction with the most fundamental laws of geometrical optics. In the same way the conception of orbits of material points seems to be, inapplicable to orbits of atomic dimensions. *It is very satisfactory, that the limit of applicability of ordinary mechanics is, by equating K (essentially) to Planck's constant* (Eq. 11),

$$[v = E/h, \tag{11]}$$

*determined to an order of magnitude, which is exactly the one to be postulated, if the new conception is to help us in our quantum difficulties.* We may add, that by Eq. (13)

$$[\lambda = h/mv. \tag{13]}$$

for a Keplerian electronic orbit of the order of magnitude of a high quantum orbit, the relation of wave-length to orbital dimensions becomes of the order of magnitude of the reciprocal of the quantum number. Hence ordinary mechanics will offer a better and better approximation in the limit of increasing quantum number (or orbital dimensions), and this is just what is to be expected from any reasonable theory.

By the fundamental equation $v = E/h$ (Eq. 11) the **phase velocity** $u$, given by Eq. (6)

$$[u = E/[2m(E − V)]^{1/2} \tag{6]}$$

proves to be dependent on the *frequency v*. Therefore, Eq. (6) is an *equation of dispersion*. By this a very interesting light is thrown on the relation of the two velocities (1) *velocity v of the moving particle*, Eq. (8)

$$[\upsilon = (\dot{x}^2 + \dot{y}^2 + \dot{z}^2)^{1/2} = [2(E - V)/m]^{1/2}; \qquad (8)]$$

(2) *phase-velocity u*, Eq. (6)

$$[u = E/[2m(E - V)]^{1/2}. \qquad (6)]$$

$\upsilon$ is easily proved to be exactly the so-called *group velocity* belonging to the dispersion formula (6).[5]

---

[5] This important theorem is due to L. de Broglie, i.e. The relation is: $\upsilon = d\nu/d(\nu/u)$.

By using this interesting result, it is possible to form an idea how ordinary mechanics is capable of giving an approximate description of our wave motion. By superposing waves of frequencies in a small interval $v; v + dv$ it is possible to construct a "*parcel of waves*", the dimensions of which are in all directions rather small, though they must be rather large in comparison to the wave-length. Now *it can be proved, that the motion of - let us say - the "center of gravity" of such a parcel will, by the laws of wave propagation, follow exactly the same orbit as the material point would by the laws of ordinary mechanics*. This equivalence is always maintained, even if the dimensions of the orbit are not large in comparison with the wave-length. But in the latter case it will have no significance, the wave parcel being spread out in all directions far over the range of the orbit. On the contrary, if the dimensions of the orbit are comparatively large, the motion of the wave parcel as a whole may afford a sufficient idea of what really happens, if we are not interested in its intrinsic constitution. As stated above this "motion as a whole" is governed by the laws of ordinary mechanics.

4. We shall not dwell on this question further, but proceed to the far more interesting applications of the theory to micro-mechanical problems. As stated above, the wave-phenomena must in this case be studied in detail. This can only be done by using an "*equation of wave propagation*." Which one is this to be? In the case of a single material point, moving in an external field of force, the simplest way is to try to use the ordinary *wave-equation*

$$\Delta\psi - \ddot{\psi}/u^2 = 0 \qquad (15)$$

and to insert for $u$ the quantity given by Eq. (6),

$$[u = E/[2m(E - V)]^{1/2}. \qquad (6)]$$

which depends on the space coordinates (through the *potential energy* V) and on the *frequency* E/h. The latter dependence restricts the use of (15) to such functions $\psi$ as depend on the time only through the factor $e^{\pm 2\pi itE/k}$. (A similar restriction is always imposed on the *wave equation*, as soon as we have dispersion.) So, we shall have

$$\ddot{\psi} = -4\pi^2 E^2 \psi/h^2 = 0,$$

Inserting this and Eq. (6) in Eq. (15) we get [the *wave equation*]

$$\Delta\psi + 8\pi^2 m(E - V)\psi/h^2 = 0, \tag{16}$$

where $\psi$ may be assumed to depend on x, y, z only. (We omit changing the notation of the dependent variable, which we really ought to do.)

Now what are we to do with Eq. (16)?

$$[\Delta\psi + 8\pi^2 m(E - V)\psi/h^2 = 0, \tag{16}]$$

At first sight this equation seems to offer ill means of solving atomic problems, e.g. of defining discrete energy-levels in the hydrogen atom. Being a partial differential equation, it offers a vast multitude of solutions, a multitude of even a higher transcendent order of magnitude than the system of solutions of the ordinary differential equations of ordinary mechanics. But the deficiency of the latter in atomic problems consisted, as is well known, by no means in that they supplied too small a number of possible orbits, but quite on the contrary, much too many. To select a discrete number of them as the "real" or "stationary" ones is, according to the view hither to adopted, the task of the "*quantum-conditions*". Our *wave equation* (16), by augmenting the possibilities indefinitely, instead of restricting them, seems to lead us from bad to worse.

*Happily, because of the very interesting character which Eq. (16) takes in actual atomic problems, this fear proves to be erroneous*. Putting for instance

$$V = - e^2/r, \tag{17}$$

(e = *electronic charge*, $r = (x^2 + y^2 + z^2)^{\frac{1}{2}}$, we get for the *simplified hydrogen atom or one body problem*:

$$\Delta\psi + 8\pi^2 m(E + e^2/r)\psi/h^2 = 0, \tag{18}$$

Now this equation for a great part of the possible *values of the energy or frequency constant E,* proves to offer no solution at all which is continuous, finite and single-valued throughout the whole space; for the E-values in question, every solution $\psi$, that satisfies the two other conditions (viz. continuity and single-valuedness) grows beyond all limits either in approaching infinity or in approaching the origin of coordinates. The only **E-values**, for which this is not the case i.e. for which *solutions* exist, that are continuous, finite and single-valued throughout the whole space are the following ones

(1)     E>0
(2)     $E = - 2\pi^2 m e^4/h^2 n^2$     $(n = 1, 2, 3, 4, ...)$ $\tag{19}$

The first set corresponds to the hyperbolic orbits in ordinary mechanics. It is the general view, that according to ordinary quantum theory the hyperbolic orbits are not submitted to quantization. In our treatment this turns out quite spontaneously from the fact that every positive value of E leads to finite solutions. ***The second set corresponds exactly to Bohr's stationary energy levels of the elliptic orbits***.

Though I cannot enter here upon the exact and rather tiresome proof of the foregoing statements, it may be interesting to describe in rough feature the solutions belonging to the second series of E-levels. The solution may be performed in three-dimensional[6] polar coordinates, by assuming $\psi$ to be a product of a function of the polar angles and a function of the radius $r$ only.

> [6] It is of course not allowed to restrict the problem to two dimensions as in ordinary mechanics since the wave-phenomenon is essentially three-dimensional.

The former is a spherical surface harmonic whose order, increased by unity, corresponds to the azimuthal quantum number. The functions of $r$, which come into play, somewhat resemble (in rough feature) the Bessel functions, though with the difference that they have but a finite number of positive roots, and this number exactly corresponds to the radial quantum number. These roots lie within a region from the origin of about the same order of magnitude as the corresponding Bohr orbit. After having passed the last root with increasing $r$ and a maximum or minimum not far away from it, the function tends to diminish exponentially as $r$ approaches infinity. So, ***the whole of the wave-phenomenon, though mathematically spreading throughout all space, is essentially restricted to a small sphere of a few Angstroms diameter which may be called "the atom" according to wave mechanics***. Any one of the above-mentioned solutions (consisting of a product of a spherical surface-harmonic and a function of $r$ only) greatly resembles a fundamental vibration of an elastic sphere, with a finite number of (1) spheres, (2) cones, (3) planes as "node surfaces." But it is surely not permissible to think that the wave-motion constituting the atom is, in general, restricted to one of these solutions, the special selection and separation of which is very much influenced by the choice of coordinates. To every one of the discrete values of E belongs a finite number of special solutions. In forming a linear aggregate of them with arbitrary constant multipliers we get the most. general solution of Eq. (18)

$$[\Delta\psi + 8\pi^2 m(E + e^2/r)\psi/h^2 = 0, \qquad\qquad (18)]$$

for the particular value of E. The number of arbitrary constants entering into this aggregate is exactly equal to what is called the "statistical weight" of this energy-level, or in other words, to the number of separate levels into which it is split up according to Bohr's theory (and, by the way, also according to the present theory) by the addition of perturbing forces, that do away with the so-called "degeneration" of the problem. It will perhaps be remembered, that in ordinary quantum theory the number of states that is supplied by the method alluded to is not exactly correct. Definite experimental evidence compels us to exclude by additional reasoning, more or less convincing from the theoretical point of view, a definite number of states, viz. those which have the equatorial quantum number zero. ***It is gratifying to be able to state, that according to the present theory the above-mentioned*** *number of arbitrary constants or, in other words, the number of separate levels or frequencies into which a degenerated E-level is split up by a perturbing potential is quite correct from the beginning.* The theory needs no supplementation since it precludes a vibrational state corresponding to a Bohr-orbit with equatorial quantum number zero.

To complete this description we may add, that to the lowest E-level, or from the wave-motion point of view its "*fundamental tone*" which corresponds to the normal state of the atom, there belongs but one mode of vibration, and this is a very simple one; the function ψ shows complete spherical symmetry and there are no node surfaces at all. Both the radial quantum number as well as the order of the spherical surface harmonic vanish.

5. I should like to discuss in a few words the question, why Eq. (18) possesses finite solutions only for certain selected values of the constant E. The whole behavior described on the foregoing pages would be quite familiar to every physicist, if the problem were a so-called "boundary condition problem," i.e. if the function ψ were required only in the interior of a given surface, let us say a sphere of given radius, and had to fulfill certain conditions on the boundary of this sphere, e.g. to vanish. Now though this is not the case, *the problem is indeed equivalent to a boundary-condition problem, the boundary being the infinite sphere.* Thus, the *selected values* (19)

$$[E = - 2\pi^2 me^4/h^2 n^2 \quad (n = 1, 2, 3, 4, …) \tag{19}]$$

are quite properly to be named "*characteristic values*" ["*eigenvalues*"] and the *solutions,* that belong to them, "*characteristic functions*" ["*eigenfunctions*"] of the problem connected with Eq. (18).

> [David Hilbert introduced the terms *Eigenwert* and *Eigenfunktion*; see Hilbert, D. (1904). *Grundzüge einer allgemeinen Theorie der linearen Integralgleichungen.*]

The mathematical reason,[7] why no boundary conditions in the proper sense of the word are neither needed nor allowed at the infinite boundary, is that a singular point of Eq. (18) is approached when we recede in any direction in space toward infinity.

> [7] See e.g. Courant, R. & Hilbert, D. (1924). *Methoden der mathemntischen Physik I.* Springer, Berlin, Chap. *5,* §9, p. 1.

This can easily be seen by splitting up the equation in the way described above, using polar coordinates. The resulting ordinary differential equation with the variable *r* has two singularities, at r = 0 and at r = ∞. It offers (for negative values of E) but one solution that remains finite at r = 0, and but one that remains finite at r = ∞. These two solutions are in general not identical, but they are for the selected values of E given by (19).

But instead of dwelling on this purely mathematical side of the subject, I should like to present an idea why Eq. (18) shows such a queer behavior so as to make the matter clear to anyone who is acquainted only with the most general principles of *wave theory*. If E is negative the bracket in Eq. (18) will be negative outside a certain sphere. Now remembering the way in which Eq. (18) was derived from Eq. (15)

$$[\Delta\psi - \ddot{\psi}/u^2 = 0, \qquad\qquad (15)]$$

we see that a negative value of the bracket in (18) clearly means a negative value of the square of **wave-velocity**, or an **imaginary value** of **wave-velocity**. What does this imply? The Laplacian operator is well known to be intimately connected with the average excess of the neighboring values over the value of the function at the point considered. Thus, the ordinary **wave-equation** (15) with a positive value of $u^2$ provides an accelerated increase (or a retarded decrease) of the function at all those points, where its value is lower than the average of the neighboring values; and, vice versa, a retarded increase (or an accelerated decrease) at those points where the function exceeds the average of its neighborhood. Thus, the ordinary **wave-equation** represents a certain tendency to smooth out again all differences between the values of the function at different points, though not at the very moment they appear and not indefinitely - as in the case of the equation for heat conduction. It will however certainly prevent the function from increasing or decreasing beyond all limit.

If the quantity $u^2$, instead of being positive, is negative which we have seen to be sometimes the case with Eq. (18),

$$[\Delta\psi + 8\pi^2 m(E + e^2/r)\psi/h^2 = 0, \qquad\qquad (18)]$$

then all the foregoing reasoning is just reversed. There is in the course of time a tendency to exaggerate infinitely all "humps" of the function and even spontaneously to form humps out of quite insignificant traces. Evidently a function which is subject to such a revolutionary sort of equation, is continually exposed to the very highest danger of increasing or decreasing beyond all limit. At any rate it is no longer astonishing, that *special conditions must be fulfilled to prevent such an occurrence.* The mathematical treatment shows that *these conditions consist exactly in E having one of the second set of characteristic values ["eigenvalues"] given by (19)*

$$[(2) \quad E = -2\pi^2 me^4/h^2 n^2 \quad (n = 1, 2, 3, 4, \ldots), \qquad\qquad (19)]$$

*whereas the first set obviously prevents all accidents by making the square of the phase velocity positive throughout the space.*

6. *I will now give an account of some of the results that have hitherto been obtained with this new mechanics.* Rather simple problems are offered by the harmonic oscillator and the rotator. The E-levels of the former prove to be

$$(n + \tfrac{1}{2}) h\nu_0; \quad n = 0, 1, 2, 3, \ldots$$

instead of $n h\nu_0$ according to the ordinary quantum theory. The E-levels of the rotator are

$$n(n + 1) h^2/8\pi^2 I$$

(I = moment of inertia), the well-known $n^2$ being replaced by $n(n + 1)$. If we are interested only in the differences of levels - as is actually the case – this amounts to the same as replacing $n^2$ by $(n + \frac{1}{2})^2$, for

$$(n + \tfrac{1}{2})^2 - n(n + 1) = \tfrac{1}{4},$$

independent of $n$. It is well known that so-called *half-quantum numbers* are actually supported by the experimental evidence on most of the simple band spectra, and are probably contradicted by none of them. Mr. Fues, whose valuable help I owe to the Rockefeller Institution (International Education Board), has worked out [8] the band theory of diatomic molecules in detail, taking into account the mutual influence of rotation and oscillation and the fact, that the latter is not of the simple harmonic type.

[8] Fues, E. (1926a). *Das Eigenschwingungsspektrum zweiatomiger Molekule in der Undulationsmechanik.* (The natural vibration spectrum of diatomic molecules in wave mechanics.) *Ann. Physik*, 385, 80, 367-396; http://dx.doi.org/10.1002/ andp.1926385 1204; another paper in press.

The result is in exact agreement with the ordinary treatment except that the quantum-numbers become half-integer also in all correction-terms. It would hardly have been possible to attack the problem just mentioned, as well as many similar ones, by direct methods, since the differential equation (16)

$$[\Delta\psi + 8\pi^2 m(E - V)\psi/h^2 = 0 \tag{16}]$$

is in general of a very difficult type. In many cases, however, this difficulty is overcome by the *theory of perturbations* which the writer has developed together with Mr. Fues. This theory, though much simpler, yet shows an interesting parallelism to the well-known *theory of perturbations* in ordinary mechanics. It allows the calculation by mere quadratures of the small modification of characteristic values and characteristic functions, that are caused by introducing an additional small term (function of the independent variables) in the coefficients of an equation whose characteristic values and characteristic functions are known.

An interesting example of the application of this mathematical theory is afforded by the *perturbation of the hydrogen-atom caused by an external homogeneous electric field (Stark-effect)*.

[The *Stark effect* is the shifting and splitting of spectral lines of atoms and molecules due to the presence of an external electric field known as the *Stark effect*; the electric-field analogue of the *Zeeman effect*.]

The discrete *Balmer levels*, shown in Eq. (19-2)

$$[(2) \quad E = -2\pi^2 m e^4/h^2 n^2 \quad (n = 1, 2, 3, 4, \ldots), \tag{19}]$$

are, as characteristic values, not simple but many fold. Each of them corresponds to $n^2$ characteristic values that coincide by chance or, more properly speaking, because of the high symmetry or simplicity of the coefficients of Eq. (18)

$$[\Delta\psi + 8\pi^2 m(E + e^2/r)\psi/h^2 = 0, \tag{18}]$$

The addition of an external electric field, small in comparison with the atomic field, does away with this symmetry and *splits up every one of the Balmer-levels into a set of near neighboring levels*, though not into as many as $n^2$, because the splitting up in this case is not thorough. The writer has been able to show that *these levels exactly coincide with those given by the well-known formula of Epstein*, a rather severe test of our wave mechanics, since experimentally no deviation whatever, at least in the limit of a weak field, would have been allowed from this famous formula.

But not only the levels, i.e. the *frequencies*, but also the *intensities* and *polarizations* of the emitted lines in the Stark effect can be calculated from wave mechanics in very fair agreement with experiment. *Hitherto we have not attached a definite physical meaning to the wave function $\psi$.* It is possible, however, to give it a certain electrodynamical meaning, which will be discussed in detail in Section 8 and *which converts our atom into a system of fluctuating charges, spread out continuously in space and generating a resultant electric moment, that changes with time with a superposition of frequencies, which exactly coincide with the differences of the vibration-frequencies E/h, i.e. coincide with the frequencies of the emitted light.* This in itself is highly satisfactory. But in addition, it is possible to calculate the *amplitudes* of the harmonic components of the *electric moment* for any direction in space, e.g., in the case of the Stark effect, parallel to the electric field or perpendicular to the field. If the theory is correct. the squares of these *amplitudes* ought to be proportional to the *intensities* of the several line components, polarized in either direction. The rather laborious calculations have been performed and the result is shown in Fig. 1 [below].[9]

**observed**


**theoretical**

...

*Fig. 1*

[9] The experimental data were taken from Stark, J. (1915). *Ann. Physik*, 48, 193. [Stark, J. (1914). *Beobachtungen über den Effekt des elektrischen Feldes auf Spektrallinien I. Quereffekt.* (Observations of the effect of the electric field on spectral lines I. Transverse effect), *Ann. Physik*. 43, 965–983. Published earlier (1913) in *Sitzungsberichten der Kgl. Preuss. Akad. d. Wiss.*] In these figures theoretical intensities that are too small to be indicated by a line of the proper length are marked by a dot. See original page 1066.

In comparing theory with experiment, it must be born in mind that the calculations have been performed only in the limit of a very weak external field and that in the region of field-strength used in experiments (about 100,000 volts/cm) a very marked influence on the intensities is found both by experiment and by theory. In particular the very weak or vanishing components are enhanced with increasing field. The sum of the intensities of all the perpendicular components of one Balmer line turns out exactly equal to the sum of the parallel components of the same line. This is in full agreement with Stark's statement that no polarization of the emitted light as a whole is produced by the field.

I ought to emphasize here, that I was led to the foregoing nearly classical calculation of *intensities* by noticing *a posteriori*, i.e. after the main features of wave mechanics had been developed, *its complete mathematical agreement with the theory of matrices put forward by Heisenberg, Born and Jordan*.[10]

[10] Heisenberg, W. (July, 1925). *Über quantentheoretische Umdeutung kinematischer und mechanischer Beziehungen.* (On the quantum-theoretical re-interpretation of kinematic and mechanical relations.) *Zeit. Phys.*, 33, 879-93; Born, M. & Jordan, P. (December, 1925). *Zur Quantenmechanik.* (On Quantum Mechanics.). *Ibid.*, 34, 858-88; Born, M., Heisenberg, W. & Jordan, P. (August, 1926). *Zur Quantenmechanik II.* (On Quantum Mechanics II.) *Ibid.*, 35, 557-615; Born, M. & Wiener, N. (1926). *Ibid.*, 36, 174; Heisenberg, W. & Jordan, P. (April, 1926). *Anwendung der Quantenmechanik auf das Problem der anomalen Zeemaneffekte.* (Application of quantum mechanics to the problem of the anomalous Zeeman effect.) *Ibid.*, 37, 263-77; Pauli, W. Jr. (1926). *Ibid.*, 36, 336; Dirac, P. A. M. (December, 1925). *The Fundamental Equations of Quantum Mechanics. Roy. Soc. Proc., A*, 109, 752, 642-53; Dirac, P. A. M. (March, 1926). *Quantum Mechanics and a preliminary investigation of the hydrogen atom. Ibid.*, 110, 755, 561-79; Dirac, P. A. M. (May, 1926). *The elimination of the nodes in quantum mechanics. Ibid.*, 111, 757, 281–305; Dirac, P. A. M. (June, 1926). *Relativity Quantum Mechanics with an Application to Compton Scattering. Ibid.*, 111, 758, 405-423.

The results shown in Fig. 1 may as well be called the results of the latter theory though they have not yet been calculated by its direct application. The connection of the two theories is a rather intricate one and is by no means to be observed at first sight.

7. It was stated in the beginning of this paper that in the present theory both the *laws of motion* and the *quantum conditions* can be deduced from one Hamiltonian principle. *To prove this, it must be shown that the wave-equation (16)*

$$[\Delta\psi + 8\pi^2 m(E - V)\psi/h^2 = 0 \tag{16}]$$

232

*can be derived from an integral-variation principle; for this equation is indeed the only fundamental equation of the theory* (in the case of a single material point, moving in a conservative field of force, the only one considered in detail on the foregoing pages).

The connection of Eq. (16) with a Hamiltonian principle is very simple and almost exactly the same as in ordinary vibration problems. Further more this connection affords the simplest means of almost cogently extending the theory to a wholly arbitrary conservative system.

Suppose, the extreme values of the following integral extending over all space were required.

$$I_1 = \iiint \{h^2[(\delta\psi/\delta x)^2 + (\delta\psi/\delta y)^2 + (\delta\psi/\delta z)^2]/8\pi^2 m + V\psi^2\} \, dxdydz, \quad (20)$$

all single-valued, finite and continuously differentiable functions $\psi$ being "admitted to concurrence" that give the following "normalizing" integral a constant value, say 1:

$$I_2 = \iiint \psi^2 \, dxdydz. \quad (21)$$

In carrying out the variation under this "accessory condition" in the well-known manner, Eq. (16) is found as the well-known necessary condition for an extreme value of (20) the constant $- E$ being the Lagrangian multiplier with which the variation of the second integral has to be multiplied and added to the first, so as to take care of the accessory condition. Thus, the normalized characteristic functions of Eq. (16) are exactly the so-called extremals of the integral (20) under the normalizing condition (21), whereas the characteristic values i.e. the values, that are admissible for the constant E are nothing else than the extreme values of integral (20). (This property of the Langrangian multiplier is well known and is easily recognized by observing, that the non-conditioned, extreme value of $I_1 - EI_2$ can be but zero, any other value being capable as well of augmentation as of diminution by simply multiplying $\psi$ by a constant.)

Now the integrand of (20) proves on closer inspection to have a very simple relation to the ordinary Hamiltonian function of our mechanical problem - in the sense of ordinary mechanics. The said function is, (cf. Section 1):

$$(1/2m) (p_x^2 + p_y^2 + p_z^2) + V(x, y, z). \quad (22)$$

Take this function to be a homogeneous quadratic function of the momenta $p_x$ etc. and of unity and replace therein $p_x$, $p_y$, $p_z$, 1 by $(h/2\pi) (\delta\psi/\delta x)$, $(h/2\pi) (\delta\psi/\delta y)$, $(h/2\pi) (\delta\psi/\delta z)$, $\psi$, respectively. There results the integrand of (20)

$$[I_1 = \iiint \{h^2[(\delta\psi/\delta x)^2 + (\delta\psi/\delta y)^2 + (\delta\psi/\delta z)^2]/8\pi^2 m + V\psi^2\} \, dxdydz. \quad (20)]$$

This immediately suggests extending our variation problem and hereby our *wave-equation* (16) to a wholly arbitrary conservative mechanical system. The Hamiltonian function of such will be of the form

$$\tfrac{1}{2} \Sigma_{l=1}^{N} \Sigma_{k=1}^{N} a_{lk} p_l p_k + V \qquad\qquad (23)$$

with $a_{lk} = a_{kl}$, the $a_{lk}$ and V being some functions of the N generalized coordinates $q_1 \dots q_N$. Take (23) to be a homogeneous quadratic function of $p_1 \dots p_N, 1$ and replace these quantities by $(h/2\pi)$ $(\delta\psi/\delta q_1)$, $\dots$ $(h/2\pi)$ $(\delta\psi/\delta q_N)$, $\psi$, respectively. Writing $\Delta_p$ for the determinant

$$\Delta_p = |\, \Sigma \pm a_{lk}\,|$$

We form the integral

$$I_1 = \int \dots \int \{(h^2/8\pi^2)\, \Sigma_{l=1}^{N} \Sigma_{k=1}^{N} a_{lk}\, (\delta\psi/\delta q_l)(\delta\psi/\delta p_k)^2$$
$$+ V\psi^2\}\Delta_p^{-\frac{1}{2}}\, dq_1 \dots dq_N, \qquad\qquad (24)$$

taken over the whole space of coordinates and seek its extreme values under the accessory condition

$$I_2 = \int \dots \int \psi^2 \Delta_p^{-\frac{1}{2}}\, dq_1 \dots dq_N, \qquad\qquad (25)$$

This leads to the generalization of Eq. (16)
$$[\Delta\psi + 8\pi^2 m(E - V)\psi/h^2 = 0, \qquad\qquad (16)]$$
viz.

$$\Delta_p^{\frac{1}{2}} \Sigma_l\, \delta/\delta q_l\, (\Delta_p^{-\frac{1}{2}} \Sigma_k a_{lk}\, \delta\psi/\delta q_k) + (8\pi^2/h^2)(E - V)\psi = 0, \qquad\qquad (26)$$

– E being, as before, the Lagrangian multiplier of (25). The double sum appearing in (26) is a sort of generalized Laplacian in the N-dimensional, non-Euclidean space of coordinates. The necessary appearance of $\Delta_p^{-\frac{1}{2}}$ in an integral like (24) or (25) is well known from Gibbs' statistical mechanics; $\Delta_p^{-\frac{1}{2}}\, dq_1 \dots dq_N$ is simply the non-Euclidean element of volume, e.g. $r^2 \sin \vartheta\, d\vartheta d\phi dr$ in the case of one material point of unit mass, whose position is fixed by three polar coordinates r, $\vartheta$, $\phi$. (In omitting the determinant, the integrals would not be invariant relative to point transformations; they would depend on the choice of generalized coordinates.) It is Eq. (26) that has been used in all problems mentioned in Section 6.

8. The question of the real ***physical meaning of the wave-function*** $\psi$ has been delayed (see Section 6) until now for the sake of discussing it but once in full generality for a wholly arbitrary system. Eq. (16) or in the more general case, Eq. (26) gives the dependence of the ***wave function*** $\psi$ on the coordinates only, the dependence on time being given for every one particular solution, corresponding to a particular characteristic value $E = E_l$, by the real part of

$$e^{[(2\pi E l t/h + \vartheta l)i]}, \quad i = \sqrt{-1}$$

the $\vartheta_l$ being phase constants. So, if $u_l$ ($l = 1, 2, 3 \dots$) be the characteristic functions the most general solution of the wave-problem will be (the real part of)

$$\psi = \Sigma_{l=1}^{\infty}\ c_l u_l\ e^{[(2\pi E_l t/h + \vartheta_l)i]}. \tag{27}$$

(For simplicity's sake we suppose the characteristic values to be all single and discrete.) The $c_l$ are real constants. Now form the square of the absolute value of the complex function $\psi$. Denoting a conjugate complex value by a *, this is

$$\psi\psi^* = 2\ \Sigma_{l,l'}\ c_l c_{l'} u_l u_{l'} \cos\ [(2\pi(E_l - E_{l'})t/h + \vartheta_l - \vartheta_{l'}]. \tag{28}$$

This of course, like $\psi$ itself, is in the general *case a function of the generalized coordinates $q_1$ ... $q_N$ and the time, - not a function of ordinary space and time* as in ordinary wave-problems. *This raises some difficulty in attaching a physical meaning to the wave-function.* In the case of the hydrogen atom (taken as a one-body problem) the difficulty disappears. In this case it has been possible to compute fairly correct values for the *intensities* e.g. of the Stark effect components (see Section 6, Fig. 1) by the following hypothesis: *the charge of the electron is not concentrated in a point, but is spread out through the whole space, proportional to the quantity $\psi\psi^*$.*

It has to be born in mind, that by this hypothesis the charge is nevertheless restricted to a domain of, say, a few Angstroms, the wave function $\psi$ practically vanishing at greater distance from the nucleus (see Section 4). The fluctuation of the charge will be governed by Eq. (28)

$$[\psi\psi^* = 2\ \Sigma_{l,l'}\ c_l c_{l'} u_l u_{l'} \cos\ [(2\pi(E_l - E_{l'})t/h + \vartheta_l - \vartheta_{l'}], \tag{28}]$$

applied to the special case of the hydrogen atom. To find the radiation, that by ordinary electrodynamics will originate from these fluctuating charges, we have simply to calculate the rectangular components of the total *electrical moment* [11] by multiplying (28) by x, y, z respectively, and integrating over the space, e.g.[12]

[11] This procedure is legitimate only because and as long as the domain to which the charge is practically restricted remains small in comparison with the optical wave-length that corresponds to the frequencies $(E_l - E_{l'})/yh$.

[12] In the sum $\Sigma_{l,l'}$, each pair of values of $l,l'$ is to be taken but once and the terms with $l' = l$ are to be halved.

$$\iiint z\psi\psi^* dxdydz = 2\ \Sigma_{l,l'}\ c_l c_{l'} \cos\ [(2\pi(E_l - E_{l'})t/h + \vartheta_l - \vartheta_{l'}].$$

$$\iiint z u_l u_{l'}\ dxdydz \tag{29}$$

Thus, the total *electric moment* is seen to be a superposition of dipoles, which are associated with the pairs of characteristic functions ["*eigenfunctions*"] which vibrate harmonically with the frequencies $(E_l - E_{l'})/h$, well known from N. Bohr's famous frequency-condition. The *intensity* of emitted radiation of a particular frequency is to be expected proportional to the square of

$$c_l c_{l'}\ \iiint z u_l u_{l'}\ dxdydz.$$

The supposition made on the $c_l$, in calculating the *intensities* of the *Stark effect* components, Fig. 1, is, that the $c$, be equal for every set of characteristic values ["eigenvalues"] derived from one Balmer level (Eq. 19-2)

[(1)  E>0
(2)  $E = -2\pi^2 me^4/h^2 n^2$  ($n = 1, 2, 3, 4, ...$)  (19)]

by the action of the electrical field. The relative *intensities* of the fine structure components will then be proportional to the square of the triple integral. This is found to be in fair agreement with experiment.

The triple integral may be shown to be equal to what in Heisenberg's theory would be called the "element of matrix $z(l, l')$". This constitutes the intimate connection between the two theories. *But the important achievement of the present theory – imperfect as it may in many respects be - seems to me to be that by a definite localization of the charge in space and time we are able from ordinary electrodynamics really to derive both the frequencies and the intensities and polarizations of the emitted light*. All so-called *selection principles* automatically result from the vanishing of the triple integral in the particular case.

Now how are these conceptions to be generalized to the case of more than one, say of $N$, electrons? Here Heisenberg's formal theory has proved most valuable. It tells us though less by physical reasoning than by its compact formal structure that Eq. (29)

[$\iiint z\psi\psi^*\mathrm{dxdydz} = 2 \Sigma_{l,l'}\, c_l c_{l'} \cos [(2\pi(E_l - E_{l'})t/h + \vartheta_l - \vartheta_{l'}]$.
$\iiint z u_l u_{l'}\, \mathrm{dxdydz}$  (29)]

giving a rectangular component of total *electric moment* has to be maintained with the only differences that (1) the integrals arc 3N-fold instead of three-fold, extending over the whole coordinate space; (2) z has to be replaced by the sum $\Sigma\, e_l z_l$ i.e. by the z-component of the total *electrical moment* which the point-charge model would have in the configuration ($x_1$, $y_1$, $z_1$; $x_2$, $y_2$, $z_2$; ...; $x_N$, $y_N$, $z_N$) that relates to the element $dx_1 ... dz_N$ of the integration.

But this amounts to making the following hypothesis as to the physical meaning of $\psi$ which of course reduces to our former hypothesis in the case of one electron only: *the real continuous partition of the charge is a sort of mean of the continuous multitude of all possible configurations of the corresponding point-charge model, the mean being taken with the quantity $\psi\psi^*$ as a sort of weight-function in the configuration space*. No very definite experimental results can be brought forward at present in favor of this generalized hypothesis. But some very general theoretical results on the quantity $\psi\psi^*$ persuade me that the hypothesis is right. For example, the value of the integral of $\psi\psi^*$ taken over the whole coordinate space proves absolutely constant (as it should, if $\psi\psi^*$ is a reasonable weight function) not only with a conservative, but also with a non-conservative system. The treatment of the latter will be roughly sketched in the following section.

9. Eq. (16)
$$[\Delta\psi + 8\pi^2 m(E - V)\psi/h^2 = 0, \tag{16}]$$
or more generally (26)
$$[\Delta_p^{1/2} \Sigma_l \delta/\delta q_l (\Delta_p^{-1/2} \Sigma_k a_{lk} \delta\psi/\delta q_k) + (8\pi^2/h^2)(E - V)\psi = 0, \tag{26}]$$
which is fundamental to all our reasoning has been arrived at under the supposition that $\psi$ depends on the time only through the factor

$$e^{\pm 2\pi i Et/h}. \tag{30}$$

But this amounts to saying, that

$$\psi = \pm 2\pi i Et/h. \tag{31}$$

From this equation and from Eq. (26) the quantity E may be eliminated and so an equation be formed that must hold in any case, whatever be the dependence of the *wave-function* $\psi$ on time:

The ambiguous sign of the last term presents no grave difficulty. Since physical meaning is attached to the product $\psi\psi^*$ only, we may postulate for $\psi$ either of the two equations (32): then $\psi^*$ will satisfy the other and their product will remain unaltered.

Eq. (32) lends itself to generalization to an arbitrary non-conservative system by simply supposing the potential function V to contain the time explicitly. A case of greatest interest is obtained by adding to the *potential energy* of a conservative system a small term; viz., the non-conservative *potential energy*, produced in the system by an incident light wave. We cannot enter here into the details, but shall only present the main features of the solution. The effect of the incident light wave is, that with each free vibration of the undisturbed system, frequency $E_l/h$, there are associated two forced vibrations with, in general, very small *amplitudes* and with the two *frequencies* $E_l/h \pm v$, $v$ being the *frequency* of the incident beam of light. Following the same principles as in the foregoing section we find every free vibration, cooperating with its associated forced vibrations to give rise to a forced light *emission* with the difference *frequency*

$$E_l/h - (E_l/h \pm v) = -/+ v$$

i.e. **with exactly the frequency of the incident beam of light**. This forced *emission* is of course to be identified with the secondary wavelets that are necessary to account for *absorption, dispersion* and *scattering*. In calculating their *amplitudes* one finds them, indeed, to increase very markedly as soon as the incident *frequency* $v$ approaches any one of the *emission frequencies* $(E_l - E_{l'})t/h$. The final formula is almost identical with the well-known **Helmholtz dispersion formula** in the form presented by Kramers.[13]

[13] Kramers, H. A. (May, 1924). *The law of dispersion and Bohr's theory of spectra.* Nature 113, 673-74; (August, 1924), 114, 310; Kramers, H. A. & Heisenberg, W. (February, 1925). *Über die Streuung von Strahlung durch Atome.* (On the scattering of radiation by atoms.) *Zeit. Physik,* 31, 681-708; http://dx.doi.org/10.1007/ BF02980624.

*The case of resonance cannot yet be treated quite satisfactorily, since a damping-term seems to be missing in our fundamental equation, even in the case of a free conservative system.* (The radiation, which according to the assumptions of Section 8 is emitted by the cooperation of every pair of free vibrations, must of course in some way alter their *amplitudes*. This it does not do according to the assumptions hitherto made.) But it is quite interesting to observe that also with that damping term still missing, *we do not encounter the accident of infinite amplitudes in the case of resonance, well known from the classical treatment of the subject.* The only thing that happens is that an incident light wave of ever so small an *amplitude* will raise the forced vibration of the system to a finite *amplitude*; and furthermore if, for example, from the beginning only one free vibration was set up, say that belonging to $E_{l'}$ and if

$$h\nu = E_{l'} - E_l,$$

then the forced vibration, raised to finite *amplitude*, is in shape and *frequency* identical with the free vibration belonging to $E_{l'}$. At the same time the *amplitude* of the vibration $E_l$ is markedly diminished. The sum of squares of all *amplitudes* remains constant under all circumstances. *This behavior seems to afford an insight (though incomplete) into the so-called transition from one stationary state to another which hitherto has been wholly inaccessible to computation.*

10. In the foregoing report *the wave theory of mechanics has been developed without reference to two very important things, viz. (1) the relativity modifications of classical mechanics, (2) the action of a magnetic field on the atom.* This may be thought rather peculiar since L. de Broglie, whose fundamental researches gave origin to the present theory, even started from the *relativistic* theory of electronic motion and from the beginning took into account a magnetic field as well as an electric one.

It is of course possible to take the same starting point also for the present theory and to carry it on fairly far in using *relativistic* mechanics instead of classical and including the action of a magnetic field. Some very interesting results are obtained in this way on the wave-length displacement, intensity and polarization of the fine structure components and of the Zeeman components of the hydrogen atom.[14]

[14] Fock, V. (1926). *Zur Schroedingerschen Wellenmechanik.* (Schroedinger's wave mechanics.) *Zeit. Physik,* 38, 242-7.

There are two reasons why I did not think it very important to enter here into this form of the theory. First, ***it has until now not been possible to extend the relativistic theory to a system of more than one electron***. But there is the region in which the solution of new problems is to be hoped from the new theory, problems that were inaccessible to the older theory.

Second, ***the relativistic theory of the hydrogen atom is apparently incomplete; the results are in grave contradiction with experiment***, since in Sommerfeld's well known formula for the displacement of the natural fine structure components the so-called azimuthal quantum number (as well as the radial quantum number) turns out as "half-integer", i.e. half of an odd number, instead of integer. So, the fine structure turns out entirely wrong.

The deficiency must be intimately connected with Uhlenbeck-Goudsmit's[15] theory of the spinning electron. But in what way the ***electron spin*** has to be taken into account in the present theory is yet unknown.

[15] Uhlenbeck, G. E. & Goudsmit, S. (November, 1925). *Ersetzung der Hypothese vom unmechanischen Zwang durch eine Forderung bezuglich des inneren Verhaltens jedes einzelnen Elektrons*. (Replacement of the hypothesis of unmechanical coercion by a requirement regarding the internal behavior of each individual electron.) *Naturw.*, 13, 47, 953-4; Uhlenbeck, G. E. & Goudsmit, S. (February, 1926). *Spinning Electrons and the Structure of Spectra. Nature*, 117, 264-265; https://doi.org/10.1038/117264a0.

## Heisenberg, W. K. (December 5, 1901 – February 1, 1976)

Heisenberg was a German theoretical physicist and one of the **key pioneers of quantum mechanics**. He published his **seminal work in 1925** in a breakthrough paper. In the subsequent series of papers with Max Born and Pascual Jordan, during the same year, this matrix formulation of quantum mechanics was substantially elaborated. He is known for the uncertainty principle, which he published in 1927. Heisenberg was awarded the 1932 Nobel Prize in Physics "*for the creation of quantum mechanics*".

Heisenberg also made important contributions to the theories of the hydrodynamics of turbulent flows, the **atomic nucleus**, **ferromagnetism**, cosmic rays, and subatomic particles. He was a principal scientist in the German nuclear weapons program during World War II. He was also instrumental in planning the first West German nuclear reactor at Karlsruhe, together with a research reactor in Munich, in 1957.

Heisenberg was born in Würzburg, Germany, to Kaspar Ernst August Heisenberg, and his wife, Annie Wecklein. His father was a secondary school teacher of classical languages who became Germany's only ordentlicher Professor (ordinarius professor) of medieval and modern Greek studies in the university system.

In his youth Heisenberg was a member and Scoutleader of the Neupfadfinder, a German Scout association and part of the German Youth Movement.

From 1920 to 1923, he studied physics and mathematics at the Ludwig Maximilian University of Munich under Arnold Sommerfeld and Wilhelm Wien; and at the Georg-August University of Göttingen with Max Born and James Franck and mathematics with David Hilbert. In June 1922, Sommerfeld took Heisenberg to Göttingen to attend the Bohr Festival, because Sommerfeld knew of Heisenberg's interest in Niels Bohr's theories on atomic physics. At the event, Bohr was a guest lecturer and gave a series of comprehensive lectures on quantum atomic physics and Heisenberg met Bohr for the first time.

Heisenberg's doctoral thesis, the topic of which was suggested by Sommerfeld, was on turbulence; the thesis discussed both the stability of laminar flow and the nature of turbulent flow. The problem of stability was investigated by the use of the Orr–Sommerfeld equation, a fourth order linear differential equation for small disturbances from laminar flow. He received his doctorate in 1923.

At Göttingen, under Born, he completed his habilitation in 1924 with a Habilitationsschrift (habilitation thesis) on the anomalous Zeeman effect.

From 1924 to 1927, Heisenberg was a Privatdozent at Göttingen, meaning he was qualified to teach and examine independently, without having a chair. From September 17, 1924 to May 1, 1925, under an International Education Board Rockefeller Foundation fellowship,

Heisenberg went to do research with Niels Bohr, director of the Institute of Theoretical Physics at the University of Copenhagen.

In Copenhagen, Heisenberg and Hans Kramers collaborated on a paper on dispersion, or the scattering from atoms of radiation whose wavelength is larger than the atoms. They showed that the successful formula Kramers had developed earlier could not be based on Bohr orbits, because the *transition frequencies* are based on level spacings which are not constant. The frequencies which occur in the Fourier transform of sharp classical orbits, by contrast, are equally spaced. But these results could be explained by a semi-classical virtual state model: the incoming radiation excites the valence, or outer, electron to a virtual state from which it decays. In a subsequent paper Heisenberg showed that this virtual oscillator model could also explain the polarization of fluorescent radiation.

These two successes, and the continuing failure of the Bohr–Sommerfeld model to explain the outstanding problem of the anomalous Zeeman effect, led Heisenberg to use the virtual oscillator model to try to calculate *spectral frequencies*. The method proved too difficult to immediately apply to realistic problems, so Heisenberg turned to a simpler example, the anharmonic oscillator.

The dipole oscillator consists of a simple harmonic oscillator, which is thought of as a charged particle on a spring, perturbed by an external force, like an external charge. The motion of the oscillating charge can be expressed as a Fourier series in the frequency of the oscillator. Heisenberg solved for the quantum behavior by two different methods. First, he treated the system with the virtual oscillator method, calculating the transitions between the levels that would be produced by the external source. He then solved the same problem by treating the anharmonic potential term as a perturbation to the harmonic oscillator and using the perturbation methods that he and Born had developed. Both methods led to the same results for the first and the very complicated second order correction terms. This suggested that behind the very complicated calculations lay a consistent scheme. Heisenberg returned to Göttingen and, with Max Born and Pascual Jordan over a period of about six months, developed the *matrix mechanics formulation of quantum mechanics*.

In his 1925 paper, which assumes that the reader is familiar with Kramers-Heisenberg transition probability calculations, Heisenberg set out to *try to construct a theory of quantum mechanics in which only relationships among observable quantities occur*. In place of assigning to the electron a point in space as a function of time he *assigned to the electron an emitted radiation;* where the observables are the *energies W(n) of the (Bohr) stationary states*, together with the associated (Einstein-Bohr) *frequencies v* and *amplitudes* which characterize the radiation emitted in the transition between the stationary states. Recognizing that quantum theory describes transitions between two stationary states he substituted two variables in place of one in the classical theory. He justified this replacement by an appeal to *Bohr's correspondence principle* and the *Pauli doctrine* that quantum mechanics must be limited to observables. In order to calculate the *energy* of a harmonic oscillator, in which the *amplitudes* do not combine in the same way as the classical harmonics, but rather in accordance with the *Ritz combination principle*, instead of reinterpreting x(t) as a *sum* over transition components, he represented the position by the *set of transition components*, thereby introducing non-commutative multiplication of matrices by physical reasoning, based on the *correspondence principle*, despite the fact that he was not then familiar with the mathematical theory of matrices.

After addressing what he referred to as the kinematic of the quantum theory, Heisenberg turned to mechanical problems aiming at the determination of the *amplitudes, frequencies* and *energies in order to construct the line spectrum of an atom from the given force on the electron*. He achieved this *by translating the old quantum condition that fixes the properties of the states to a new condition that fixes the properties of the transitions between states*, replacing the differential in the equation for the *phase* integral by a difference, resulting in an equation that has a simple quantum theoretical connection to the *Kramer's dispersion theory*.

On July 9, Heisenberg gave Born his paper to review and submit for publication. Heisenberg's seminal paper was published in September 1925. [Heisenberg, W. (July, 1925). *Über quantentheoretische Umdeutung kinematischer und mechanischer Beziehungen*. (On the Quantum-Theoretical Re-interpretation of Kinematic and Mechanical Relations.) *Zeit. Physik*, 33, 879-93.] *This is the first paper in the famous trilogy which launched the matrix mechanics formulation of quantum mechanics*.

When Born read the paper, he recognized the formulation as one which could be transcribed and extended to the systematic language of matrices, which he had learned from his study under Jakob Rosanes at Breslau University. Up until this time, matrices were seldom used by physicists; they were considered to belong to the realm of pure mathematics. Gustav Mie had used them in a paper on electrodynamics in 1912; and Born had used them in his work on the lattice theory of crystals in 1921. While matrices were used in these cases, the algebra of matrices with their multiplication did not enter the picture as they did in the matrix formulation of quantum mechanics.

Born, with the help of his assistant and former student Pascual Jordan, began immediately to make the transcription and extension, and they submitted their results for publication; the paper was received for publication just 60 days after Heisenberg's paper. [Born, M. & Jordan, P. (December, 1925). *Zur Quantenmechanik*. (On Quantum Mechanics.) *Zeit. Phys.*, 34, 858-88.] A follow-on paper was submitted for publication before the end of the year by all three authors. [Born, M., Heisenberg, W. & Jordan, P. (August, 1926). *Zur Quantenmechanik II*. (On Quantum Mechanics II.) *Zeit. Phys.*, 35, 557-615.]

On May 1, 1926, Heisenberg began his appointment as a university lecturer and assistant to Bohr in Copenhagen. It was in Copenhagen, in 1927, that Heisenberg developed his *uncertainty principle*, while working on the mathematical foundations of quantum mechanics. On February 23, Heisenberg wrote a letter to fellow physicist Wolfgang Pauli, in which he first described his new principle. In his paper on the principle, Heisenberg used the word "Ungenauigkeit" (imprecision), not "uncertainty", to describe it.

In 1928, Heisenberg was appointed ordentlicher Professor (professor ordinarius) of theoretical physics and head of the department of physics at the University of Leipzig; he gave his inaugural lecture there on 1 February 1928. In his first paper published from Leipzig, Heisenberg used the *Pauli exclusion principle* to solve the mystery of *ferromagnetism*.

In 1928, the British mathematical physicist Paul Dirac had derived his *relativistic wave equation of quantum mechanics*, which implied the existence of positive electrons, later to be named *positrons*. In early 1929, Heisenberg and Pauli submitted the first of two papers laying the foundation for *relativistic quantum field theory*.

In 1929, Heisenberg went on a lecture tour of China, Japan, India, and the United States. In the spring of 1929, he was a visiting lecturer at the University of Chicago, where he lectured on quantum mechanics.

In 1932, from a cloud chamber photograph of cosmic rays, the American physicist Carl David Anderson identified a track as having been made by a *positron*. In mid-1933, Heisenberg presented his *theory of the positron*. His thinking on Dirac's theory and further development of the theory were set forth in two papers. The first, [Heisenberg, W. (March, 1934). *Bemerkungen zur Diracschen Theorie des Positrons.* (Remarks on the Dirac theory of positron.) *Zeit. Phys.*, 90, 3-4, 209-31] was published in 1934, and the second, Heisenberg, W., Euler, H. (1936). *Folgerungen aus der Diracschen Theorie des Positrons. Zeit. Physik,* 98, 714-32; https://doi.org/10.1007/BF01343663], was published in 1936.

In these papers Heisenberg was the first to *reinterpret the Dirac equation as a "classical" field equation for any point particle of spin ħ/2, itself subject to quantization conditions involving anti-commutators*. Thus, reinterpreting it as a quantum field equation accurately describing electrons, Heisenberg put matter on the same footing as electromagnetism: as being described by *relativistic quantum field equations* which allowed the possibility of particle creation and destruction. (Hermann Weyl had already described this in a 1929 letter to Albert Einstein.)

In 1928, Albert Einstein nominated Heisenberg, Born, and Jordan for the Nobel Prize in Physics. The announcement of the Nobel Prize in Physics for 1932 was delayed until November 1933. It was at that time that it was announced Heisenberg had won the Prize for 1932 "for the creation of quantum mechanics, the application of which has, inter alia, led to the discovery of the allotropic forms of hydrogen".

Heisenberg enjoyed classical music and was an accomplished pianist. His interest in music led to meeting his future wife. In January 1937, Heisenberg met Elisabeth Schumacher (1914–1998) at a private music recital. Elisabeth was the daughter of a well-known Berlin economics professor, and her brother was the economist E. F. Schumacher, author of *Small Is Beautiful*. Heisenberg married her on April 29. Fraternal twins Maria and Wolfgang were born in January 1938, whereupon Wolfgang Pauli congratulated Heisenberg on his "pair creation"—a word play on a process from elementary particle physics, pair production. They had five more children over the next 12 years: Barbara, Christine, Jochen, Martin and Verena. In 1936 he bought a summer home for his family in Urfeld am Walchensee, in southern Germany.

Heisenberg was involved in the German nuclear weapons program, known as *Uranverein*, which was formed on September 1, 1939, the day World War II began, The Kaiser-Wilhelm Institut für Physik (KWIP, Kaiser Wilhelm Institute for Physics) in Berlin-Dahlem, was placed under the authority of the Heereswaffenamt (HWA, Army Ordnance Office), and the military control of the nuclear research commenced. In February 1942, at a scientific conference called by the Army Weapons Office, Heisenberg presented a lecture to Reichs officials on energy acquisition from nuclear fission entitled *Die theoretischen Grundlagen für die Energiegewinning aus der Uranspaltung* (The theoretical basis for energy generation from uranium fission). He lectured on the enormous energy potential of nuclear fission, stating that 250 million electron volts could be released through the fission of an atomic nucleus. Heisenberg stressed that pure U-235 had to be obtained to achieve a chain reaction; and explored various ways of obtaining isotope $^{235}_{92}U$ in its pure form, including uranium enrichment and an alternative layered method of normal uranium and a moderator in a machine.

In April 1942, Reichs Minister Rust decided to move the nuclear project from the Physics Institute to the Reichs Research Council; returning the Physics Institute to the Kaiser Wilhelm Society, and naming Heisenberg as Director at the Institute. With this appointment, Heisenberg obtained his first professorship. Heisenberg still also had his position in the department of physics at the University of Leipzig.

In February 1943, Heisenberg was appointed to the Chair for Theoretical Physics at the Friedrich-Wilhelms-Universität (today, the Humboldt-Universität zu Berlin). In April, his election to the Preußische Akademie der Wissenschaften (Prussian Academy of Sciences) was approved. That same month, he moved his family to their retreat in Urfeld as Allied bombing increased in Berlin.

The Alsos Mission, an Allied effort to determine if the Germans had an atomic bomb program and to exploit German atomic related facilities, research, material resources, and scientific personnel for the benefit of the US, generally moved into areas which had just come under control of the Allied military forces, but sometimes they operated in areas still under control by German forces. The Kaiser-Wilhelm-Institut für Physik (KWIP, Kaiser Wilhelm Institute for Physics) had been bombed so it had mostly been moved in 1943 and 1944 to Hechingen and its neighboring town of Haigerloch, on the edge of the Black Forest, which eventually became included in the French occupation zone. This allowed the American task force of the Alsos Mission to take into custody a large number of German scientists associated with nuclear research. In January 1945, Heisenberg, with most of the rest of his staff, moved from the Kaiser-Wilhelm Institut für Physik to the facilities in the Black Forest.

On 30 March, 1945, the Alsos Mission reached Heidelberg, where important scientists were captured. Their interrogation revealed that Otto Hahn was at his laboratory in Tailfingen, while Heisenberg and Max von Laue were at Heisenberg's laboratory in Hechingen, and that the experimental natural uranium reactor that Heisenberg's team had built in Berlin had been moved to Haigerloch. Thereafter, the main focus of the Alsos Mission was on these nuclear facilities in the Württemberg area. Heisenberg was captured and arrested in Urfeld, on May 3, 1945, in an alpine operation in territory still under control by German forces.

Germany surrendered on May 7. Heisenberg would not see his family again for eight months, as he was moved across France and Belgium and flown to England on July 3, 1945. Nine prominent German scientists, including Heisenberg, who were members of the *Uranverein* were captured and incarcerated at Farm Hall in England. The facility had been a safe house of the British foreign intelligence MI6. During their detention, their conversations were recorded. Conversations thought to be of intelligence value were transcribed and translated into English. The Farm Hall transcripts reveal that Heisenberg, along with other physicists interned at Farm Hall including Otto Hahn and Carl Friedrich von Weizsäcker, were glad the Allies had won the war. Heisenberg told other scientists that he had never contemplated a bomb, only an atomic pile to produce energy.

On 3 January 1946, the Operation Epsilon detainees were transported to Alswede in Germany. Heisenberg settled in Göttingen, which was in the British zone of Allied-occupied Germany. Following the Kaiser Wilhelm Society's obliteration by the Allied Control Council, and the establishment of the Max Planck Society in the British zone, the Kaiser Wilhelm Institute for Physics was renamed, and Heisenberg became the director of the Max Planck Institute for Physics.

In 1951, Heisenberg agreed to become the scientific representative of the Federal Republic of Germany at the UNESCO conference, with the aim of establishing a European laboratory for nuclear physics. Heisenberg's aim was to build a large particle accelerator, drawing on the resources and technical skills of scientists across the Western Bloc. On 1 July 1953 Heisenberg signed the convention that established CERN on behalf of the Federal Republic of Germany. Although he was asked to become CERN's founding scientific director, he declined. Instead, he was appointed chair of CERN's science policy committee and went on to determine the scientific program at CERN.

In 1958, the Max Planck Institute for Physics was moved to Munich and renamed Max Planck Institute for Physics und Astrophysics, of which Heisenberg was a co-director, and then sole director until he resigned his directorship on December 31, 1970.

[Heisenberg gave a joint lecture with Dirac at the old Cavendish Laboratory on May 22, 1963, which I attended. [Underwood, T. G. (1962-3). *Cambridge University lecture notebook 6.*]

Heisenberg died age 74 of kidney cancer at his home, on February 1, 1976. Heisenberg is buried in Munich Waldfriedhof.

# Heisenberg, W. (December, 1922). Zur Quantentheorie der Linienstruktur und der anomalen Zeemaneflekte. (On the quantum theory of line structure and anomalous Zeeman effect.)

[*Zeit. Physik*, 8, 273-97; https://doi.org/10.1007/BF01329602.]

Received December 17, 1921.

Institut für theoretisehe Physik, München.

[G. Breit. (September, 1923). *The Heisenberg Theory of the Anomalous Zeeman Effect. Nature*, 112, 396; https://doi.org/10.1038/112396a0.]

## Abstract

In his theory of doublets Heisenberg (1922. *Zeit. Physik*, 8, 273) assumes that the atom may be looked at as made of two parts: (1) the shell and (2) the valence electron. Expressing angular momenta in multiples of $h/2\pi$ and choosing the direction of the angular momentum of the shell as positive, the electron is allowed to have angular momenta $I = \frac{1}{2}, \pm 3/2, \pm 5/2, \ldots$ in the s, p, d, … states respectively, and the shell has in all of the states the angular momentum $\frac{1}{2}$. The observed Zeeman patterns show that $I = 3/2$ in $2p_1$ and $I = -3/2$ in $2p_2$. The observed energy levels show that the energy in $2p_1$ is higher than in $2p_2$. The writer experienced the following difficulty in accounting for this relative position of energy levels.]

[Cassidy, D. C. (1978). *Physics Today*, 31, 7, 23; https://doi.org/10.1063/ 1.2995102: Heisenberg had just celebrated his twentieth birthday when he presented his first paper for publication in 1921. This paper, a long and complex study entitled "*On the Quantum Theory of Line Structure and of the Anomalous Zeeman Effects*" immediately placed its young author on the forefront of theoretical spectroscopy. "He understands everything," Niels Bohr remarked. But, as often happens with brilliant first papers, its unique proposals were as controversial and perplexing as the phenomena they purported to explain.]

## Heisenberg, W. (July 25, 1925). Über quantentheoretische Umdeutung kinematischer und mechanischer Beziehungen. (Quantum-theoretical re-interpretation of kinematic and mechanical relations.)

[*Zeit. Phys.*, 33, 879-93; https://doi.org/10.1007/ BF01328377; (translation (2014) by Luca Doria, Institute of Theoretical Physics, Göttingen; translation by D. H. Delphenich; https://neo-classical-physics.info/ electromagnetism. html); and translation in van der Waerden, B. L., ed. (1968). *Sources of Quantum Mechanics*, 12, 261-76. Dover, New York.]

Submitted July 29, 1925.

Göttingen, Institut fur theoretische Physik.

[This translation is based largely on the December 2014 translation by Luca Doria, though *the notation has been restored to that used by Heisenberg and equation numbering is provided for both versions.* [Heisenberg, W. (July, 1925). *On the quantum reinterpretation of kinematical and mechanical relationships*. Werner Heisenberg Institute of Theoretical Physics, Göttingen.]

Most of the comments on this paper are based on Aitchison, A. J. R., MacManus, D. A. & Snyder, T. M. (2004). *Understanding Heisenberg's 'Magical' Paper of July 1925: a New Look at the Calculational Details. arXiv*:quant-ph/0404009; also in (November, 2004). *Am. J. Phys.*, 72, 11, 1370-9. Other comments, as indicated, are from the Appendix to Fedak, W. A. & Prentis, J. J. (2009). The 1925 Born and Jordan paper "*On quantum mechanics*". *Am. J. Phys.*, 77, 2, 128-139.]

Heisenberg proposes a quantum mechanics in which only relationships among observable quantities occur, not possible to assign to the electron a point in space as a function of time, builds on Kramer's dispersion theory and instead assigns to the electron an *emitted radiation*, substitutes *frequencies* and *amplitudes* of Fourier components of emitted radiation of electron, instead of reinterpreting x(t) as a *sum* over transition components represents position by *set* of transition components, assigns *transition frequencies* and *transition amplitudes* as observables, replaces classical component by *transition* component corresponding to the quantum jump from state *n* to state *n − α*, translates the old *quantum condition* that fixes the properties of the *states* to a new condition to calculate the amplitude of a *transition* between two states by replacing the differential by a difference, in quantum case *frequencies* do not combine in same way as classical harmonics but in accordance with the *Ritz combination principle* under which spectral lines of any element include frequencies that are either the sum or the difference of the frequencies of two other lines, in quantum case frequencies combine by multiplying *transition amplitudes* (equivalent to matrix multiplication), results in non-commutativity of kinematical quantities, shows simple quantum theoretical connection to Kramers' dispersion theory, the *equation of motion* $\ddot{x} + f(x) = 0$ and the *quantum condition* $h = 4\pi m \sum_{\alpha=-\infty}^{+\infty} \{|a(n, n+\alpha)|^2 \omega(n, n+\alpha) - |a(n, n-\alpha)|^2 \omega(n, n-\alpha)\}$ together contain if solvable *a complete determination not only of the frequencies and energies but also of the quantum theoretical transition probabilities*.

[Fedak, W. A. & Prentis, J. J. (2009), p. 128: "The name *"quantum mechanics"* appeared for the first time in the literature in Born, M. (December, 1924). *Über Quantenmechanik. Z. Phys.* 26, 379–395." ... For Born and others, *quantum mechanics* denoted a canonical theory of atomic and electronic motion of the same level of generality and consistency as classical mechanics. The transition from classical mechanics to a true quantum mechanics remained an elusive goal prior to 1925.

Heisenberg made the breakthrough in his historic 1925 paper.[2]

[2] Heisenberg, W. (July, 1925). *Über quantentheoretische Umdeutung kinematischer und mechanischer Beziehungen.* (On the quantum-theoretical reinterpretation of kinematic and mechanical relations.) *Zeit. Phys.*, 33, 879-93.

*Heisenberg's bold idea was to retain the classical equations of Newton but to replace the classical position coordinate with a "quantum-theoretical quantity."* The new position quantity contains information about the measurable line spectrum of an atom rather than the unobservable orbit of the electron."]

## Abstract

In this work we will try to obtain the basis for a quantum mechanics theory which is *based uniquely on relationships between in principle observable quantities*.

## Introduction

It is known that against the formal rules of the quantum theory used for the calculation of the observable quantities (for example the energy levels of the hydrogen atom) the serious objection can be raised that 1) those calculational rules contain as essential components relationships between quantities that seemingly in principle cannot be observed (like for example the electron position and period of revolution of the electron) and 2) also those rules apparently lack every clear physical basis unless one does not want to remain attached to the hope that those until now unobserved quantities will be made experimentally accessible in the future. This hope might be regarded as justified if the above-mentioned rules were internally consistent and applicable to a clearly defined range of quantum theoretical problems.

Anyway, experience shows that 1) *only the hydrogen atom and its Stark effect fit into those formal rules of quantum theory*, 2) already in the "crossed fields" problem (hydrogen atom in electric and magnetic fields in different directions) fundamental difficulties arise, 3) the reaction of atoms to periodically varying fields surely cannot be described by the mentioned rules and 4) finally an expansion of the quantum rules for the treatment of many-electrons atoms has been proved unfeasible.

It became customary to characterize the failure of the quantum rules (that were already essentially characterized through the application of classical mechanics) as a deviation from classical mechanics. However, this description can hardly be viewed as logical when one considers that already the *Einstein-Bohr frequency condition* represents such a complete departure from classical mechanics or better, from the point of view the wave theory, from the underlying kinematics of this mechanics, that it is absolutely not possible even for the simplest quantum theoretical problem to maintain the validity of classical mechanics.

In this situation, *it is advisable to completely give up any hope about the observation of hitherto unobserved quantities (like the electrons' position and period)* and at the same time acknowledge that 1) the partial agreement with experience of the mentioned quantum rules is more or less an accident and 2) to *try to construct a theory of quantum mechanics in which only relationships among observable quantities occur*. As the most important first steps toward such a theory of quantum mechanics one can refer to the *dispersion theory of Kramers* (1) and following works based on it (2).

> 1) Kramers, H. A. (May, 1924). *The law of dispersion and Bohr's theory of spectra. Nature* 113, 673-74 [; derives dispersion (scattering) formula for electromagnetic radiation incident on an atom by assuming incident radiation characterized by train of polarized harmonic waves, positive virtual oscillators correspond to absorption frequencies (same as Ladenburg). Kramers' formula includes an addition term representing negative virtual oscillators that corresponds to emission frequencies].
> 2) Born, M. (December, 1924). *Über Quantenmechanik*, (About quantum mechanics.) *Zeit. Phys.*, 26, 379-95 [; the thesis contains an attempt to establish the first step towards quantum mechanics of coupling, which gives an account of the most important properties of atoms (stability, resonance for the jump frequencies, correspondence principle) and arises naturally from the classical laws. This theory contains Kramers' dispersion formula and shows a close relationship to Heisenberg's formulation of the rules of the anomalous Zeeman effect]; Kramers, H. A. & Heisenberg, W. (1925). *Über die Streuung von Strahlung durch Atome.* (On the scattering of radiation by atoms.) *Zeit. Physik*, 31, 681-708; http://dx.doi.org/10.1007/BF02980624; Born, M. & P. Jordan, P. (1925). *Zur Quantenmechanik.* (On Quantum Mechanics.) *Zeit. Physik.* (Forthcoming.)

In the following, *we shall try to present some new quantum mechanical relationships and apply them to the detailed treatment of some special problems*. We shall limit ourselves to problems with one degree of freedom.

§ 1. In the classical theory, *the radiation of a moving electron* (in the wave-zone $\mathbf{E} \sim \mathbf{H} \sim 1/r$) is not completely given by the expressions

$$\mathbf{E} = e/r^3c^2 \, [\mathbf{r}(\mathbf{r}\mathbf{v}^{\cdot})] \tag{1}$$
$$\mathbf{H} = e/r^2c^2 \, (\mathbf{v}^{\cdot}\mathbf{r}) \tag{2}$$

[where $\mathbf{E}$ and $\mathbf{H}$ are the fields strengths at a point for the electric field and magnetic field respectively, e is the *electron charge*, $\mathbf{r}$ is the *distance of the electron from the field point*, and $\mathbf{v}$ the *electron velocity*)]

but we have other terms at the next order, e.g. of the form

$$e/rc^3 \, (\mathbf{v}^{\cdot}\mathbf{v}) \tag{3}$$

that we can denote as quadrupole radiation, and at the next higher order we have terms of the form

$$e/rc^4 \, (\mathbf{v}^{\cdot}\mathbf{v}^2) \tag{4}$$

and in this way the approximation can be carried out at any desired order.

One can ask himself *how the higher terms look like in the quantum theory*.

Since in the classical theory *the higher orders can be easily calculated when the motion of the electron or its Fourier representation are given respectively, one can expect the same in the quantum theory*. This question does not have to do with electrodynamics but this is - and this seems particularly important to us - of *pure kinematical nature*. We can pose this question as follows: *given instead of the classical quantity x(t) a quantum theoretical one, which quantum theoretical quantity enters in the place of x(t)²*?

Before being able to answer this question, we have to remember that *in the quantum theory it was not possible to assign to the electron a point in space as a function of time through observable quantities*. However surely also in the quantum theory *one can assign to the electron an emitted radiation*. First, *this radiation will be described by frequencies* [v] which quantum theoretically arise as function of two variables in the form:

$$\nu \, (n, n - \alpha) = 1/h \, \{W(n) - W(n - \alpha)\} \tag{5}$$

and in the classical theory in the form:

$$\nu \, (n, \alpha) = \alpha \cdot \nu \, (n,) = \alpha \, 1/h \, dW/dn. \tag{6}$$

[where the observables are the *energies W(n) of the (Bohr) stationary states*, together with the associated (Einstein-Bohr) *frequencies v*, which characterize radiation emitted in the transition $n \rightarrow n - \alpha$.]

(From here onwards, we define $nh = J$ where J is one of the *canonical constants*).

As characteristic for the comparisons of the classical mechanics to the quantum theory, with regard to the *frequencies* one can write the "***combination relations***"

*Classically*:

$$\nu(n, \alpha) + \nu(n, \beta) = \nu(n, \alpha + \beta) \qquad [7]$$

[where $\nu(n, \alpha) = \alpha\, 1/h\, dW/dn$]

*Quantum theoretically*:

$$\nu(n, n - \alpha) + \nu(n - \alpha, n - \alpha - \beta) = \nu(n, n - \alpha - \beta) \qquad [8]$$
$$\nu(n - \beta, n - \alpha - \beta) + \nu(n, n - \beta) = \nu(n, n - \alpha - \beta) \qquad [9]$$

[where $\nu(n, n - \alpha) = 1/h\, \{W(n) - W(n - \alpha)\}$]

Secondly, besides the *frequencies*, the **amplitudes** are necessary for the description of radiation. The **amplitudes** can be written as complex vectors (each with six independent components) and determine *polarization* and *phase*. They are also function of the two variables $n$ and $\alpha$ so that the corresponding part of the radiation will be represented with

*Quantum theoretically*:

$$\mathbf{R}\{\mathbf{A}(n, n - \alpha)\, e^{i\omega(n,n-\alpha)t}\} \qquad [10](1)$$

*Classically*:

$$\mathbf{R}\{\mathbf{A}_\alpha(n)\, e^{i\omega(n)\alpha t}\} \qquad [11](2)$$

[Heisenberg brings complex numbers and the square root of $-1$ into quantum theory in the same way as in the classical theory, by expressing the **amplitudes** as ***Fourier series*** in terms of ***exponentials***.

Fourier's original formulation of the ***Fourier transform*** did not use complex numbers, but rather sines and cosines. Statisticians and others still use this form. An absolutely integrable function $f$ for which Fourier inversion holds can be expanded in terms of genuine frequencies $\lambda$ (avoiding negative frequencies, which are sometimes considered hard to interpret physically) by
$$f(t) = \int_0^\infty \{a(\lambda) \cos(2\pi\lambda t) + b(\lambda) \sin(2\pi\lambda t)\}.]$$

First of all, the *phase* (contained in **A**) appears to have no meaning in the quantum theory since in this theory the *frequencies* are not in general commensurable with their harmonics. However, *we will immediately see that also in the quantum theory the phase has a precise meaning which has an analog in the classical theory*.

Let us consider now a particular quantity x(t) in the classical theory such that it can be regarded as represented by the totality of quantities of the form

$$\mathbf{A}_\alpha\,(n)\,e^{i\omega(n)\alpha t} \qquad\qquad [12]$$

which depending on the motion being periodic or not, represents x(t) with a sum or an integral

$$x(t) = \Sigma_{\alpha=-\infty}^{+\infty}\,\mathbf{A}_\alpha\,(n)\,e^{i\omega(n)\alpha t} \qquad\qquad [13]$$

or $\qquad x(t) = \int_{\alpha=-\infty}^{+\infty}\,\mathbf{A}_\alpha\,(n)\,e^{i\omega(n)\alpha t}\,d\alpha \qquad\qquad [14](2a)$

A similar combination of the corresponding quantum-theoretical quantities seems to be impossible in a unique manner and therefore not meaningful in view of the equal weight of the quantities $n$ and $n-\alpha$ [i.e. in the *amplitude* A(n, n − α) and *frequency* ω(n, n − α)]. *However, one may readily regard the ensemble of quantities*

$$A(n,\,n-\alpha)\,e^{i\omega(n,n-\alpha)t} \qquad\qquad [15]$$

*as a representation of the quantity x(t)* and then try to answer the question posed before: *how would the quantity x(t)² be represented?*

> [Aitchison, MacManus & Snyder. (2004). *Understanding Heisenberg's 'Magical' Paper of July 1925: a New Look at the Calculational Details*, pages 3-4: "An example of something he wishes to exclude from the new theory is the time-dependent *position* coordinate x(t). In considering what might replace it, he turns to the *probabilities for transitions between stationary states*. Consider a simple one-dimensional model of an atom consisting of an electron undergoing periodic motion. For a state characterized by the label *n*, fundamental *frequency* ω(*n*) and *coordinate* x(*n*, t), one can represent x(*n*, t) as a Fourier series
>
> $$x(n,\,t) = \Sigma_{\alpha=-\infty}^{+\infty}\,A_\alpha(n)\,e^{i\omega(n)\alpha t},$$
>
> where $A_\alpha$ is the *amplitude* of the α th harmonic.
>
> According to classical theory, the *energy emitted per unit time* (that is, the power) in a transition corresponding to the α th harmonic v(*n*)α is
>
> $$-\,(dE/dt)_\alpha = e^2/3\pi\varepsilon_0 c^3\,[v(n)\alpha]^4\,|A_\alpha(n)|^2\,.$$
>
> In the quantum theory, however, the *transition frequency* corresponding to the classical 'v(*n*)α' is in general not a simple multiple of a fundamental frequency, but is given by ω (*n*, *n* − α) = 1/h {E(*n*) − E (*n* − α)}, thus v(*n*)α is replaced by ω (*n*, *n* − α). Correspondingly, Heisenberg introduces the quantum analogue of Aα(*n*), which he writes as **A**(*n*, *n* − α). Further, − (dE/dt)α has, in the quantum theory, to be replaced by the *product of the transition probability per unit time*, P(*n*, *n* − α), and the *emitted energy* hω(*n*, *n* − α); resulting in
>
> $$P(n,\,n-\alpha) = e^2/3\pi\varepsilon_0 hc^3\,[\omega(n,\,n-\alpha)]^3\,|A(n,\,n-\alpha)|^2.$$

This equation refers, however, to only one specific transition. For a full description of atomic dynamics (as then conceived), one will need to consider all the quantities $A(n, n - \alpha) \, e^{i\omega(n, n - \alpha)t}$. In the classical case, the terms $A\alpha(n) \, e^{i\omega(n)\alpha t}$ may be combined to yield x(t) using $x(n, t) = \Sigma_{\alpha = -\infty}^{+\infty} A_\alpha(n) \, e^{i\omega(n)\alpha t}$.

It is the **transition amplitudes** $\mathbf{A}(n, n - \alpha)$ which Heisenberg fastens upon as being satisfactorily 'observable'; like the **transition frequencies**, they depend on two discrete variables.

… This is the first of Heisenberg's 'magical jumps' - and certainly a very large one. Representing x(t) in this way seems to be the sense in which he considered himself to be 're-interpreting the kinematics'."]

[Fedak, W. A. & Prentis, J. J. (2009), *Appendix,* p. 136: *"Reinterpretation 1: Position.* Heisenberg considered one-dimensional periodic systems. The classical motion of the system (in a stationary state labeled *n*) is described by the time-dependent position x(*n*, t). Heisenberg represents this periodic function by the Fourier series

$$x(n, t) = \Sigma_{\alpha = -\infty}^{+\infty} a_\alpha(n) \, e^{i\omega(n)\alpha t} \qquad\qquad \text{(A1)} \qquad\qquad [14](2a)$$

The $\alpha$th Fourier component related to the *n*th stationary state has amplitude $a_\alpha(n)$ and frequency $\alpha\omega(n)$. According to the **correspondence principle**, the $\alpha$th Fourier component of the classical motion in the state *n* corresponds to the quantum jump from state *n* to state *n* − $\alpha$. Motivated by this principle, Heisenberg replaced the classical component $a(n) \, e^{i\omega(n)\alpha t}$ by the transition component $a(n, n - \alpha) \, e^{i\omega(n, n-\alpha)t}$. … Unlike the sum over the classical components in Eq. (A1), Heisenberg realized that a similar sum over the transition components is meaningless. Such a quantum *Fourier series* could not describe the electron motion in one stationary state (*n*) because each term in the sum describes a transition process associated with two states (*n* and *n* − $\alpha$).

Heisenberg's next step was bold and ingenious. *Instead of reinterpreting x(t) as a sum over transition components, he represented the position by the set of transition components*. We symbolically denote Heisenberg's reinterpretation as

$$x \rightarrow \{a(n, n - \alpha) \, e^{i\omega(n, n-\alpha)t}. \qquad\qquad \text{(A2)}$$

*Equation (A2) is the first breakthrough relation*."]

Classically, the answer is obviously

$$\mathbf{B}_\beta(n) \, e^{i\omega(n)\beta t} = \Sigma_{-\infty}^{+\infty} \mathbf{A}_\alpha \, \mathbf{A}_{\beta-\alpha} \, e^{i\omega(n)(\alpha+\beta-\alpha)t} \qquad\qquad [16](3)$$

or
$$= \int_{-\infty}^{+\infty} \mathbf{A}_\alpha \, \mathbf{A}_{\beta-\alpha} \, e^{i\omega(n)(\alpha+\beta-\alpha)t} \, d\alpha \qquad\qquad [17](4)$$

so that

$$x(t)^2 = \Sigma_{\beta=-\infty}^{+\infty} \mathbf{B}_\beta \, e^{i\omega(n)\beta t} \qquad\qquad [18](5)$$

or, respectively

$$= \int_{-\infty}^{+\infty} \mathbf{B}_\beta \, e^{i\omega(n)\beta t} d\beta \qquad\qquad [19](6)$$

[from $x(t) = \Sigma_{\alpha=-\infty}^{+\infty} \mathbf{A}_\alpha \, (n) \, e^{i\omega(n)\alpha t}$,

$\qquad x(t)^2 = \Sigma_{\alpha=-\infty}^{+\infty} \Sigma_{\beta-\alpha=-\infty}^{+\infty} \mathbf{A}_\alpha \, \mathbf{A}_{\beta-\alpha} \, e^{i\omega(n)(\alpha+\beta-\alpha)t}$,

$\qquad x(t)^2 = \Sigma_{\alpha=-\infty}^{+\infty} \Sigma_{\beta-\alpha=-\infty}^{+\infty} \mathbf{A}_\alpha \, \mathbf{A}_{\beta-\alpha} \, e^{i\omega(n)\beta t}$,

or $\qquad x(t)^2 = \Sigma_{\beta=-\infty}^{+\infty} B_\beta \, e^{i\omega(n)\beta t}$

where $\quad B_\beta(n) \, e^{i\omega(n)\beta t} = \Sigma_{-\infty}^{+\infty} A_\alpha \, A_{\beta-\alpha} \, e^{i\omega(n)(\alpha+\beta-\alpha)t}$.]

[Aitchison, MacManus & Snyder. (2004), p. 4-5: "The crucial difference in the quantum case is that the *frequencies* do not combine in the same way as the classical harmonics, but rather in accordance with the *Ritz combination principle*:

$$\omega(n, n-\alpha) + \omega(n-\alpha, n-\beta) = \omega(n, n-\beta),$$

which is of course consistent with $\omega \, (n, n-\alpha) = 1/h \, \{W(n) - W(n-\alpha)\}$.

> [The *Rydberg–Ritz combination principle* is an empirical generalization proposed by Walther Ritz in 1908 to describe the relationship of the spectral lines for all atoms. The principle states that the spectral lines of any element include frequencies that are either the sum or the difference of the frequencies of two other lines.]

Thus in order to end up with the particular *frequency* $\omega(n, n-\beta)$, it seems 'almost necessary' (in Heisenberg's words) to combine the quantum *amplitudes* in such a way as to ensure the *frequency* combination above; that is, as

$B(n, n-\beta) \, e^{i\omega(n, n-\beta)t} = \Sigma_\alpha A(n, n-\alpha) \, e^{i\omega(n, n-\alpha)t} \, A(n-\alpha, n-\beta) \, e^{i\omega(n-\alpha, n-\beta)t}$.

Cancelling the exponentials on both sides, we are left with

$$B(n, n-\beta) = \Sigma_\alpha A(n, n-\alpha) \, A(n-\alpha, n-\beta),$$

which is Heisenberg's law for multiplying *transition amplitudes* together."]

[Fedak, W. A. & Prentis, J. J. (2009), *Appendix*, p. 136: "*Reinterpretation 2: Multiplication.* To calculate the energy of a harmonic oscillator, Heisenberg needed to know the quantity $x^2$. ***How do you square a set of transition components?*** Heisenberg posed this fundamental question twice in his paper. His answer gave birth to the algebraic structure of quantum mechanics. … Heisenberg answered this question by reinterpreting the square of a Fourier series with the help of the *Ritz principle*. He evidently was convinced that quantum multiplication, whatever it looked like, must reduce to Fourier-series multiplication in the classical limit. ... In the new quantum theory Heisenberg replaced

$$x^2(n, t) = \Sigma_{\beta = -\infty}^{+\infty} \, b_\beta(n) \, e^{i\omega(n)\beta t} \tag{A3}$$

... with

$$x^2 \rightarrow \{b(n, n - \beta) \, e^{i\omega(n, \, n-\beta)t}, \tag{A5}$$

where the n→n − β **transition amplitude** is

$$b(n, n - \beta) = \Sigma_{\alpha = -\infty}^{+\infty} \, a(n, n - \alpha) \, a(n - \alpha, n - \beta) \tag{A6}$$

In constructing Eq. (A6) Heisenberg uncovered the symbolic algebra of atomic processes. ... Eq. (A6) allowed Heisenberg to algebraically manipulate the transition components."]

It seems that quantum theoretically the easiest and most natural assumption is to replace Eqs. (3), (4)

$$[\mathbf{B}_\beta(n) \, e^{i\omega(n)\beta t} = \Sigma_{-\infty}^{+\infty} \, \mathbf{A}_\alpha \, \mathbf{A}_{\beta-\alpha} \, e^{i\omega(n)(\alpha+\beta-\alpha)t} \tag{[16](3)}$$

$$\text{or} \qquad = \int_{-\infty}^{+\infty} \, \mathbf{A}_\alpha \, \mathbf{A}_{\beta-\alpha} \, e^{i\omega(n)(\alpha+\beta-\alpha)t} \, d\alpha \tag{[17](4)]}$$

with

$$\mathbf{B}(n, n - \beta) \, e^{i\omega(n, n-\beta)t} = \Sigma_{\alpha = -\infty}^{+\infty} \, \mathbf{A}(n, n - \alpha) \, \mathbf{A}(n - \alpha, n - \beta) \, e^{i\omega(n, n-\beta)t} \tag{[20](7)}$$

$$\text{or} \qquad = \int_{-\infty}^{+\infty} \, \mathbf{A}(n, n - \alpha) \, \mathbf{A}(n - \alpha, n - \beta) \, e^{i\omega(n, n-\beta)t} \, d\alpha \tag{[21](8)}$$

and indeed, this way of combination follows almost inevitably from the **frequency combination relation**. If we accept the assumptions (7), (8) one recognizes also that the *phases* of the quantum theoretical **A** have the same relevant physical significance as in the classical theory: only the beginning time and hence a phase constant common to all the **A** is arbitrary and without physical meaning but the **phase** of every single A enters in the quantity **B** [1].

[1] Compare also to Kramers, H.A., & Heisenberg, W. (1925). [*Über die Streuung von Strahlung durch Atome.* (On the scattering of radiation by atoms.) *Zeit. Physik,* 31, 681-708; https://doi.org/ 10.1007/BF02980624.] In the expressions used there for the induced scattering momentum, the phases are essentially contained.

A geometric interpretation of these quantum theoretic *phase* relationships in analogy to the classical theory seems at first not possible.

We ask now about how to represent the quantity $x(t)^3$ and we find without difficulty:

*Classically*:

$$\mathbf{C}(n, \gamma) = \Sigma_{\alpha = -\infty}^{+\infty} \, \Sigma_{\beta = -\infty}^{+\infty} \, \mathbf{A}_\alpha(n) \, \mathbf{A}_\beta(n) \, \mathbf{A}_{\gamma-\alpha-\beta}(n) \tag{[22](9)}$$

*Quantum theoretically*:

$$\mathbf{C}(n, n{-}\gamma) = \Sigma_{\alpha = -\infty}^{+\infty} \, \Sigma_{\beta = -\infty}^{+\infty} \, \mathbf{A}(n, n{-}\alpha) \, \mathbf{A}(n{-}\alpha, n{-}\alpha{-}\beta) \, \mathbf{A}(n{-}\alpha{-}\beta, n{-}\gamma) \tag{[23](10)}$$

or the corresponding formulae with integrals.

In a similar way, all the quantities of the form x(t) n can be expressed quantum theoretically and when a function f[x(t)] is given, one can always obviously find the quantum theoretical analog if it is possible to expand this function in powers of x. *A substantial difficulty arises when we consider two quantities x(t), y(t) and we ask about the product x(t)y(t).*

Let be x(t) characterized with **A** and y(t) with **B** so the representation of x(t)y(t) results:

*Classically*:

$$\mathbf{C}_\beta = \Sigma_{\alpha=-\infty}^{+\infty} \mathbf{A}_\alpha(n) \, \mathbf{B}_{\beta-\alpha}(n) \qquad\qquad [24]$$

*Quantum theoretically*:

$$\mathbf{C}(n, n-\beta) = \Sigma_{\alpha=-\infty}^{+\infty} \mathbf{A}(n, n-\alpha) \, \mathbf{B}(n-\alpha, n-\beta). \qquad\qquad [25]$$

[Aitchison, MacManus & Snyder. (2004), p. 5: "Born recognized
$$C(n, n-\beta) = \Sigma_{\alpha=-\infty}^{+\infty} A(n, n-\alpha) \, B(n-\alpha, n-\beta)$$
as matrix multiplication (something unknown to Heisenberg in July 1925), and he and Jordan rapidly produced the first paper to state the fundamental commutation relation (in modern notation …)
$$\mathbf{xp} - \mathbf{px} = i\hbar. \qquad\qquad (11)$$

Dirac's paper followed soon after, and then the 'three-man' paper of Born, Heisenberg and Jordan. [Born, M., Heisenberg, W. & Jordan, P. (August, 1926). *Zur Quantenmechanik II.* (On Quantum Mechanics II.) *Zeit. Phys.*, 35, 557-615."]

[Aitchison, MacManus & Snyder. (2004), pp. 6-7: "It took Born only a few days to show that Heisenberg's *quantum condition* (16) was in fact the diagonal matrix element of
$$\Sigma_{\alpha=-\infty}^{+\infty} \mathbf{A}(n, n-\alpha) \, \mathbf{B}(n-\alpha, n-\beta). \qquad\qquad (7)$$

or in modern notation, of $\mathbf{xp} - \mathbf{px}$, and to guess that the off-diagonal elements of $\mathbf{xp} - \mathbf{px}$ were zero, a result which was shown to be compatible with the *equations of motion* in Born and Jordan's paper. [Born, M. & Jordan, P. (December, 1925). *Zur Quantenmechanik.* (On Quantum Mechanics.) *Zeit. Phys.*, 34, 858-88."]

… Heisenberg's *transition amplitude* $\mathbf{A}(n, n-\alpha)$ is indeed precisely the same as the quantum-mechanical matrix element $(n-\alpha|x|n)$, where $|n)$ is the exact eigenstate with energy W(n). The relation of (16)
$$[h = 4\pi m \, \Sigma_{\alpha=0}^{+\infty} \{|a(n, n+\alpha)|^2 \omega(n, n+\alpha)$$
$$- |a(n, n-\alpha)|^2 \omega(n, n-\alpha)\} \qquad\qquad [34](16)]$$
to the *fundamental commutator* (11)
$$[x^{\cdot\cdot} + f(x) = 0 \qquad\qquad [26](11)]$$
is briefly recalled in Appendix A.

…

*Appendix A: The quantum condition and* **$xp - px$** $= i\hbar$.

Consider the $(n, n)$ element of $(x\dot{x} - \dot{x}x)$. This is

$$\Sigma_\alpha\, a(n, n - \alpha)\, i\omega(n - \alpha, n)\, a(n - \alpha, n)$$
$$- \Sigma_\alpha\, i\omega(n, n - \alpha)\, a(n, n - \alpha)\, a(n - \alpha, n).$$

$$[\Sigma_\alpha\, a(n, n - \alpha) \cdot d/dt\, a(n - \alpha, n)$$
$$- \Sigma_\alpha\, d/dt\, a(n, n - \alpha) \cdot a(n - \alpha, n),$$
$$\text{where } a(n, n - \alpha) = a_\alpha(n, n - \alpha)\, e^{i\alpha\omega(n, n - \alpha)t}]$$

[The square root of $-1$ (i) is introduced into **$x\dot{x} - \dot{x}x$** by differentiating the Fourier series $x = \Sigma_{\alpha = -\infty}^{+\infty} a_\alpha(n)\, e^{i\alpha\omega(n)t}$ expressed in terms of exponentials with respect to t in $\dot{x}$.]

In the first term, the sum over $\alpha > 0$ may be re-written as

$$- i\, \Sigma_{\alpha>0}\, \omega(n, n - \alpha)\, |a(n, n - \alpha)|^2$$

using $\omega(n, n-\alpha) = - \omega(n-\alpha, n)$ from $\nu(n, n - \alpha) = 1/h\, \{W(n) - W(n - \alpha)\}$ in paragraph 1(original page 881), and $a_\alpha(n - \alpha, n) = a_\alpha(n, n - \alpha)$ from the quantum-theoretical analogue of $a_{-\alpha}(n) = a_\alpha(n)$ on original page 885, assuming as Heisenberg did that the $a_\alpha$'s are chosen to be real, while the sum over $\alpha < 0$ becomes, similarly,

$$i\, \Sigma_{\alpha>0}\, \omega(n + \alpha, n)|a(n + \alpha, n)|^2$$

on changing $\alpha$ to $- \alpha$.

Similar steps in the second term led to the result

$$(\mathbf{x\dot{x}} - \mathbf{\dot{x}x})\, (n, n) = 2i\, \Sigma_{\alpha>0}\, \{\omega(n + \alpha, n)|a(n + \alpha, n)|^2$$
$$- \omega(n, n - \alpha)|a(n, n - \alpha)|^2\,\} = 2ih/(4\pi m),$$

where the last step follows from the '***quantum condition***' (16). Setting **p** $= m\mathbf{\dot{x}}$ [in $m(x\dot{x} - \dot{x}x)\, (n, n) = ih/4\pi$] we find

$$(\mathbf{xp} - \mathbf{px})(n, n) = i\hbar$$

for all values of n, [which is the modern formulation of (16).] This is the result which Born found shortly after reading Heisenberg's paper. ***In the further development of the theory the value of the 'fundamental commutator' xp − px, namely iℏ times the unit matrix, was taken to be a basic postulate***. The sum rule (16) is then derived by taking the $(n, n)$ matrix element of the relation $[\mathbf{x}, [\mathbf{H}, \mathbf{x}]] = \hbar^2/m$."]

While classically $x(t)y(t)$ always equal to $y(t)x(t)$ is, in general it must not be the case in the quantum theory. In special cases, for example when one considers $x(t)x(t)^2$, the difficulty does not arise.

[Aitchison, MacManus & Snyder. (2004), p. 5: "This 'difficulty' clearly unsettled Heisenberg: but ***it very quickly became clear that the non-commutativity (in general) of kinematical quantities in quantum theory was the really essential new technical idea in the paper***."]

258

As in the question posed at the beginning of this paragraph, when one considers a form like

$$v(t)v^{\cdot}(t)$$

one has to substitute $vv^{\cdot}$ quantum theoretically with $(vv^{\cdot} + v^{\cdot}v)/2$ so that $vv^{\cdot}$ becomes the derivative of $v^2/2$. In a similar way, natural mass quantum-theoretic mean values can always be given, which, however, are hypothetical to an even higher degree than the formulas (7) and (8). *

$$[\mathbf{B}(n, n - \beta)\, e^{i\omega(n,n-\beta)t} = \Sigma_{\alpha = -\infty}^{+\infty}\ \mathbf{A}(n, n - \alpha)\, \mathbf{A}(n - \alpha, n - \beta)\, e^{i\omega(n,n-\beta)t} \qquad [20](7)$$

$$\text{or} \qquad\qquad\qquad = \int_{-\infty}^{+\infty} \mathbf{A}(n, n - \alpha)\, \mathbf{A}(n - \alpha, n - \beta)\, e^{i\omega(n,n-\beta)t}\, d\alpha. \qquad [21](8)]$$

Apart from the difficulty just described, formulas of type (7), (8) were general enough to express the interaction of the electrons in an atom through the characteristic **amplitudes** of the electrons. *

[* My translation. These two sentences were omitted from Luca Dora's translation:

"In ahnlicher Weise lassen sich wohl stets naturgemasse quanten-theoretisehe Mittelwerte angeben, die allerdings in noch hoherem Grade hypothetiseh sind als die Formela (7) und (8).

Abgesehen von der eben geschilderten Schwierigkeit durften Formeln vom Typus (7), (8) allgemein genugen, um aueh die Wechselwirkung der Elektronen in einem Atom durch die eharakteristischen Amplituden der Elektronen auszudrucken."

Original page 884.]

§ 2. After these considerations which subject was the kinematic of the quantum theory, we will turn to **mechanical problems aiming at the determination of A, v, W from the given forces of the system**.

[Aitchison, MacManus & Snyder. (2004), p. 5: "Having identified the **transition amplitudes** $X(n, n-\alpha)$ and **frequencies** $\omega(n, n-\alpha)$ as the 'observables' with which the new theory should deal, Heisenberg now turns his attention to how they may be determined 'from the given forces of the system' - that is, by the dynamics."]

In the previously presented theory [the Old Quantum Mechanics], this problem is solved in two steps:

1. Integration of the **equations of motion**

$$x^{\cdot\cdot} + f(x) = 0 \qquad\qquad\qquad\qquad\qquad [26](11)$$

[where f(x) is the **force per mass** function.]

2. Determination of the **constants** arising from periodic motion [through the quantum condition] with

$$\int pdq = \int mx^{\cdot}dx = J\ (= nh), \qquad\qquad\qquad\qquad [27](12)$$

[where m is the **mass** and the integral is to be evaluated over the period of one period of the motion].

> [Fedak, W. A. & Prentis, J. J. (2009), *Appendix*, pp 136-7: "*Reinterpretation 3: Motion.* Equations (A2), (A5), and (A6) represent the new "kinematics" of quantum theory—the new meaning of the position x. Heisenberg next turned his attention to the new "mechanics." ***The goal of Heisenberg's mechanics is to determine the amplitudes, frequencies, and energies from the given forces***. Heisenberg noted that in the old quantum theory $a_\alpha(n)$ and $\varpi(n)$ are determined by solving the classical ***equation of motion***
>
> $$x\ddot{} + f(x) = 0 \qquad\qquad (A7) \qquad\qquad [26](11)$$
>
> and quantizing the classical solution—making it depend on *n* - via the quantum condition
>
> $$\int m\dot{x}\,dx = nh. \qquad\qquad (A8) \qquad\qquad [27](12)$$
>
> Heisenberg assumed that Newton's second law in Eq. (A7) is valid in the new quantum theory provided that the classical quantity x is replaced by the set of quantities in Eq. (A2),
>
> $$x \rightarrow \{a(n, n-\alpha)\,e^{i\omega(n, n-\alpha)t}. \qquad\qquad (A2)$$
>
> and f(x) is calculated according to the new rules of amplitude algebra. ***Keeping the same form of Newton's law of dynamics, but adopting the new kinematic meaning of x is the third Heisenberg breakthrough.***"]

If one wants to construct a quantum theoretical mechanics which is the possible classical analog, it is probably very close to bring the ***equation of motion*** Eq. (11)

$$[x\ddot{} + f(x) = 0 \qquad\qquad (A7) \qquad\qquad [26](11)]$$

directly into the quantum theory ***where it is only necessary to take over***, for not abandoning the foundation of in principle observable quantities, ***instead of the quantities x¨, f(x), their quantum theoretic representations known from § 1***.

> [Aitchison, MacManus & Snyder. (2004), p. 5: "or, as we would say today, by taking matrix elements of the operator equation of motion x¨ + f(x) = 0."]

In the classical theory, it is possible to search for a solution of Eq. (11) by first expressing x[(t)] in Fourier series or Fourier integrals with undetermined coefficients (and frequencies); although in general we obtain infinitely many equations with infinitely many unknowns (or integral equations) which can be solved only in special cases with simple recursion formulae for [the Fourier coefficients] X.

However, in the quantum theory, we are dependent on this kind of solution for Eq. (11)

$$[x\ddot{} + f(x) = 0 \qquad\qquad\qquad [26](11)]$$

which, as discussed before, prevents the definition of direct [quantum-theoretical] analogues of the function x(n, t).

This has as consequence that the quantum theoretical solution of Eq. (11) is feasible at first only in the simplest cases. Before going over these simple examples, we would like to derive quantum theoretically the value of the constant in Eq. (12).

$$[\int pdq = \int m\dot{x}\,dx = J\ (= nh),\qquad\qquad [27](12)]$$

We assume also that the (classical) motion is periodic:

$$x = \Sigma_{\alpha=-\infty}^{+\infty}\ a_\alpha(n)\ e^{i\alpha\omega(n)t}\qquad\qquad [28](13)$$

[Note: the complex amplitude vectors for absorption $\mathbf{A}_a$ and emission $\mathbf{A}_e$ transitions are replaced by $a_\alpha$, which are no longer vectors but amplitudes in the Fourier expansion of the coordinate x of the electron.]

then

$$m\dot{x} = m\,\Sigma_{\alpha=-\infty}^{+\infty}\ a_\alpha(n)\ .\ i\alpha\omega(n)\ e^{\alpha\omega(n)t}\qquad\qquad [29]$$

and

$$\int m\dot{x}\ dx = \int m\dot{x}^2\ dt = 2\pi m\,\Sigma_{\alpha=-\infty}^{+\infty}\ a_\alpha(n)a_{-\alpha}(n)\alpha^2\omega(n).\qquad\qquad [30]$$

[where the integral is evaluated over one period of the motion].

Further, since $a_{-\alpha}(n) = a_\alpha(n)$ (x must be real), it follows

$$\int m\dot{x}^2\ dt = 2\pi m\,\Sigma_{\alpha=-\infty}^{+\infty}\ |a_\alpha(n)|^2\alpha^2\omega(n).\qquad\qquad [31](14)$$

[Aitchison, MacManus & Snyder. (2004), p. 6: "Heisenberg argues that (14) does not sit well with the **Correspondence Principle**, since the latter should only determine J up to an additive constant (times h). Setting (14) equal to $n$h, he converts it to the form

$$h = 2\pi m\,\Sigma_{\alpha=-\infty}^{+\infty}\ \alpha\ d/dn\ \{\alpha\omega(n)\cdot|a_\alpha(n)|^2\}$$

which determines the $a_\alpha(n)$'s only to within a constant."]

Until now, this **phase** integral was set to a multiple of h ($n$h); such a condition is not only forced into the classical calculation but it looks arbitrary also from the previous point of view of the **correspondence principle** because correspondence-wise the J is set only up to an additive constant as a multiple integer of h and instead of Eq. (14)

$$[\int m\dot{x}^2\ dt = 2\pi m\,\Sigma_{\alpha=-\infty}^{+\infty}\ |a_\alpha(n)|^2\alpha^2\omega(n).\qquad\qquad [31](14)]$$

one should have had

$$d/dn\ (n\text{h}) = d/dn\cdot\int m\dot{x}^2\ dt\qquad\qquad [32]$$

which means

$$h = 2\pi m\,\Sigma_{\alpha=-\infty}^{+\infty}\ \alpha\ d/dn\ \{\alpha\omega(n)\cdot|a_\alpha(n)|^2\}\qquad\qquad [33](15).$$

[Bohr's **Correspondence Principle** states that *the behavior of systems described by the theory of quantum mechanics (or by the old quantum theory) reproduces*

261

*classical physics in the limit of large quantum numbers*. In other words, it says that for large orbits and for large energies, quantum calculations must agree with classical calculations. The principle was formulated by Niels Bohr in 1920, though he had previously made use of it as early as 1913 in developing his model of the atom. [Bohr, N. (October, 1920), *Über die Serienspektra der Elemente*. (About the serial spectra of the elements.), *Zeit. Physik*, 2, 5, 423-78; https://doi.org/10.1007/BF01329978; translation in *Niels Bohr Collected Works* (1976). Edited by L. Rosenfeld, J. Rud Nielsen. Volume 3, 241-282.]]

Such a relation though fixes the $a_\alpha$ only up to a constant and this indetermination led empirically to the difficulty of half-integer quantum numbers.

[Aitchison, MacManus & Snyder. (2004), p. 6: "... the summation can alternatively be written as over positive values of α, replacing 2πm by 4πm. *In another crucial jump, Heisenberg now replaces the differential in (15) by a difference*, giving
$$h = 4\pi m \sum_{\alpha=0}^{+\infty} \{|a(n, n+\alpha)|^2 \omega(n, n+\alpha) - |a(n, n-\alpha)|^2 \omega(n, n-\alpha)\}.\text{"}]$$

If we ask for a quantum theoretical relation between observable quantities according to Eq. (14) and (15), the missing unambiguity comes out by itself again.

Indeed *only Eq. (15) has a simple quantum theoretical connection to the Kramer's dispersion theory*:

$$h = 4\pi m \sum_{\alpha=0}^{+\infty} \{|a(n, n+\alpha)|^2 \omega(n, n+\alpha) - |a(n, n-\alpha)|^2 \omega(n, n-\alpha)\} \qquad [34](16)$$

[Aitchison, MacManus & Snyder. (2004), p. 6: "As Heisenberg later recalled, he had noticed that 'if I wrote down this [presumably (15) above] and tried to translate it according to the scheme of dispersion theory, I got the Thomas-Kuhn sum rule [which is equation (16)]. And that is the point. Then I thought, 'That is apparently how it is done'.

By 'the scheme of dispersion theory', Heisenberg is referring to what Jammer called *Born's Correspondence Rule*, [Born, M. (December, 1924). *Über Quantenmechanik*, (About quantum mechanics.) *Zeit. Phys.*, 26, 379-95] namely
$$\alpha \, \partial\Phi(n)/\partial n \leftrightarrow \Phi(n) - \Phi(n-\alpha),$$
or rather to its iteration in the form
$$\alpha \, \partial\Phi(n, \alpha)/\partial n \leftrightarrow \Phi(n+\alpha, n) - \Phi(n, n-\alpha)$$
as used in the *Kramers-Heisenberg theory of dispersion*. [Kramers, H. A. & Heisenberg, W. (February, 1925). *Über die Streuung von Strahlung durch Atome*. (On the scattering of radiation by atoms.) *Zeit. Physik*, 31, 681-708; http://dx.doi.org/10.1007/BF02980624]".]

[*The Kramers-Heisenberg dispersion formula is an expression for the cross section for scattering of a photon by an atomic electron*. It was derived before the advent of quantum mechanics by Hendrik Kramers and Werner Heisenberg in 1925, based on the **Correspondence Principle** applied to the classical dispersion formula for light.

The quantum mechanical derivation was given by Paul Dirac in 1927. [Dirac, P. A. M. (March, 1927). *The quantum theory of the emission and absorption of radiation. Roy. Soc. Proc., A*, 114, 767, 243-65.]]

[Fedak, W. A. & Prentis, J. J. (2009), *Appendix: "Reinterpretation 4: Quantization.* "How did Heisenberg reinterpret the old quantization condition in Eq. (A8)?

$$\int mx\dot{}\,dx = nh. \qquad\qquad (A8) \qquad\qquad [27](12)$$

Given the Fourier series in Eq. (A1),

$$x(n, t) = \Sigma_{\alpha = -\infty}^{+\infty}\, a_\alpha\,(n)\, e^{i\omega(n)\alpha t} \qquad\qquad (A1) \qquad\qquad [14](2a)$$

the quantization condition, $nh = \int mx\dot{}^2\, dt$, can be expressed in terms of the Fourier parameters $a_\alpha\,(n)$ and $\omega(n)$ as

$$nh = \int mx\dot{}^2\, dt = 2\pi m\,\Sigma_{\alpha = -\infty}^{+\infty}\, |a_\alpha(n)|^2\alpha^2\omega(n). \qquad\qquad (A9) \qquad\qquad [31](14)$$

For Heisenberg, setting $\int pdx$ (= $\int mx\dot{}\,dx$) equal to an integer multiple of h was an arbitrary rule that did not fit naturally into the dynamical scheme. Because his theory focuses exclusively on transition quantities, Heisenberg needed to translate the old quantum condition that fixes the properties of the **states** to a new condition that fixes the properties of the **transitions between states**. Heisenberg believed that what matters is the difference between $\int pdx$ evaluated for neighboring states: $[\int pdx]_n - [\int pdx]_{n-1}$. He therefore took the derivative of Eq. (A9) with respect to *n* to eliminate the forced *n* dependence and to produce a differential relation that can be reinterpreted as a difference relation between transition quantities. In short, Heisenberg converted

$$h = 2\pi m\,\Sigma_{\alpha = -\infty}^{+\infty}\, \alpha\, d/dn\, (|a_\alpha(n)|^2\alpha\omega(n)) \qquad\qquad (A10)$$

$$\text{to} \qquad h = 4\pi m\,\Sigma_{\alpha = 0}^{+\infty}\, \{|a(n, n + \alpha)|^2\omega(n, n + \alpha)$$

$$- |a(n, n - \alpha)|^2\omega(n, n - \alpha)\} \qquad (A11) \qquad\qquad [34](16)$$

In a sense Heisenberg's "**amplitude condition**" in Eq. (A11) (16) is the counterpart to Bohr's **frequency condition** (Ritz's **frequency combination rule**). Heisenberg's condition relates the **amplitudes** of different lines within an atomic spectrum and Bohr's condition relates the **frequencies**.

***Equation (A11) is the fourth Heisenberg breakthrough.***

Equations (A7) and (A11)

$$[x\ddot{} + f(x) = 0 \qquad\qquad\qquad\qquad [26](11)$$

$$h = 4\pi m\,\Sigma_{\alpha = 0}^{+\infty}\, \{|a(n, n + \alpha)|^2\omega(n, n + \alpha)$$

$$- |a(n, n - \alpha)|^2\omega(n, n - \alpha)\} \qquad (A11) \qquad\qquad [34](16)$$

263

constitute Heisenberg's new mechanics. In principle, these two equations can be solved to find $a(n, n - \alpha)$ and $\omega(n, n - \alpha)$. ***No one before Heisenberg knew how to calculate the amplitude of a quantum jump***. Equations (A2), (A6), (A7), and (A11) define Heisenberg's program for constructing the line spectrum of an atom from the given force on the electron.]

Indeed, this relationship

$$[h = 4\pi m \, \Sigma_{\alpha = 0}{}^{+\infty} \, \{|a(n, n + \alpha)|^2 \omega(n, n + \alpha)$$
$$- |a(n, n - \alpha)|^2 \omega(n, n - \alpha)\} \qquad [34](16)]$$

is sufficient for a unique determination of the $a$'s because the initially undetermined constant in the quantities $a$ will be fixed by itself by the condition which should give a normal state where no more radiation is present. Let the normal state be described by $n_0$, then

$$a(n_0, n_0 - \alpha) = 0 \text{ for } \alpha > 0. \qquad [35]$$

The question about half-integer or integer quantization cannot be present in a quantum mechanics where only relations between observable quantities are used.

Eqs. (11) and (16)

$$[\ddot{x} + f(x) = 0 \qquad [26](11)$$
$$h = 4\pi m \, \Sigma_{\alpha = -\infty}{}^{+\infty} \, \{|a(n, n + \alpha)|^2 \omega(n, n + \alpha)$$
$$- |a(n, n - \alpha)|^2 \omega(n, n - \alpha)\} \qquad [34](16)]$$

together contain, if solvable, ***a complete determination not only of the frequencies and energies, but also of the quantum theoretical transition probabilities***. However, the actual mathematical procedure succeeds only in the easiest cases. A particular complication comes also from systems like the hydrogen atom: since the solutions represent partly periodic and partly aperiodic motions, it has the consequence that the quantum theoretic series (7), (8)

$$[B(n, n - \beta) \, e^{i\omega(n, n - \beta)t} = \Sigma_{\alpha = -\infty}{}^{+\infty} \, A(n, n - \alpha) \, A(n - \alpha, n - \beta) \, e^{i\omega(n, n - \beta)t} \qquad [20](7)$$
$$\text{or} \qquad = \int_{-\infty}{}^{+\infty} \, A(n, n - \alpha) \, A(n - \alpha, n - \beta) \, e^{i\omega(n, n - \beta)t} \, d\alpha. \qquad [21](8)]$$

and Eq. (16) always fall in both the sum and the integral case. ***Quantum mechanically, it is not possible to divide "periodic and aperiodic motions"***.

Despite that, one might see Eq. (11) and Eq. (16) at least in principle as a satisfactory solution of the mechanical problem, if it is possible to show that this solution coincides (or is not in contradiction) with the until now known quantum mechanical relationships and that a small perturbation of a mechanical problem gives rise to additional orders in the energies or frequencies respectively which correspond to the expressions found by Kramers and Born (in contrast to which would have led to the classical theory). Further, one must investigate if in general Eq. (11) in the suggested quantum theoretical interpretation corresponds an energy integral $m \, \dot{x}^2/2 + U(x) = \text{const.}$ and if such obtained energy (analogously as classically holds $\nu = \partial W/\partial J$) the relation $\Delta W = h\nu$ is sufficient. A general answer to these questions might demonstrate the coherence of the present experiments and lead to a quantum mechanics which operates only with observable quantities. Apart from a general relationship between the **Kramer's dispersion formula** and Eq. (11) and (16), we can only answer the above stated questions in very special solvable cases through simple recursion.

The general connection between **Kramer's dispersion theory** and our Eqs. (11) and (16)

$$[x'' + f(x) = 0 \qquad\qquad\qquad [26](11)$$

$$h = 4\pi m \sum_{\alpha=-\infty}^{+\infty} \{|a(n, n+\alpha)|^2 \omega(n, n+\alpha)$$
$$- |a(n, n-\alpha)|^2 \omega(n, n-\alpha)\} \qquad\qquad [34](16)]$$

consists in the fact that in Eq. (11) (more precisely from its quantum-theoretical analog) one finds, just as in the classical theory, that the oscillating electron behaves like a free electron when acted upon by light, of much higher frequency than any eigenfrequency of the system. This result follows also from Kramers' dispersion theory when Eq. (16) is taken into account. Indeed, Kramers finds for **moment** induced by a wave of the form $E \cos 2\pi vt$

$$M = 2e^2 E \cos(2\pi vt)/h \sum_{\alpha=0}^{+\infty} [|a(n, n+\alpha)|^2 v(n, n+\alpha)/\{v^2 (n, n+\alpha) - v^2\}$$
$$- |a(n, n-\alpha)|^2 v(n, n-\alpha)/\{v^2 (n, n-\alpha) - v^2\}] \qquad [36]$$

So that for $v > v(n, n+\alpha)$

$$M = 2e^2 E \cos(2\pi vt)/v^2 h \sum_{\alpha=0}^{+\infty} [|a(n, n+\alpha)|^2 v(n, n+\alpha)$$
$$- |a(n, n-\alpha)|^2 v(n, n-\alpha)] \qquad\qquad [37]$$

which using Eq. (16) becomes

$$M = - e^2 E \cos(2\pi vt)/v^2 4\pi^2 m. \qquad\qquad [38]$$

§ 3. In the following, as the simplest example, the anharmonic oscillator will be treated:

$$x'' + \omega_0^2 x + \lambda x^2 = 0 \qquad\qquad [39](17)$$

Classically, this equation can be satisfied by an Anzatz for the solution of the form:

$$x = \lambda a_0 + a_1 \cos \omega t + \lambda a_2 \cos 2\omega t + \lambda^2 a3 \cos 3\omega t + \ldots + \lambda^{\tau-1} a_\tau \cos \tau \omega t \quad [40]$$

where the $a$ are power series in $\lambda$, the first terms of which are independent from $\lambda$.

[Aitchison, MacManus & Snyder. (2004), pp. 8-9: "… Heisenberg proposes to seek a solution analogous to this, using the 'representation' of x(t) in terms of the quantities $\{a(n, n - \alpha) e^{i\omega(n, n - \alpha)t}\}$. It seems reasonable to assume that, as the index $\alpha$ increases away from zero, in integer steps, each successive amplitude will (to leading order in $\lambda$) be suppressed by an additional power of $\lambda$, as in the classical case. Thus, Heisenberg suggests that, in the quantum case, x(t) should be represented by terms of the form

265

$\lambda a(n, n)$ ; $a(n, n-1) \cos \omega(n, n-1)t$ ; $\lambda a(n, n-2)t$ ; ... $\lambda^{\tau-1} a(n, n-\tau)$ $\cos \omega(n, n-\tau)t$ ...

where,

$a(n, n) = a^{(0)}(n, \text{n}) + \lambda a^{(1)}(n, n) + \lambda^2 a^{(2)}(n, n) + \ldots$

$a(n, n-1) = a^{(0)}(n, n-1) + \lambda a^{(1)}(n, n-1) + \lambda^2 a^{(2)}(n, n-1) + \ldots$

and so on, and

$\omega(n, n-\alpha) = \omega^{(0)}(n, n-\alpha) + \lambda\omega^{(1)}(n, n-\alpha) + \lambda^2\omega^{(2)}(n, n-\alpha) + \ldots$

As Born and Jordan pointed out*, some use of 'correspondence' arguments has been made here, in assuming that, as $\lambda \to 0$, only transitions between adjacent states are possible.

* Born, M. & Jordan, P. (December, 1925). *Zur Quantenmechanik.* (On Quantum Mechanics.) *Zeit. Phys.*, 34, 858-88.

***Heisenberg now simply writes down what he asserts to be the quantum versions.*"]**

[For derivation relating the amplitudes $a(n, n-\alpha)$ to the corresponding quantities $\lambda^{\tau-1}a(n, n-\tau)$ see Aitchison, MacManus & Snyder. (2004), pp. 10-11.]

Quantum theoretically, we try to find an analogous expression representing x with terms of the form

$\lambda a(n, n)$ ; $a(n, n-1) \cos \omega(n, n-1)t$ ; $\lambda a(n, n-2)t$ ;

$\ldots \lambda^{\tau-1}a(n, n-\tau) \cos \omega(n, n-\tau)t \ldots$      [41]

The recursion formulae for the determination of $a$ and $\omega$ (up to order $\lambda$) according to Eq. (3), (4) or Eq. (7), (8) are:

*Classically*

... [Original page 888.]

[Aitchison, MacManus & Snyder. (2004). *Understanding Heisenberg's 'Magical' Paper of July 1925: a New Look at the Calculational Details.* § 4. *Conclusion*, pages 18-19: "The fact is, Heisenberg's 'amplitude calculus' works: at least for the simple one-dimensional problems on which he tried it out, it is an eminently practical procedure, requiring no sophisticated mathematical knowledge to implement. Since it uses the correct equations of motion, and incorporates the fundamental commutator (11)

    $[\mathbf{xp} - \mathbf{px}](n, n) = ih$           (11)]

via the '***quantum condition***' (16)

    $[h = 4\pi m \Sigma_{\alpha=-\infty}^{+\infty} \{|a(n, n+\alpha)|^2\omega(n, n+\alpha)$

          $- |a(n, n-\alpha)|^2\omega(n, n-\alpha)\}$      [34][16]

the answers obtained are completely correct, in the sense of agreeing with conventional quantum mechanics.

... The multiplication law (10)

$$[B(n, n - \beta) = \Sigma_\alpha A(n, n - \alpha) A(n - \alpha, n - \beta) \ (10)]$$

has a convincing physical rationale, even for those who (like Heisenberg) do not recognize it as matrix multiplication ... . The simple examples of this introduce the fundamental quantum idea that a transition from one state to the other occurs via all possible intermediate states, something which can take time to emerge in the traditional wave-mechanical approach. ... Finally, the type of perturbation theory employed here ... [is] more easily related to the classical analysis than is conventional quantum-mechanical perturbation theory ... .It is of course true that many important problems in quantum mechanics are much more conveniently handled in the wave-mechanical formalism ... [in which] the ... 'matrix elements' are the elements of Heisenberg's matrices."]

...

[Original page 890 ff; translation page 13 ff:]

... Furthermore, the energy calculated from Eq. 27 satisfies the relation (cf. Eq. 24):

$$\omega(n, n - 1)/2\pi = 1/h \ . \ [W(n) - W(n - 1)], \tag{62}$$

which can be regarded as a necessary condition for the possibility of a determination of the transition probabilities according to Eqs. 11 and 16

$$[\ddot{x} + f(x) = 0 \tag{[26](11)}$$

$$h = 4\pi m \ \Sigma_{\alpha = -\infty}^{+\infty} \ \{|a(n, n + \alpha)|^2 \omega(n, n + \alpha)$$
$$- |a(n, n - \alpha)|^2 \omega(n, n - \alpha)\} \tag{[34](16)}$$

...

Whether a method to determine quantum-theoretical data using relations between observable quantities as proposed here, can be regarded as satisfactory in principle, or whether this method indeed after all represents a too rough approach to the physical problem of constructing a theoretical quantum mechanics, an obviously very involved problem at the moment, can be decided only by a deeper mathematical investigation of the method which has been very superficially employed here.

## Two new quantum experiments just proved Einstein wrong.

Darren Orf, January 8, 2026.

In 1927, Albert Einstein and Neils Bohr debated the nature of what's known as *complementarity*—the idea that a photon's dual wave-like and particle natures can't be measured at the same time.

Now, *two unrelated-but-similar experiments confirm that, in a double-slit experiment, detection of a photon's path (its particle nature) would wash out its interference pattern (its wave nature), proving Bohr right*.

Using these experimental set-ups could help scientists probe other questions of the quantum realm, including the nitty-gritty details of the interaction between decoherence and entanglement.

The 1920s were an incredible time for science. The decade—prefaced by the stunning confirmation of Albert Einstein's General Theory of Relativity, thanks to the solar eclipse in May of 1919—saw the confirmation of the existence of other galaxies, the discovery of penicillin, and the birth of modern quantum mechanics. A complete banger of a decade, scientifically speaking.

However, this plethora of science knowledge came with some understandable disagreements. For example, it was becoming increasingly clear that Einstein's elegant theories that appeared to govern the universe on a macro scale didn't perfectly square with the quantum weirdness being discovered at the subatomic scale. One of the most famous debates of the decade involved arguably two of the greatest physicists of the century: Einstein himself and Neils Bohr, the world-famous Danish theoretical physicist who is a foundational figure in quantum mechanics. The matter in question was the idea of "*complementarity*," which stated (in simple terms) *that a photon's particle and wave nature can't be measured at the same time*.

Einstein believed that, by using a complicated *double-slit experiment* involving a spring that recoils when a photon enters, one could learn about that photon's particle nature based on which slit it entered and its wave nature based on the interference pattern. Bohr disagreed, stating that the quantum mechanical uncertainty principle essentially made this detection impossible.

Now, in two back-to-back experiments—both published in the journal *Physical Review Letters*—scientists at MIT and the University of Science and Technology of China (USTC) have seemingly settled the matter in favor of Bohr. Earlier this year, Wolfgang Ketterle (along with colleagues at MIT) created what the study authors call an "idealized version of the double slit experiment," in which they used individual atoms as slits and weak beams of light so that those atoms scattered only one photon. Ketterle and his team essentially discovered an inverse relationship between the photon's two states: the more information that was obtained about the photon's path (i.e. particle nature), the lower the visibility of the interference pattern (i.e. wave nature).

The USTC experiment, which probes the same debate through different means, used a rubidium atom held in place by optical tweezers. The team then used lasers and electromagnetic forces to control the atom's quantum properties and scattered light in two directions. They also found that Bohr's assertion held out.

"Seeing quantum mechanics 'in action' at this fundamental level is simply breathtaking," Chao-Yang Lu told New Scientist. "Bohr's counterargument was brilliant. But the thought experiment remained theoretical for almost a century."

Lu and his team hope to use their experimental set-up to explore more, lesser-known aspects of the quantum realm, including the nature of the relationship between decoherence and entanglement.

## Chadwick, J. (October 20, 1891 – July 24, 1974)

Chadwick was a British experimental physicist who received the Nobel Prize in Physics in 1935 for his ***discovery of the neutron***. In 1941, he wrote the final draft of the MAUD Report, which inspired the U.S. government to begin serious atomic bomb research efforts. He was the head of the British team that worked on the Manhattan Project during World War II. He was knighted in Britain in 1945 for his achievements in nuclear physics.

Chadwick was born on October 20, 1891 in Cheshire, England, the first child of John Joseph Chadwick, a cotton spinner, and Anne Mary Knowles, a domestic servant. He was named James after his paternal grandfather. In 1895, his parents moved to Manchester, leaving him in the care of his maternal grandparents. He went to Bollington Cross Primary School, and was offered a scholarship to Manchester Grammar School, which his family had to turn down as they could not afford the small fees that still had to be paid. Instead, he attended the Central Grammar School for Boys in Manchester, rejoining his parents there. He now had two younger brothers, Harry and Hubert; a sister had died in infancy. At the age of 16, he sat two examinations for university scholarships, and won both of them.

Chadwick chose to attend Victoria University of Manchester, which he entered in 1908. He meant to study mathematics, but enrolled in physics by mistake. Like most students, he lived at home, walking the 4 miles (6.4 km) to the university and back each day. At the end of his first year, he was awarded a Heginbottom Scholarship to study physics. ***The physics department was headed by Ernest Rutherford***, who assigned research projects to final-year students, and he instructed Chadwick to devise a means of comparing the amount of radioactive energy of two different sources. The idea was that they could be measured in terms of the activity of 1 gram (0.035 oz) of radium, a unit of measurement which would become known as the curie. Rutherford's suggested approach was unworkable—something Chadwick knew but was afraid to tell Rutherford—so Chadwick pressed on, and eventually devised the required method. The results became Chadwick's first paper, which, co-authored with Rutherford, was published in 1912. [Rutherford, E & Chadwick, J. (1912). *"A Balance Method for Comparison of Quantities of Radium and Some of its Applications"*. *Proceedings of the Physical Society*. 24 (1): 141–151; doi:10.1088/1478-7814/24/1/320] He graduated in 1911.

Chadwick graduated with First Class Honors from the Victoria University of Manchester in 1911, and continued to study under Rutherford.  Having devised a means of measuring gamma radiation, Chadwick proceeded to measure the absorption of gamma rays by various gases and liquids. This time the resulting paper was published under his name alone. He was awarded his M.Sc. in 1912, and was appointed a Beyer Fellow. The following year, he was awarded an 1851 Exhibition Scholarship, which allowed him to study and research at a university in continental Europe. In 1913, he chose to go to the Physikalisch-Technische Reichsanstalt in Berlin to study beta radiation under Hans Geiger. Using Geiger's recently developed Geiger counter, which provided more accuracy than the earlier photographic techniques, he was able to demonstrate that beta radiation did not produce discrete lines, as has been previously thought, but rather a continuous spectrum with peaks in certain regions. On a visit to Geiger's laboratory, Albert Einstein told Chadwick that: "I can explain either of these things, but I can't explain them both at the same time." The continuous spectrum would remain an unexplained phenomenon for many years. Still in Germany when World War I broke out in Europe, he spent the next four years in the Ruhleben internment camp, where he was allowed to set up a laboratory in the stables and conduct scientific experiments using improvised materials such as radioactive toothpaste. With the help of Charles Drummond Ellis, he worked on the ionization of phosphorus, and the photochemical reaction of carbon monoxide and chlorine. He was released after the Armistice with Germany came into effect in November 1918, and returned to his parents' home in Manchester, where he wrote up his findings over the previous four years for the 1851 Exhibition commissioners.

Rutherford gave Chadwick a part-time teaching position at Manchester, allowing him to continue research. He looked at the nuclear charge of platinum, silver, and copper, and experimentally found that this was the same as the atomic number within an error of less than 1.5 per cent. In April 1919, **Rutherford became director of the Cavendish Laboratory at the University of Cambridge**, and Chadwick joined him there a few months later. Chadwick was awarded a Clerk Maxwell Studentship in 1920, and enrolled as a Ph.D. student at Gonville and Caius College, Cambridge, where Chadwick earned his Doctor of Philosophy degree under Rutherford's supervision. The first half of his thesis was his work with atomic numbers; in the second, he looked at the forces inside the nucleus. He was awarded his degree in June 1921, and became a Fellow of Gonville and Caius College in November.

In 1921, while working with Niels Bohr, Rutherford theorized about the existence of **neutrons**, which could somehow compensate for the repelling effect of the positive charges of protons by causing an attractive nuclear force and thus keep the nuclei from flying apart, due to the repulsion between protons. Rutherford's conjecture and the hypothetical "**neutron**" were not widely accepted. In his 1931 monograph on the Constitution of Atomic Nuclei and Radioactivity, George Gamow, then at the Institute for Theoretical Physics in Copenhagen, did not mention the neutron. At the time of their 1932 measurements in Paris that would lead to the discovery of the neutron, Irène Joliot-Curie and Frédéric Joliot were unaware of the conjecture.

Rutherford and Chadwick immediately began an experimental program at the Cavendish Laboratory in Cambridge to search for the **neutron**. The experiments continued throughout the 1920s without success.

Chadwick's Clerk Maxwell Studentship expired in 1923, and he was succeeded by the Russian physicist Pyotr Kapitza. The Chairman of the Advisory Council of the Department of Scientific and Industrial Research, Sir William McCormick, arranged for Chadwick to become Rutherford's assistant director of research. He was Rutherford's assistant director of research at the Cavendish Laboratory for over a decade at a time when it was one of the world's foremost centers for the study of physics. In this role, he helped Rutherford select Ph.D. students. Over the next few years these would include John Cockcroft, Norman Feather, and Mark Oliphant, who would become firm friends with Chadwick. As many students had no idea what they wanted to research, Rutherford and Chadwick would suggest topics. He edited all the papers produced by the laboratory.

Throughout the 1920s, physicists assumed that the atomic nucleus was composed of protons and "nuclear electrons". Under this hypothesis, the nitrogen-14 ($^{14}$N) nucleus would be composed of 14 protons and 7 electrons, so that it would have a net charge of +7 elementary charge units and a mass of 14 atomic mass units. This nucleus would also be orbited by another 7 electrons, termed "external electrons" by Rutherford, to complete the $^{14}$N atom. However, problems with the hypothesis soon became apparent.

In 1925, Chadwick met Aileen Stewart-Brown, the daughter of a Liverpool stockbroker. The two were married in August 1925, with Kapitza as best man. The couple had twin daughters, Joanna and Judith, who were born in February 1927.

In his research, Chadwick continued to probe the nucleus. In 1925, the concept of spin had allowed physicists to explain the Zeeman effect, but it also created unexplained anomalies. At the time it was believed that the nucleus consisted of protons and electrons, so nitrogen's nucleus, for example, with a mass number of 14, was assumed to contain 14 protons and 7 electrons. This gave it the right mass and charge, but the wrong spin.

Ralph Kronig pointed out in 1926 that the observed hyperfine structure of atomic spectra was inconsistent with the proton–electron hypothesis. This structure is caused by the influence of the nucleus on the dynamics of orbiting electrons. The magnetic moments of supposed "nuclear electrons" should produce hyperfine spectral line splittings similar to the Zeeman effect, but no such effects were observed. It seemed that the magnetic moment of the electron vanished when it was within the nucleus.

In 1927, Charles Ellis and W. Wooster at the Cavendish Laboratory measured the energies of β-decay electrons. They found that the distribution of energies from any particular radioactive nuclei was broad and continuous, a result that contrasted notably with the distinct energy values observed in alpha and gamma decay. Furthermore, the continuous energy distribution seemed to indicate that energy was not conserved by this "***nuclear electrons***" process.

The Klein paradox, discovered by Oskar Klein in 1928, presented further quantum mechanical objections to the notion of an electron confined within a nucleus. Derived from the Dirac equation, this clear and precise paradox suggested that an electron approaching a high potential barrier has a high probability of passing through the barrier by a pair creation process. Apparently, an electron could not be confined within a nucleus by any potential well. The meaning of this paradox was widely debated at the time.

At a conference at Cambridge on beta particles and gamma rays in 1928, Chadwick met Geiger again. Geiger had brought with him a new model of his Geiger counter, which had been improved by his student, Walther Müller. Chadwick had not used one since the war, and the new Geiger–Müller counter was potentially a major improvement over the scintillation techniques then in use at Cambridge, which relied on the human eye for observation. The major drawback with it was that it detected alpha, beta and gamma radiation, and radium, which the Cavendish Laboratory normally used in its experiments, emitted all three, and was therefore unsuitable for what Chadwick had in mind. However, polonium is an alpha emitter, and Lise Meitner sent Chadwick about 2 millicuries (about 0.5 µg) from Germany.

While on a visit to Utrecht University in 1928, Kronig learned of a surprising aspect of the rotational spectrum of $N_2+$. The precision measurement made by Leonard Ornstein, the director of Utrecht's Physical Laboratory, showed that the spin of a nitrogen nucleus must be equal to one. However, if the nitrogen-14 ($^{14}N$) nucleus was composed of 14 protons and 7 electrons, an odd number of spin-1/2 particles, then the resultant nuclear spin should be half-integer. Kronig therefore suggested the possibility that "protons and electrons do not retain their identity to the extent they do outside the nucleus".

Observations of the rotational energy levels of diatomic molecules using Raman spectroscopy by Franco Rasetti in 1929 were inconsistent with the statistics expected from the proton–electron hypothesis. Rasetti obtained band spectra for $H_2$ and $N_2$ molecules. While the lines for both diatomic molecules showed alternation in intensity between light and dark, the pattern of alternation for $H_2$ is opposite to that of the $N_2$. After carefully analyzing these experimental results, German physicists Walter Heitler and Gerhard Herzberg showed that the hydrogen nuclei obey Fermi statistics and the nitrogen nuclei obey Bose statistics. However, a then unpublished result of Eugene Wigner showed that a composite system with an odd number of spin-1/2 particles must obey Fermi statistics; a system with an even number of spin-1/2 particle obeys Bose statistics. If the nitrogen nucleus had 21 particles, it should obey Fermi statistics, contrary to fact. Thus, Heitler and Herzberg concluded: "the electron in the nucleus ... loses its ability to determine the statistics of the nucleus".

In 1929, Bohr proposed to modify the law of energy conservation to account for the continuous energy distribution, a proposal that earned the support of Werner Heisenberg. Such considerations were apparently reasonable, inasmuch as the laws of quantum mechanics had so recently overturned the laws of classical mechanics.

*By about 1930, it was generally recognized that it was difficult to reconcile the proton–electron model for nuclei with the Heisenberg uncertainty relation of quantum mechanics*. This relation, $\Delta x \cdot \Delta p \geq 1/2\hbar$, implies that an electron confined to a region the size of an atomic nucleus typically has a kinetic energy of about 40 MeV, which is larger than the observed energy of beta particles emitted from the nucleus. Such energy is also much larger than the binding energy of nucleons, which Aston and others had shown to be less than 9 MeV per nucleon.

While all of these considerations did not "prove" an *electron* could not exist in the nucleus, they were confusing and challenging for physicists to interpret. Many theories were invented to explain how the above arguments could be wrong. In his 1931 monograph, Gamow summarized all of these contradictions, marking the statements regarding electrons in the nucleus with warning symbols.

**Discovery of the neutron**

In 1930, Walther Bothe and his collaborator Herbert Becker in Giessen, Germany found that if the energetic alpha particles emitted from polonium fell on certain light elements, specifically beryllium ($^{9}_{4}Be$), boron ($^{11}_{5}B$), or lithium ($^{7}_{3}Li$), an unusually penetrating radiation was produced. Beryllium produced the most intense radiation. Polonium is highly radioactive, producing energetic alpha radiation, and it was commonly used for scattering experiments at the time. Alpha radiation can be influenced by an electric field because it is composed of charged particles. The observed penetrating radiation was not influenced by an electric field, however, so it was thought to be gamma radiation. The radiation was more penetrating than any gamma rays known, and the details of experimental results were difficult to interpret.

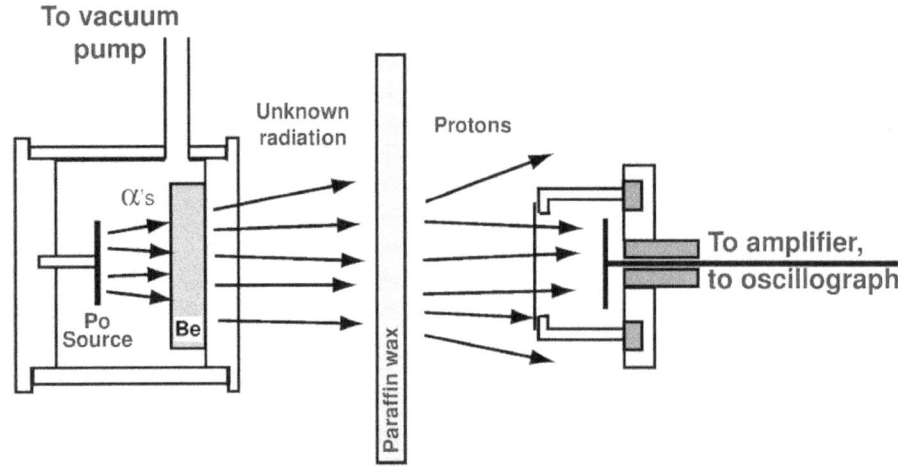

A schematic diagram of the experiment used to discover the neutron in 1932. At left, a polonium source was used to irradiate beryllium with alpha particles, which induced an uncharged radiation. When this radiation struck paraffin wax, protons were ejected. The protons were observed using a small ionization chamber. Adapted from Chadwick, J. (1932). *"Existence of a Neutron".* *Proceedings of the Royal Society A.* 136 (830): 692–708; doi:10.1098/rspa.1932.0112.

In Germany, Walther Bothe and his student, Herbert Becker, had used polonium to bombard beryllium with alpha particles, producing an unusual form of radiation. Chadwick had his Australian 1851 Exhibition scholar, Hugh Webster, duplicate their results. ***To Chadwick, this was evidence of something that he and Rutherford had been hypothesizing for years: the neutron, a theoretical nuclear particle with no electric charge***.

Then in January 1932, Feather drew Chadwick's attention to another surprising result. Frédéric and Irène Joliot-Curie in Paris had shown that if what they thought was gamma radiation using polonium and beryllium as a source fell on paraffin wax, or any other hydrogen-containing compound, it ejected protons of very high energy (5 MeV). This observation was not in itself inconsistent with the assumed gamma ray nature of the new radiation, but that interpretation (Compton scattering) had a logical problem. From energy and momentum considerations, a gamma ray would have to have impossibly high energy (50 MeV) to scatter a massive proton. In Rome, the young physicist Ettore Majorana declared that the manner in which the new radiation interacted with protons required a neutral particle as heavy as a proton, but declined to publish his result despite the encouragement of Enrico Fermi.

Rutherford and Chadwick also disagreed; protons were too heavy for that. But ***neutrons*** would need only a small amount of energy to achieve the same effect. In Rome, Ettore

Majorana came to the same conclusion: the Joliot-Curies had discovered the neutron but did not know it.

On hearing of the Paris results in 1932, Rutherford and James Chadwick at the Cavendish Laboratory also did not believe the gamma ray hypothesis since it failed to conserve energy. Chadwick dropped all his other responsibilities to concentrate on proving the existence of the **neutron**, assisted by Feather and frequently working late at night. The previous year, Chadwick, J.E.R. Constable, and E.C. Pollard had already conducted experiments on disintegrating light elements using alpha radiation from polonium using a simple apparatus that consisted of a cylinder containing a polonium source and the target. The resulting radiation could then be directed at a material such as paraffin wax. They had also developed more accurate and efficient methods for detecting, counting, and recording the ejected protons. The displaced particles would go into a small ionization chamber where they could be detected with an oscilloscope. Chadwick repeated the creation of the radiation using beryllium to absorb the alpha particles:

$$^9Be + {}^4He\ (\alpha) \rightarrow {}^{12}C + 1n.$$

Assisted by Norman Feather, Chadwick quickly performed a series of experiments showing that the gamma ray hypothesis was untenable. Then, following the Paris experiment, he aimed the radiation at paraffin wax, a hydrocarbon high in hydrogen content, hence offering a target dense with protons. As in the Paris experiment, the radiation energetically scattered some of the protons. Chadwick measured the range of these protons and also measured how the new radiation impacted the atoms of various gases. Measurements of the recoil energy showed that the mass of the radiation particles must be similar to the mass of the proton: the new radiation could not consist of gamma rays. *Uncharged particles with about the same mass to the proton matched the properties Rutherford described in 1920 and which had later been called neutrons.*

In February 1932, after only about two weeks of experimentation with **neutrons**, Chadwick sent a letter to Nature titled "*Possible Existence of a Neutron*". He communicated his findings in detail in an article sent to *Proceedings of the Royal Society A* titled "*The Existence of a Neutron*" in May. His **discovery of the neutron was a milestone in understanding the nucleus**. Reading Chadwick's paper, Robert Bacher and Edward Condon realized that anomalies in the then-current theory, like the spin of nitrogen, would be resolved if the **neutron** has a spin of 1/2 and that a nitrogen nucleus consisted of seven protons and seven neutrons.

*Chadwick won the Nobel Prize in Physics in 1935 for his discovery of the neutron in 1932.*

The year 1932 was later referred to as the "annus mirabilis" for nuclear physics in the Cavendish Laboratory, with discoveries of the *neutron*, artificial nuclear disintegration by the Cockcroft–Walton particle accelerator, and the positron.

The theoretical physicists Niels Bohr and Werner Heisenberg considered whether the *neutron* could be a fundamental nuclear particle like the proton and electron, rather than a proton–electron pair. *Heisenberg showed that the neutron was best described as a new nuclear particle*, but its exact nature remained unclear.

> [Heisenberg, W. (1932). *About the construction of atomic nuclei. I. Z. Physik*, 77, 1–11. https://doi.org/10.1007/BF01342433. The consequences of the assumption that the atomic nuclei are made up of protons and neutrons without the participation of electrons are discussed. § 1. The Hamiltonian function of the nucleus. § 2. The relationship between charge and mass and the special stability of the He nucleus. § 3 to 5; Stability of the nuclei and radioactive decay series. § 6. Discussion of the basic physical assumptions.]

In his 1933 Bakerian Lecture, Chadwick estimated that a *neutron* had a mass of about 1.0067 Da. Since a proton and an electron had a combined mass of 1.0078 u, this implied the neutron as a proton–electron composite had a binding energy of about 2 MeV, which sounded reasonable, *although it was hard to understand how a particle with so little binding energy could be stable*.

Chadwick *followed his discovery of the neutron by measuring its mass*.

Estimating such a small mass difference required challenging precise measurements, however, and several conflicting results were obtained in 1933–4. By bombarding boron with alpha particles, Frédéric and Irène Joliot-Curie obtained a large value for the *mass of a neutron*, but Ernest Lawrence's team at the University of California produced a small one. Then Maurice Goldhaber, a refugee from Nazi Germany and a graduate student at the Cavendish Laboratory, suggested to Chadwick that *deuterons* could be photo-disintegrated by the 2.6 MeV gamma rays of $^{208}$Tl (then known as thorium C"):

$$^{2}_{1}D \quad + \quad \gamma \quad \rightarrow \quad ^{1}_{1}H \quad + \quad n$$

An accurate value for the *mass of the neutron* could be determined from this process. Chadwick and Goldhaber tried this and found that it worked. They measured the kinetic energy of the proton produced as 1.05 MeV, leaving the mass of the neutron as the unknown in the equation. Chadwick and Goldhaber calculated that it was either 1.0084 or 1.0090 atomic units, depending on the values used for the masses of the proton and *deuteron*. (The modern accepted value for the mass of the neutron is 1.00866 Da.) The mass of the neutron was too large to be a proton–electron pair.

*In 1935 Chadwick and his doctoral student Maurice Goldhaber reported the first accurate measurement of the mass of the neutron.* They used the 2.6 MeV gamma rays of Thallium-208 ($^{208}$Tl) (then known as "thorium C") to photo-disintegrate the *deuteron*.

$$^{2}_{1}D \quad + \quad \gamma \quad \rightarrow \quad ^{1}_{1}H \quad + \quad n$$

In this reaction, the resulting proton and neutron have about equal kinetic energy, since their masses are about equal. The kinetic energy of the resulting proton could be measured (0.24 MeV), and therefore the deuteron's binding energy could be determined (2.6 MeV − 2(0.24 MeV) = 2.1 MeV, or 0.0023 Da). The neutron's mass could then be determined by the simple mass balance

$$m_d \quad + \quad b.e. \quad = \quad m_p \quad + \quad m_n$$

where $m_{d,p,n}$ refer to the deuteron, proton, or neutron mass, and "b.e." is the binding energy. The masses of the deuteron and proton were known; Chadwick and Goldhaber used values 2.0142 Da and 1.0081 Da, respectively. They found that *the neutron's mass was slightly greater than the mass of the proton* 1.0084 Da or 1.0090 Da, depending on the precise value used for the deuteron mass. *The mass of the neutron was too large to be a proton–electron composite, and the neutron was therefore identified as an elementary particle.* Chadwick and Goldhaber predicted that a free neutron would be able to decay into a proton, electron, and neutrino (free neutron decay).

For his discovery of the neutron, Chadwick was awarded the Hughes Medal in 1932, the Nobel Prize in Physics in 1935, the Copley Medal in 1950, and the Franklin Medal in 1951. His discovery of the neutron made it possible to produce elements heavier than uranium in the laboratory by the capture of slow neutrons followed by beta decay. Unlike the positively charged alpha particles, which are repelled by the electrical forces present in the nuclei of other atoms, neutrons do not need to overcome any Coulomb barrier, and can therefore penetrate and enter the nuclei of even the heaviest elements such as uranium. This inspired Enrico Fermi to investigate the nuclear reactions brought about by collisions of nuclei with slow neutrons, work for which Fermi would receive the Nobel Prize in 1938.

Wolfgang Pauli proposed another kind of particle on 4 December 4, 1930 to explain the continuous spectrum of beta radiation that Chadwick had reported in 1914. Since not all of the energy of beta radiation could be accounted for, the law of conservation of energy appeared to be violated, but Pauli argued that this could be redressed if another, undiscovered, particle was involved. Pauli also called this particle a neutron, but it was clearly not the same particle as Chadwick's neutron. Fermi renamed it the **neutrino** (Italian for "little neutron"). In 1934, Fermi proposed his theory of beta decay which explained that the electrons emitted from the nucleus were created by the decay of a neutron into a proton, an electron, and a neutrino. The neutrino could account for the missing energy, but a particle with little mass and no electric charge was difficult to observe. Rudolf Peierls and Hans Bethe calculated that neutrinos could easily pass through the Earth, so the chances of detecting them were slim. [Frederick Reines and Clyde Cowan would confirm the neutrino on June 14, 1956 by placing a detector within a large antineutrino flux from a nearby nuclear reactor.

With the onset of the Great Depression in the United Kingdom, the government became more parsimonious with funding for science. At the same time, Lawrence's recent invention, the **cyclotron**, promised to revolutionize experimental nuclear physics, and Chadwick felt that the Cavendish Laboratory would fall behind unless it also acquired one. He therefore chafed under Rutherford, who clung to the belief that good nuclear physics could still be done without large, expensive equipment, and turned down the request for a cyclotron.

Chadwick left the Cavendish Laboratory in 1935 to become a professor of physics at the University of Liverpool, where he overhauled an antiquated laboratory and, by installing a cyclotron, made it an important center for the study of nuclear physics. Chadwick was a critic of Big Science in general, and Lawrence in particular, whose approach he considered careless and focused on technology at the expense of science. When Lawrence proposed the existence of a new and hitherto unknown particle that he claimed was a possible source of limitless energy at the Solvay Conference in 1933, Chadwick responded that the results were more likely attributable to contamination of the equipment. While Lawrence rechecked his results at Berkeley only to find that Chadwick was correct, Rutherford and Oliphant conducted an investigation at the Cavendish that found that deuterium fuses to form helium-3, thereby causing the effect that Lawrence had observed. This was another major discovery, but the Oliphant–Rutherford particle accelerator was an expensive state-of-the-art piece of equipment.

In March 1935, Chadwick received an offer of the Lyon Jones Chair of Physics at the University of Liverpool, in his wife's home town, to succeed Lionel Wilberforce. The laboratory was so antiquated that it still ran on direct current electricity, but Chadwick seized the opportunity, assuming the chair on October 1 1935. The university's prestige was soon bolstered by Chadwick's Nobel Prize, which was announced in November 1935. Chadwick set about acquiring a cyclotron for Liverpool. He started by spending £700 to refurbish the antiquated laboratories at Liverpool, so some components could be made in-house. He was able to persuade the university to provide £2,000 and obtained a grant for another £2,000 from the Royal Society. To build his cyclotron, Chadwick brought in two young experts, Bernard Kinsey and Harold Walke, who had worked with Lawrence at the University of California. A local cable manufacturer donated the copper conductor for the coils. The cyclotron's 50-ton magnet was manufactured in Trafford Park by Metropolitan-Vickers, which also made the vacuum chamber. The cyclotron was completely installed and running in July 1939. The total cost of £5,184 was more than Chadwick had received from the university and the Royal Society, so Chadwick paid the rest from his 159,917 kr (£8,243) Nobel Prize money.

During the Second World War, Chadwick carried out research as part of the Tube Alloys project to build an atom bomb, while his Manchester lab and environs were harassed by Luftwaffe bombing. When the Quebec Agreement merged his project with the American Manhattan Project, he became part of the British Mission, and worked at the Los Alamos Laboratory and in Washington, D.C. He surprised everyone by earning the almost-complete trust of project director Leslie R. Groves, Jr.

For his efforts, Chadwick received a knighthood in the New Year Honors on 1 January 1,1945. In July 1945, he viewed the Trinity nuclear test. After this, he served as the British scientific advisor to the United Nations Atomic Energy Commission. Uncomfortable with the trend toward Big Science, he returned to Cambridge where he became Master of Gonville and Caius College in 1948.

Chadwick did not believe that there was any likelihood of another war with Germany in 1939, and took his family for a holiday on a remote lake in northern Sweden. The news of the outbreak of the Second World War therefore came as a shock. Determined not to spend another war in an internment camp, Chadwick made his way to Stockholm as fast as he could, but when he arrived there with his family, he found that all air traffic between Stockholm and London had been suspended. They made their way back to England on a tramp steamer. When he reached Liverpool, Chadwick found Joseph Rotblat, a Polish postdoctoral fellow who had come to work with the cyclotron, was now destitute, as he was cut off from funds from Poland. Chadwick promptly hired Rotblat as a lecturer, despite his poor grasp of English.

In October 1939, Chadwick received a letter from Edward Appleton, the Secretary of the Department of Scientific and Industrial Research, asking for his opinion on the feasibility of an atomic bomb. Chadwick responded cautiously. He did not dismiss the possibility, but carefully went over the many theoretical and practical difficulties involved. Chadwick decided to investigate the properties of uranium oxide further with Rotblat. In March 1940, Otto Frisch and Rudolf Peierls at the University of Birmingham re-examined the theoretical issues involved in a paper that became known as the Frisch–Peierls memorandum. Instead of looking at unenriched uranium oxide, they considered what would happen to a sphere of pure uranium-235, and found that not only could a chain reaction occur, but that it might require as little as 1 kilogram (2.2 lb) of uranium-235, and unleash the energy of tons of dynamite.

A special subcommittee of the Committee for the Scientific Survey of Air Warfare (CSSAW), known as the MAUD Committee, was created to investigate the matter further. It was chaired by George Paget Thomson, and its original membership included Chadwick, along with Mark Oliphant, John Cockcroft, and Philip Moon. While other teams investigated uranium enrichment techniques, Chadwick's team at Liverpool concentrated on determining the nuclear cross section of uranium-235. By April 1941, it had been experimentally confirmed that the critical mass of uranium-235 might be 8 kilograms (18 lb) or less. His research into such matters was complicated by all-but-incessant Luftwaffe bombings of the environs of his Liverpool lab; the windows were blown out so often that they were replaced by cardboard.

In July 1941, Chadwick was chosen to write the final draft of the MAUD Report, which, when presented by Vannevar Bush to President Franklin D. Roosevelt in October 1941, inspired the U.S. government to pour millions of dollars into the pursuit of an atom bomb. When George B. Pegram and Harold Urey visited Britain to see how the project, now known as Tube Alloys, was going, Chadwick was able to tell them: "I wish I could tell you that the bomb is not going to work, but I am 90 per cent sure that it will".

In a recent book about the Bomb project, Graham Farmelo wrote that "Chadwick did more than any other scientist to give Churchill the Bomb. ... Chadwick was tested almost to the breaking point."[86] So worried that he could not sleep, Chadwick resorted to sleeping pills, which he continued to take for most of his remaining years. Chadwick later said that he realized that "a nuclear bomb was not only possible—it was inevitable. Sooner or later these ideas could not be peculiar to us. Everybody would think about them before long, and some country would put them into action". Sir Hermann Bondi suggested that it was fortunate that Chadwick, not Rutherford, was the doyen of UK physics at the time, as the latter's prestige might otherwise have overpowered Chadwick's interest in "looking forward" to the Bomb's prospects.

## Manhattan Project

Owing to the danger from aerial bombardment, the Chadwicks sent their twins to Canada as part of a government evacuation scheme. Chadwick was reluctant to move Tube Alloys there, believing that the United Kingdom was a better location for the isotope separation plant. The enormous scope of the effort became more apparent in 1942: even a pilot separation plant would cost over £1 million and strain Britain's resources, to say nothing of a full-scale plant, which was estimated to cost somewhere in the vicinity of £25 million. It would have to be built in America. At the same time that the British became convinced that a joint project was necessary, the progress of the American Manhattan Project was such that British cooperation seemed less essential, although the Americans were still eager to use Chadwick's talents.

The matter of cooperation had to be taken up at the highest level. In September 1943, Prime Minister Winston Churchill and President Roosevelt negotiated the Quebec Agreement, which reinstated cooperation between Britain, the United States, and Canada. Chadwick, Oliphant, Peierls and Simon were summoned to the United States by the director of Tube Alloys, Sir Wallace Akers, to work with the Manhattan Project. The Quebec Agreement established a new Combined Policy Committee to direct the joint project. The Americans disliked Akers, so Chadwick was appointed technical advisor to the Combined Policy Committee, and the head of the British Mission.

Leaving Rotblat in charge in Liverpool, Chadwick began a tour of the Manhattan Project facilities in November 1943, except for the Hanford Site where plutonium was produced, which he was not allowed to see. He became the only man, apart from Groves and his second in command, to have access to all the American research and production facilities for the uranium bomb. Observing the work on the K-25 gaseous diffusion facility at Oak Ridge, Tennessee, Chadwick realized how wrong he had been about building the plant in wartime Britain. The enormous structure could never have been concealed from the Luftwaffe. In early 1944, he moved to Los Alamos, New Mexico, with his wife and their twins, who now spoke with Canadian accents. For security reasons, he was given the cover name of James Chaffee.

Chadwick accepted that the Americans did not need British help, but that it could still be useful in bringing the project to an early and successful conclusion. Working closely with the director of the Manhattan Project, Major General Leslie R. Groves, Jr., he attempted to do everything he could to support the effort. He also endeavored to place British scientists in as many parts of the project as possible to facilitate a post-war British nuclear weapons project to which Chadwick was committed. Requests from Groves via Chadwick for particular scientists tended to be met with an immediate rejection by the company, ministry or university currently employing them, only to be overcome by the overriding priority accorded to Tube Alloys. As a result, the British team was critical to the Project's success.

Although he had more knowledge of the project than anyone else from Britain, Chadwick had no access to the Hanford site. Lord Portal was offered a tour of Hanford in 1946. "This was the only plant to which Chadwick had been denied access in wartime, and now he asked Groves if he could accompany Portal. Groves replied that he could, but if he did then 'Portal will not see very much'".
For his efforts, Chadwick received a knighthood in the New Year Honors on 1 January 1945. He considered this to be a recognition of the work of the whole Tube Alloys project.

By early 1945, Chadwick was spending most of his time in Washington, D.C., and his family relocated from Los Alamos to a house on Washington's Dupont Circle in April 1945. He was present at the meeting of the Combined Policy Committee on 4 July when Field Marshal Sir Henry Maitland Wilson gave Britain's agreement to use the atomic bomb against Japan, and at the Trinity nuclear test on 16 July, when the first atom bomb was detonated. Inside its pit was a polonium-beryllium modulated neutron initiator, a development of the technique that Chadwick had used to discover the neutron over a decade before. William L. Laurence, a reporter from The New York Times attached to the Manhattan Project, wrote that "never before in history had any man lived to see his own discovery materialize itself with such telling effect on the destiny of man".

Shortly after the war ended, Chadwick was appointed to the Advisory Committee on Atomic Energy (ACAE). He was also appointed as the British scientific advisor to the United Nations Atomic Energy Commission. He clashed with fellow ACAE member Patrick Blackett, who disagreed with Chadwick's conviction that Britain needed to acquire its own nuclear weapons; but it was Chadwick's position that was ultimately adopted. He returned to Britain in 1946, to find a country still beset by wartime rationing and shortages.

At this time, Sir James Mountford, the Vice Chancellor of the University of Liverpool, wrote in his diary "he had never seen a man 'so physically, mentally and spiritually tired" as Chadwick, for he "had plumbed such depths of moral decision as more fortunate men are never called upon even to peer into ... [and suffered] ... almost insupportable agonies of responsibility arising from his scientific work".

In September 1949 Edward Teller was in England to discuss nuclear power and safety, and was dining with Sir James Chadwick and his wife at their home in Cambridge. His wife was a charming conversationalist but Sir James was (typically) taciturn. When Teller made an unflattering comment on Groves, he became very talkative, saying that the project would never have succeeded without Groves: while acknowledging his dislike of the British.

In 1948, Chadwick accepted an offer to become Master of Gonville and Caius College, Cambridge. The job was prestigious but ill-defined; the Master was the titular head of the college, but authority actually resided in a council of 13 fellows, of whom one was the Master. As Master, Chadwick strove to improve the academic reputation of the college. He increased the number of research fellowships from 31 to 49, and sought to bring talent into the college. This involved controversial decisions, such as hiring in 1951 the Chinese biochemist Tien-chin Tsao and the Hungarian-born economist Peter Bauer. In what became known as the Peasants' Revolt, fellows led by Patrick Hadley voted an old friend of Chadwick's off the council and replaced him with Bauer. More friends of Chadwick's were removed over the following years, and he retired in November 1958. It was during his mastership that Francis Crick, a Ph.D. student at Gonville and Caius College, Rosalind Franklin and James Watson discovered the structure of DNA.

***By the 1970s Chadwick became more frail, and seldom left his flat***, although he travelled to Liverpool for celebrations of his eightieth birthday. A lifelong atheist, he saw no reason to adopt religious faith in later life. ***He died in his sleep on July 24, 1974 in Cambridge at the age of 82***. [*sic* the author]

Chadwick's papers are held at the Churchill Archives Centre in Cambridge, and are accessible to the public.

# Chadwick, J. (February, 1932). Possible Existence of a Neutron.

*Nature*, 129, 312; https://doi.org/10.1038/129312a0.

## Abstract

It has been shown by Bothe and others that beryllium when bombarded by α-particles of polonium emits a radiation of great penetrating power, which has an absorption coefficient in lead of about 0.3 $(cm)^{-1}$. Recently Mme. Curie-Joliot and M. Joliot found, when measuring the ionization produced by this beryllium radiation in a vessel with a thin window, that the ionization increased when matter containing hydrogen was placed in front of the window. The effect appeared to be due to the ejection of protons with velocities up to a maximum of nearly $3 \times 10^9$ cm. per sec. They suggested that the transference of energy to the proton was by a process similar to the Compton effect, and estimated that the beryllium radiation had a quantum energy of $50 \times 10^6$ electron volts.

# Chadwick, J. (1932). The Existence of a Neutron.

*Proceedings of the Royal Society A*. 136 (830): 692–708; doi:10.1098/rspa.1932.0112.

Received May 10, 1932.

§ 1. It was shown by Bothe and Becker* that some light elements when bombarded by α-particles of polonium emit radiations which appear to be of the γ-ray type.

> * (1930). *Z. Physik*, 66, p. 289.

The element beryllium gave a particularly marked effect of this kind, and later observations by Bothe, by Mme. Curie-Joliot** and by Webster J ***showed that the radiation excited in beryllium possessed a penetrating power distinctly greater than that of any γ-radiation yet found from the radioactive elements.

> ** Curie, I, (1931). *C. R. Acad. Sci. Paris*, 193, p. 1412.
> *** (1932). *Proc. Roy. Soc. A*, 136, p. 428.

In Webster's experiments the intensity of the radiation was measured both by means of the Geiger-Muller tube counter and in a high-pressure ionization chamber. He found that the beryllium radiation had an absorption coefficient in lead of about 0.22 cm$^{-1}$ as measured under his experimental conditions. Making the necessary corrections for these conditions, and using the results of Gray and Tarrant to estimate the relative contributions of scattering, photoelectric absorption, and nuclear absorption in the absorption of such penetrating radiation, Webster concluded that the radiation had a quantum energy of about $7 \times 10^6$ electron volts. Similarly, he found that the radiation from boron bombarded by α-particles of polonium consisted in part of a radiation rather more penetrating than that from beryllium, and he estimated the quantum energy of this component as about $10 \times 10^6$ electron volts. These conclusions agree quite well with the supposition that the radiations arise by the capture of the α-particle into the beryllium (or boron) nucleus and the emission of the surplus energy as a quantum of radiation.

The radiations showed, however, certain peculiarities, and at my request the beryllium radiation was passed into an expansion chamber and several photographs were taken. No unexpected phenomena were observed though, as will be seen later, similar experiments have now revealed some rather striking events. The failure of these early experiments was partly due to the weakness of the available source of polonium, and partly to the experimental arrangement, which, as it now appears, was not very suitable.

Quite recently, Mme. Curie-Joliot and M. Joliot* made the very striking observation that these radiations from beryllium and from boron were able to eject protons with considerable velocities from matter containing hydrogen.

* Curie & Joliot, (1932), *C. It. Acad. Sci. Paris*, 194, p. 273.

In their experiments the radiation from beryllium was passed through a thin window into an ionization vessel containing air at room pressure. When paraffin wax, or other matter containing hydrogen, was placed in front of the window, the ionization in the vessel was increased, in some cases as much as doubled. The effect appeared to be due to the ejection of protons, and from further experiment they showed that the protons had ranges in air up to about 26 cm., corresponding to a velocity of nearly $3 \times 10^9$ cm. per second. They suggested that energy was transferred from the beryllium radiation to the proton by a process similar to the Compton effect with electrons, and they estimated that the beryllium radiation had a quantum energy of about $50 \times 10^6$ electron volts. The range of the protons ejected by the boron radiation was estimated to be about 8 cm. in air, giving on a Compton process an energy of about $35 \times 10^6$ electron volts for the effective quantum. **

** Many of the arguments of the subsequent discussion apply equally to both radiations, and the term 'beryllium radiation' may often be taken to include the boron radiation

There are two grave difficulties in such an explanation of this phenomenon. Firstly, it is now well established that the frequency of scattering of high energy quanta by electrons is given with fair accuracy by the Klein-Nishina formula, and this formula should also apply to the scattering of quanta by a proton. The observed frequency of the proton scattering is, however, many thousand times greater than that predicted by this formula. Secondly, it is difficult to account for the production of a quantum of $50 \times 10^6$ electron volts from the interaction of a beryllium nucleus and an $\alpha$-particle of kinetic energy of $5 \times 10^6$ electron volts. The process which will give the greatest amount of energy available for radiation is the capture of the $\alpha$-particle by the beryllium nucleus, $Be^9$, and its incorporation in the nuclear structure to form a carbon nucleus $C^{13}$. The mass defect of the $C^{13}$ nucleus is known both from data supplied by measurements of the artificial disintegration of boron $B^{10}$ and from observations of the band spectrum of carbon; it is about $10 \times 10^6$ electron volts. The mass defect of $Be^9$ is not known, but the assumption that it is zero will give a maximum value for the possible change of energy in the reaction $Be^9 + \alpha \rightarrow C^{13} +$ quantum. On this assumption it follows that the energy of the quantum emitted in such a reaction cannot be greater than about $14 \times 10^6$ electron volts. It must, of course, be admitted that this argument from mass defects is based on the hypothesis that the nuclei are made as far as possible of $\alpha$-particles; that the $Be^9$ nucleus consists of 2 $\alpha$-particles + 1 proton + 1 electron and the

$C^{13}$ nucleus of 3 α-particles + 1 proton + 1 electron. So far as the lighter nuclei are concerned, this assumption is supported by the evidence from experiments on artificial disintegration, but there is no general proof.

Accordingly, I made further experiments to examine the properties of the radiation excited in beryllium. It was found that the radiation ejects particles not only from hydrogen but from all other light elements which were examined. The experimental results were very difficult to explain on the hypothesis that the beryllium radiation was a quantum radiation, *but followed immediately if it were supposed that the radiation consisted of particles of mass nearly equal to that of a proton and with no net charge, or neutrons.* A short statement of some of these observations was published in Nature. *

> * Chadwick, J. (February, 1932). *Possible Existence of a Neutron. Nature*, 129, 312; https://doi.org/10.1038/129312a0 [above].

This paper contains a fuller description of the experiments, *which suggest the existence of neutrons* and from which some of the properties of these particles can be inferred. In the succeeding paper Dr. Feather will give an account of some observations by means of the expansion chamber of the collisions between the beryllium radiation and nitrogen nuclei, and this is followed by an account by Mr. Dee of experiments to observe the collisions with electrons.

**§ 2. Observations of Recoil Atoms.** —The properties of the beryllium radiation were first examined by means of the valve counter used in the work** on the artificial disintegration by α-particles and described fully there.

> ** Chadwick, Constable & Pollard, (1931). *Proc. Roy. Soc. A*, 130, p. 463.

Briefly, it consists of a small ionization chamber connected to a valve amplifier. The sudden production of ions in the chamber by the entry of an ionizing particle is detected by means of an oscillograph connected in the output circuit of the amplifier. The deflections of the oscillograph were recorded photographically on a film of bromide paper. The source of polonium was prepared from a solution of radium (D+E+F) *** by deposition on a disc of silver.

> *** The radium D was obtained from old radon tubes generously presented by Dr. C. F. Burnam and Dr. F. West, of the Kelly Hospital, Baltimore.

The disc had a diameter of 1 cm. and was placed close to a disc of pure beryllium of 2 cm. diameter, and both were enclosed in a small vessel which could be evacuated, fig. 1. The first ionization chamber used had an opening of 13 mm. covered with aluminum foil of 4.5 cm. air equivalent, and a depth of 15 mm. This chamber had a very low natural effect, giving on the average only about 7 deflections per hour.

When the source vessel was placed in front of the ionization chamber, the number of deflections immediately increased. For a distance of 3 cm. between the beryllium and the counter the number of deflections was nearly 4 per minute. Since the number of deflections remained sensibly the same when thick metal sheets, even as much as 2 cm. of lead, were interposed between the source vessel and the counter, it was clear that these deflections were due to a penetrating radiation emitted from the beryllium. It will be shown later that the deflections were due to atoms of nitrogen set in motion by the impact of the beryllium radiation. When a sheet of paraffin wax about 2 mm. thick was interposed in the path of the radiation just in front of the counter, the number of deflections recorded by the oscillograph increased markedly. This increase was due to particles ejected from the paraffin wax so as to pass into the counter. By placing absorbing screens of aluminum between the wax and the counter the absorption curve shown in fig. 2, curve A, was obtained. From this curve it appears that the particles have a maximum range of just over 40 cm. of air, assuming that an Al foil of 1.64 mg. per square centimeter is equivalent to 1 cm. of air. By comparing the sizes of the deflections (proportional to the number of ions produced in the chamber) due to these particles with those due to protons of about the same range it was obvious that the particles were protons. From the range-velocity curve for protons we deduce therefore that the maximum velocity imparted to a proton by the beryllium radiation is about $3.3 \times 10^9$ cm. per second, corresponding to an energy of about $5.7 \times 10^6$ electron volts.

The effect of exposing other elements to the beryllium radiation was then investigated. An ionization chamber was used with an opening covered with a gold foil of 0-5 mm. air equivalent. The element to be examined was fixed on a clean brass plate and placed very close to the counter opening. In this way lithium, beryllium, boron, carbon and nitrogen, as para-cyanogen, were tested. … For a given position of the beryllium source relative to the counter the number of recoil atoms was roughly the same for each gas. This point will he referred to later. It appears then that the beryllium radiation can impart energy to the atoms of matter through which it passes and that the chance of an energy transfer does not vary widely from one element to another.

It has been shown that protons are ejected from paraffin wax with energies up to a maximum of about $5.7 \times 10^6$ electron volts. If the ejection be ascribed to a Compton recoil from a quantum of radiation, then the energy of the quantum must be about $55 \times 10^6$ electron volts, for the maximum 2 energy which can be given to a mass $m$ by a quantum hv is $2/(2 + mc^2/hv) . hv$.

The energies of the recoil atoms produced by this radiation by the same process in other elements can be readily calculated. For example, the nitrogen recoil atoms should have energies up to a maximum of 450,000 electron volts. Taking the energy necessary to form a pair of ions in air as 35 electron volts, the recoil atoms of nitrogen should produce not more than about 13,000 pairs of ions. Many of the deflections observed with nitrogen, however, corresponded to far more ions than this; some of the recoil atoms produced from 30,000 to 40,000 ion pairs. In the case of the other elements a similar discrepancy was noted between the observed energies and ranges of the recoil atoms and the values calculated on the assumption that the atoms were set in motion by recoil from a quantum of $55 \times 10^6$ electron volts. The energies of the recoil atoms were estimated from the number of ions produced in the counter, as given by the size of the oscillograph deflections. A sufficiently good measurement of the ranges could be made either by varying the distance between the element and the counter or by interposing thin screens of gold between the element and the counter.

The nitrogen recoil atoms were also examined, in collaboration with Dr. N. Feather, by means of the expansion chamber. The source vessel was placed immediately above an expansion chamber of the Shimizu type, so that a large proportion of the beryllium radiation traversed the chamber. A large number of recoil tracks was observed in the course of a few hours. Their range, estimated by eye, was sometimes as much as 5 or 6 mm. in the chamber, or, correcting for the expansion, about 3 mm. in standard air. These visual estimates were confirmed by a preliminary series of experiments by Dr. Feather with a large automatic expansion chamber, in which photographs of the recoil tracks in nitrogen were obtained. Now the ranges of recoil atoms of nitrogen of different velocities have been measured by Blackett and Lees. Using their results we find that the nitrogen recoil atoms produced by the beryllium radiation may have a velocity of at least $4 \times 10^8$ cm. per second, corresponding to an energy of about $1.2 \times 10^6$ electron volts. In order that the nitrogen nucleus should acquire such an energy in a collision with a quantum of radiation, it is necessary to assume that the energy of the quantum should be about $90 \times 10^6$ electron volts, if energy and momentum are conserved in the collision. It has been shown that a quantum of $55 \times 10^6$ electron volts is sufficient to explain the hydrogen collisions. In general, the experimental results show that if the recoil atoms are to be explained by collision with a quantum, we must assume a larger and larger energy for the quantum as the mass of the struck atom increases.

## § 3. The Neutron Hypothesis.

—It is evident that we must either relinquish the application of the conservation of energy and momentum in these collisions or adopt another hypothesis about the nature of the radiation. *If we suppose that the radiation is not a quantum radiation, but consists of particles of mass very nearly equal to that of the proton, all the difficulties connected with the collisions disappear*, both with regard to their frequency and to the energy transfer to different masses. In order to explain the great penetrating power of the radiation *we must further assume that the particle has no net charge*. We may suppose it to consist of a proton and an electron in close combination, the "*neutron*" discussed by Rutherford* in his Bakerian Lecture of 1920.

> * Rutherford, (1920). *Proc. Roy. Soc., A*, 97, p. 374. Experiments to detect the formation of *neutrons* in a hydrogen discharge tube were made by Glasson, J. L. (1921), *Phil. Mag.*, 42, p. 596, and by Roberts, J. K. (1922), *Proc. Roy. Soc.*, A, 102, p. 72. Since 1920 many experiments in search of these *neutrons* have been made in this laboratory.

When such *neutrons* pass through matter, they suffer occasionally close collisions with the atomic nuclei and so give rise to the recoil atoms which are observed. Since the mass of the *neutron* is equal to that of the proton, the recoil atoms produced when the *neutrons* pass through matter containing hydrogen will have all velocities up to a maximum which is the same as the maximum velocity of the *neutrons*. The experiments showed that the maximum velocity of the protons ejected from paraffin wax was about $3.3 \times 10^9$ cm. per second. This is therefore the maximum velocity of the *neutrons* emitted from beryllium bombarded by α-particles of polonium. From this we can now calculate the maximum energy which can be given by a colliding *neutron* to other atoms, and we find that the results are in fair agreement with the energies observed in the experiments. For example, a nitrogen atom will acquire in a head-on collision with the *neutron* of mass 1 and velocity $3.3 \times 10^9$ cm per second a velocity of $4.4 \times 10^8$ cm. per second, corresponding to an energy of $1.4 \times 10^6$ electron volts, a range of about 3.3 mm. in air, and a production of ions of about 40,000 pairs. Similarly, an argon atom may acquire an energy of $0.54 \times 10^6$ electron volts, and produce about 15,000 ion pairs. Both these values are in good accord with experiment. *

> * It was noted that a few of the nitrogen recoil atoms produced about 50 to 60,000 ion pairs. These probably correspond to the cases of disintegration found by Feather and described in his paper.

It is possible to prove that the *mass of the neutron* is roughly equal to that of the proton, by combining the evidence from the hydrogen collisions with that from the nitrogen collisions. In the succeeding paper, Feather records experiments in which about 100 tracks of nitrogen recoil atoms have been photographed in the expansion chamber. The measurement of the tracks shows that the maximum range of the recoil atoms is 3.5 mm in air at 15° C. and 760 mm pressure, corresponding to a velocity of $4.7 \times 10^8$ cm per second according to Blackett and Lees.

If M, V be the *mass and velocity of the neutron* then the *maximum velocity* given to a *hydrogen atom* is

$$u_p = 2M/(M + 1) \cdot V,$$

and the *maximum velocity* given to a *nitrogen atom* is

$$u_n = 2M/(M + 1) \cdot V,$$

whence

$$(M + 14)/(M + 1) = u_p/u_n = 3.3 \times 10^9 / 4.7 \times 10^8,$$

and

$$M = 1.15.$$

The total error in the estimation of the velocity of the nitrogen recoil atom may easily be about 10 per cent., and *it is legitimate to conclude that the mass of the neutron is very nearly the same as the mass of the proton*.

*We have now to consider the production of the neutrons from beryllium by the bombardment of the α-particles*. We must suppose that an α-particle is captured by a $Be^9$ nucleus with the formation of a carbon $C^{12}$ nucleus and the emission of a *neutron*. The process is analogous to the well-known artificial disintegrations, but a *neutron* is emitted instead of a proton. The energy relations of this process cannot be exactly deduced, for the *masses* of the $Be^9$ nucleus and the *neutron* are not known accurately. It is, however, easy to show that such a process fits the experimental facts. We have

$$Be^9 + He^4 + \text{kinetic energy of } \alpha$$
$$= C^{12} + n^1 + \text{kinetic energy of } C^{12} + \text{kinetic energy of } n^1.$$

If we assume that the beryllium nucleus consists of two α-particles and a neutron, then its mass cannot be greater than the sum of the masses of these particles, for the binding energy corresponds to a defect of mass. The energy equation becomes

$$(8.00212 + n^1) + 4.00106 + \text{K.E. of } \alpha > 12.0003 + n^1 + \text{K.E. of } C^{12} + \text{K.E. of } n^1$$

or

$$\text{K.E. of } n^1 < \text{K.E. of } \alpha + 0.003 - \text{K.E. of } C^{12}.$$

Since the kinetic energy of the α-particle of polonium is $5.25 \times 10^6$ electron volts, it follows that the energy of emission of the *neutron* cannot be greater than about $8 \times 10^6$ electron volts. The velocity of the *neutron* must therefore be less than $3.9 \times 10^9$ cm per second. We have seen that the actual maximum velocity of the **neutron** is about $3.3 \times 10^9$ cm per second, so that the proposed disintegration process is compatible with observation.

A further test of the **neutron** hypothesis was obtained by examining the radiation emitted from beryllium in the opposite direction to the bombarding α-particles. The source vessel, fig. 1, was reversed so that a sheet of paraffin wax in front of the counter was exposed to the "backward" radiation from the beryllium. The maximum range of the protons ejected from the wax was determined as before, by counting the numbers of protons observed through different thicknesses of aluminum interposed between the wax and the counter.

The absorption curve obtained is shown in curve B, fig. 2. The maximum range of the protons was about 22 cm. in air, corresponding to a velocity of about $2.74 \times 10^9$ cm per second. Since the polonium source was only about 2 mm. away from the beryllium, this velocity should be compared with that of the neutrons emitted not at 180° but at an angle not much greater than 90° to the direction of the incident α-particles. A simple calculation shows that the velocity of the **neutron** emitted at 90° when an α-particle of full range is captured by a beryllium nucleus should be $2.77 \times 10^9$ cm. per second, taking the velocity of the **neutron** emitted at 0° in the same process as $3.3 \times 10^9$ cm. per second. The velocity found in the above experiment should be less than this, for the angle of emission is slightly greater than 90°. The agreement with calculation is as good as can be expected from such measurements.

**§ 4. The Nature of the Neutron.** —It has been shown that the origin of the radiation from beryllium bombarded by α-particles and the behavior of the radiation, so far as its interaction with atomic nuclei is concerned, receive a simple explanation ***on the assumption that the radiation consists of particles of mass nearly equal to that of the proton which have no charge***. The simplest hypothesis one can make about the nature of the particle is to suppose that it consists of a proton and an electron in close combination, giving a net charge 0 and a mass which should be slightly less than the mass of the hydrogen atom. This hypothesis is supported by an examination of the evidence which can be obtained about the ***mass of the neutron***.

As we have seen, a rough estimate of the ***mass of the neutron*** was obtained from measurements of its collisions with hydrogen and nitrogen atoms, but such measurements cannot be made with sufficient accuracy for the present purpose. We must turn to a consideration of the energy relations in a process in which a **neutron** is liberated from an atomic nucleus; if the masses of the atomic nuclei concerned in the process are accurately known, a good estimate of the ***mass of the neutron*** can be deduced. The mass of the beryllium nucleus has, however, not yet been measured, and, as was shown in § 3, only general conclusions can be drawn from this reaction. Fortunately, there remains the case of boron. It was stated in § 1 that boron bombarded by α-particles of polonium also emits a radiation which ejects protons from materials containing hydrogen. Further examination showed that this radiation behaves in all respects like that from beryllium, and it must therefore be assumed to consist of **neutrons**. It is probable that, the **neutrons** are emitted from the isotope $B^{11}$, for we know that the isotope $B^{10}$ disintegrates with the emission of a proton. *

\* Chadwick, Constable & Pollard, *loc. cit.*

The process of disintegration will then be

$$B^{11} + He^4 \rightarrow N^{14} + n^1.$$

The masses of $B^{11}$ and $N^{14}$ are known from Aston's measurements, and the further data required for the deduction of the **mass of the neutron** can be obtained by experiment.

In the source vessel of fig. 1 the beryllium was replaced by a target of powdered boron, deposited on a graphite plate. The range of the protons ejected by the boron radiation was measured in the same way as with the beryllium radiation. The effects observed were much smaller than with beryllium, and it was difficult to measure the range of the protons accurately. The maximum range was about 16 cm in air, corresponding to a velocity of 2.5 x $10^9$ cm per second. This then is the maximum velocity of the **neutron** liberated from boron by an α-particle of polonium of velocity 1.59 x $10^9$ cm per second. Assuming that momentum is conserved in the collision, the velocity of the recoiling $N^{14}$ nucleus can be calculated, and we then know the kinetic energies of all the particles concerned in the disintegration process. The energy equation of the process is

Mass of $B^{11}$ + mass of $He^4$ + K.E. of $He^4$
$\qquad$ = mass of $N^{14}$ + mass of $n^1$ + K.E. of $N^{14}$ + K.E. of $n^1$.

The masses are $B^{11}$ = 11.00825 ± 0.0016; $He^4$ = 4.00106 ± 0.0006; $N^{14}$ = 14.0042 ± 0.0028. The kinetic energies in mass units are α-particle = 0.00565; **neutron** = 0.0035; and nitrogen nucleus = 0.00061. We find therefore that the **mass of the neutron** is 1.0067. The errors quoted for the mass measurements are those given by Aston. They are the maximum errors which can be allowed in his measurements, and the probable error may be taken as about one-quarter of these. *

\* The mass of $B^{11}$ relative to $B^{10}$ has been checked by optical methods by Jenkins & McKellar (1932) *Phys. Rev.*, 39, p. 549. Their value agrees with Aston's to 1 part in $10^5$. This suggests that great confidence may be put in Aston's measurements.

Allowing for the errors in the mass measurements it appears that *the mass of the neutron cannot be less than 1.003, and that it probably lies between 1.005 and 1.008*.

Such a value for the *mass of the neutron* is to be expected if the *neutron* consists of a proton and an electron, and it lends strong support to this view. Since the sum of the masses of the proton and electron is 1.0078, the binding energy, or mass defect, of the neutron is about 1 to 2 million electron volts. This is quite a reasonable value. We may suppose that the proton and electron form a small dipole, or we may take the more attractive picture of a proton embedded in an electron. On either view, we may expect the *"radius" of the neutron* to be a few times $10^{-13}$ cm.

**§ 5. The Passage of the Neutron through Matter.** —The electrical field of a *neutron* of this kind will clearly be extremely small except at very small distances of the order of $10^{-12}$ cm. In its passage through matter the *neutron* will not be deflected unless it suffers an intimate collision with a nucleus. The potential of a *neutron* in the field of a nucleus may be represented roughly by fig. 3. The radius of the collision area for sensible deflection of the *neutron* will be little greater than the *radius of the nucleus*. Further, the *neutron* should be able to penetrate the nucleus easily, and it may be that the scattering of the *neutrons* will be largely due to the internal field of the nucleus, or, in other words, that the scattered *neutrons* are mainly those which have penetrated the potential barrier. On these views we should expect the collisions of a *neutron* with a nucleus to occur very seldom, and that the scattering will be roughly equal in all directions, at least as compared with the Coulomb scattering of a charged particle.

These conclusions were confirmed in the following way. The source vessel, with Be target, was placed rather more than 1 inch from the face of a closed counter filled with air, fig. 1. The number of deflections, or the number of nitrogen recoil atoms produced in the chamber, was observed for a certain time. The number observed was 190 per hour, after allowing for the natural effect. A block of lead 1 inch thick was then introduced between the source vessel and the counter. The number of deflections fell to 166 per hour. Since the number of recoil atoms produced must be proportional to the number of *neutrons* passing through the counter, these observations show that 13 per cent, of the *neutrons* had been absorbed or scattered in passing through 1 inch of lead.

Suppose that a *neutron* which passes within a distance from the center of the lead nucleus is scattered and removed from the beam. Then the fraction removed from the beam in passing through a thickness t of lead will be $\pi p^2 n t$, where n is the number of lead atoms per unit volume. Hence $\pi p^2 n t = 0.13$, and $p = 7 \times 10^{-13}$ cm. This value for the collision radius with lead seems perhaps rather small, but it is not unreasonable. We may compare it with the radii of the radioactive nuclei calculated from the disintegration constants by Gamow and Houtermans, * viz., about 7 x $10^{-13}$ cm.

* *Z. Physik*, (1928), 52, p. 453.

Similar experiments were made in which the **neutron** radiation was passed through blocks of brass and carbon. The values of p deduced in the same way were 6 x $10^{-13}$ cm. and 3.5 x $10^{-13}$ cm. respectively.

***The target areas for collision for some light elements were compared by another method.*** The second ionization chamber was used, which could be filled with different gases by circulation. The position of the source vessel was kept fixed relative to the counter, and the number of deflections was observed when the counter was filled in turn with **hydrogen, nitrogen, oxygen, and argon**. Since the number of **neutrons** passing through the counter was the same in each case, the number of deflections should be proportional to the **target area for collision**, neglecting the effect of the material of the counter, and allowing for the fact that argon is monatomic. It was found that nitrogen, oxygen, and argon gave about the same number of deflections; **the target areas of nitrogen and oxygen are thus roughly equal**, and the target area of argon is nearly twice that of these. With hydrogen the measurements were very difficult, for many of the deflections were very small owing to the low ionizing power of the proton and the low density of the gas. **It seems probable from the results that the target area of hydrogen is about two-thirds that of nitrogen or oxygen**, but it may be rather greater than this.

There is as yet little information about the **angular distribution of the scattered neutrons**. In some experiments kindly made for me by Dr. Gray and Mr. Lea, the scattering by lead was compared in the backward and forward directions, using the ionization in a high-pressure chamber to measure the **neutrons**. They found that the amount of scattering was about that to be expected from the measurements quoted above, and that the intensity per unit solid angle was about the same between 30° to. 90° in the forward direction as between 90° to 150° in the backward direction. The scattering by lead is therefore not markedly anisotropic.

***Two types of collision may prove to be of peculiar interest, the collision of a neutron with a proton and the collision with an electron***. A detailed study of these collisions with an elementary particle is of special interest, for it should provide information about the structure and field of the **neutron**, whereas the other collisions will depend mainly on the structure of the atomic nuclei. Some preliminary experiments by Mr. Lea, using the pressure chamber to measure the scattering of **neutrons** by paraffin wax and by liquid hydrogen, suggest that the collision with a proton is more frequent than with other light atoms. This is not in accord with the experiments described above, but the results are at present indecisive. These collisions can be more directly investigated by means of the expansion chamber or by counting methods, and it is hoped to do so shortly.

***The collision of a neutron with an electron*** has been examined in two ways, by the expansion chamber and by the counter. An account of the expansion chamber experiments is given by Mr. Dee in the third paper of this series. Mr. Dee has looked for the general ionization produced by a large number of ***neutrons*** in passing through the expansion chamber, and also for the short electron tracks which should be the result of a very close collision between a ***neutron*** and an electron. His results show that collisions with electrons are extremely rare compared even with those with nitrogen nuclei, and he estimates that a ***neutron*** can produce on the average not more than 1 ion pair in passing through 3 meters of air.

In the counter experiments a beam of ***neutrons*** was passed through a block of brass, 1 inch thick, and the maximum range of the protons ejected from paraffin wax by the emergent beam was measured. From this range the ***maximum velocity of the neutrons*** after travelling through the brass is obtained and it can be compared with the maximum velocity in the incident beam. ***No change in the velocity of the neutrons due to their passage through the brass could be detected***. The accuracy of the experiment is not high, for the estimation of the end of the range of the protons was rather difficult. The results show that the ***loss of energy of a neutron*** in passing through 1 inch of brass is not more than about $0.4 \times 10^6$ electron volts. A path of 1 inch in brass corresponds as regards electron collisions to a path of nearly $2 \times 10^4$ cm of air, so that this result would suggest that a ***neutron*** loses less than 20 volts per centimeter path in air in electron collisions. This experiment thus lends general support to those with the expansion chamber, though it is of far inferior accuracy. We conclude that the transfer of energy from the ***neutron*** to electrons is of very rare occurrence. This is not unexpected. Bohr* has shown on quite general ideas that collisions of a neutron with an electron should be very few compared with nuclear collisions.

* Bohr, *Copenhagen discussions*, unpublished.

Massey, ** on plausible assumptions about the field of the ***neutron***, has made a detailed calculation of(the loss of energy to electrons, and finds also that it should be small, not more than 1 ion pair per meter in air.
** Massey, (1932), *Nature*, 129, p. 469, corrected p. 691.

**General Remarks.**

*It is of interest to examine whether other elements, besides beryllium and boron, emit neutrons when bombarded by α-particles*. So far as experiments have been made, no case comparable with these two has been found. Some evidence was obtained of the emission of *neutrons* from fluorine and magnesium, but the effects were very small, rather less than 1 per cent, of the effect obtained from beryllium under the same conditions. There is also the possibility that some elements may emit *neutrons* spontaneously, e.g., potassium, which is known to emit a nuclear *β-radiation* accompanied by a more penetrating radiation. Again, no evidence was found of the presence of *neutrons*, and it seems fairly certain that the penetrating type is, as has been assumed, a *γ-radiation*.

Although there is certain evidence for the emission of *neutrons* only in two cases of nuclear transformations, *we must nevertheless suppose that the neutron is a common constituent of atomic nuclei*. We may then proceed to *build up nuclei out of α-particles, neutrons and protons*, and *we are able to avoid the presence of uncombined electrons in a nucleus*. This has certain advantages for, as is well known, the electrons in a nucleus have lost some of the properties which they have outside, e.g., their spin and magnetic moment. If the α-particle, the neutron, and the proton are the only units of nuclear structure, we can proceed to calculate the mass defect or binding energy of a nucleus as the difference between the mass of the nucleus and the sum of the masses of the constituent particles. *It is, however, by no means certain that the α-particle and the neutron are the only complex particles in the nuclear structure*, and therefore the mass defects calculated in this way may not be the true binding energies of the nuclei. In this connection it may be noted that the examples of disintegration discussed by Dr. Feather in the next paper are not all of one type, and he suggests that in some cases a particle of mass 2 and charge 1, the hydrogen isotope recently reported by Urey, Brickwedde and Murphy, may be emitted. It is indeed possible that this particle also occurs as a unit of nuclear structure.

*It has so far been assumed that the neutron is a complex particle consisting of a proton and an electron*. This is the simplest assumption and it is supported by the evidence that the mass of the neutron is about 1.006, just a little less than the sum of the masses of a proton and an electron. Such a neutron would appear to be the first step in the combination of the elementary particles towards the formation of a nucleus. It is obvious that this neutron may help us to visualize the building up of more complex structures, but the discussion of these matters will not be pursued further for such speculations, though not idle, are not at the moment very fruitful. *It is, of course, possible to suppose that the neutron may be an elementary particle*. This view has little to recommend it at present, except the possibility of explaining the statistics of such nuclei as $N^{14}$.

There remains to discuss the transformations which take place when an α-particle is captured by a beryllium nucleus, $Be^9$. The evidence given here indicates that the main type of transformation is the formation of a $C^{12}$ nucleus and the emission of a **neutron**. The experiments of Curie-Joliot and Joliot, * of Auger, ** and of Dee show quite definitely that there is some radiation emitted by beryllium which is able to eject fast electrons in passing through matter.

* (1932*). C. R. Acad. Sci. Paris*, 194, p. 708 and p. 876.
** (1932). *C. R. Acad. Sci. Paris*, 194, p. 877.

I have made experiments using the Geiger point counter to investigate this radiation and the results suggest that the electrons are produced by a γ-radiation. ***There are two distinct processes which may give rise to such a radiation. In the first place***, we may suppose that the transformation of $Be^9$ to $C^{12}$ takes place sometimes with the formation of an excited $C^{12}$ nucleus which goes to the ground state with the emission of γ-radiation. This is similar to the transformations which are supposed to occur in some cases of disintegration with proton emission, e.g., $B^{10}$, $F^{19}$, $Al^{27}$; the majority of transformations occur with the formation of an excited nucleus, only in about one-quarter is the final state of the residual nucleus reached in one step. We should then have two groups of **neutrons** of different energies and a γ-radiation of quantum energy equal to the difference in energy of the **neutron** groups. The quantum energy of this radiation must be less than the maximum energy of the **neutrons** emitted, about 5.7 x $10^6$ electron volts. ***In the second place,*** we may suppose that occasionally the beryllium nucleus changes to a $C^{13}$ nucleus and that all the surplus energy is emitted as radiation. In this case the quantum energy of the radiation may be about 10 x $10^6$ electron volts.

It is of interest to note that Webster has observed a soft radiation from beryllium bombarded by polonium α-particles, of energy about 5 x $10^5$ electron volts. This radiation may well be ascribed to the first of the two processes just discussed, and its intensity is of the right order. On the other hand, some of the electrons observed by Curie-Joliot and Joliot had energies of the order of 2 to 10 x $10^6$ volts, and Auger recorded one example of an electron of energy about 6,5 X $10^6$ volts. These electrons may be due to a hard γ-radiation produced by the second type of transformation. *

* Although the presence of fast electrons can be easily explained in this way, the possibility that some may be due to secondary effects of the **neutrons** must not be lost sight of.

It may be remarked that no electrons of greater energy than the above appear to be present. This is confirmed by an experiment** made in this laboratory by Dr. Occhialini.

** Cf. also Rasetti, (1932), *Naturwiss.*, 20, p. 252.

Two tube counters were placed in a horizontal plane and the number of coincidences recorded by them was observed by means of the method devised by Rossi. The beryllium source was then brought up in the plane of the counters so that the radiation passed through both counters in turn. No increase in the number of coincidences could be detected. It follows that there are few, if any, β-rays produced with energies sufficient to pass through the walls of both counters, a total of 4 mm brass; that is, with energies greater than about $6 \times 10^6$ volts. This experiment further shows that the **neutrons** very rarely produce coincidences in tube counters under the usual conditions of experiment.

In conclusion, I may restate briefly the case for supposing that ***the radiation the effects of which have been examined in this paper consists of neutral particles rather than of radiation quanta***. Firstly, there is no evidence from electron collisions of the presence of a radiation of such a quantum energy as is necessary to account for the nuclear collisions. Secondly, ***the quantum hypothesis can be sustained only by relinquishing the conservation of energy and momentum***. On the other hand, ***the neutron hypothesis gives an immediate and simple explanation of the experimental facts***; it is consistent in itself and it throws new light on the problem of nuclear structure.

## Summary.

The properties of the penetrating radiation emitted from beryllium (and boron) when bombarded by the α-particles of polonium have been examined. ***It is concluded that the radiation consists, not of quanta as hitherto supposed, but of neutrons, particles of mass 1, and charge 0.*** Evidence is given to show that ***the mass of the neutron is probably between 1.005 and 1.008.*** This suggests that the neutron consists of a proton and an electron in close combination, the binding energy being about 1 to $2 \times 10^6$ electron volts. From experiments on the passage of the **neutrons** through matter the frequency of their collisions with atomic nuclei and with electrons is discussed.

I wish to express my thanks to Mr. H. Nutt for his help in carrying out the experiments.

## Heisenberg, W. (January, 1932). Über den Bau der Atomkerne. (About the construction of atomic nuclei.)

*Z. Physik*, 77, 1–11; https://doi.org/10.1007/BF01342433.

The consequences of the assumption that the atomic nuclei are made up of protons and neutrons without the participation of electrons are discussed. § 1. The Hamiltonian function of the nucleus. § 2. The relationship between charge and mass and the special stability of the He nucleus. § 3 to 5; Stability of the nuclei and radioactive decay series. § 6. Discussion of the basic physical assumptions.

## Heisenberg, W. (March, 1932). Über den Bau der Atomkerne. II. (About the construction of atomic nuclei, II.)

*Z. Physik*, 78, 156–64; doi:10.1007/BF01337585.

§ 1. Stability of even and odd neutron nuclei. § 2. Scattering of γ rays at the atomic nucleus. § 3. The properties of the neutron.

## Heisenberg, W. (September, 1933). Über den Bau der Atomkerne. III. (About the construction of atomic nuclei, III.)

*Z. Physik*, 80, 587–96; doi:10.1007/BF01335696.

Heisenberg's landmark papers approached the description of protons and neutrons in the nucleus through quantum mechanics. While Heisenberg's theory for protons and neutrons in the nucleus was a "major step toward understanding the nucleus as a quantum mechanical system", he still assumed the presence of nuclear electrons. In particular, *Heisenberg assumed the neutron was a proton–electron composite, for which there is no quantum mechanical explanation*. Heisenberg had no explanation for how lightweight electrons could be bound within the nucleus.

Heisenberg introduced the first theory of *nuclear exchange forces* that bind the nucleons. He considered *protons and neutrons to be different quantum states of the same particle*, i.e., nucleons distinguished by the value of their nuclear isospin quantum numbers.

The nature of the neutron was a primary topic of discussion at the 7th Solvay Conference held in October 1933, attended by Heisenberg, Niels Bohr, Lise Meitner, Ernest Lawrence, Fermi, Chadwick, and others. As posed by Chadwick in his Bakerian Lecture in 1933, the primary question was the mass of the neutron relative to the proton. If the neutron's mass was less than the combined masses of a proton and an electron (1.0078 Da), then the neutron could be a proton-electron composite because of the mass defect from the nuclear binding energy. ***If*** [the neutron's mass was] ***greater than the combined masses, then the neutron was elementary like the proton***. The question was challenging to answer because the electron's mass is only 0.05% of the proton's, hence exceptionally precise measurements were required.

## Heisenberg, W. (1932). The development of quantum mechanics.

Nobel Lecture, December 11, 1933;
https://www.nobelprize.org/prizes/physics/1932/heisenberg/lecture/.

Heisenberg was awarded the 1932 Nobel Prize "for the creation of quantum mechanics, the application of which has, inter alia, led to the discovery of the allotropic forms of hydrogen". The Nobel Prize in Physics 1932. NobelPrize.org. https://www.nobelprize.org/prizes/physics/ 1932/summary/. In Niels Bohr's theory of the atom, electrons absorb and emit radiation of fixed wavelengths when jumping between fixed orbits around a nucleus. The theory provided a good description of the spectrum created by the hydrogen atom, but needed to be developed to suit more complicated atoms and molecules. In 1925, Werner Heisenberg formulated a type of quantum mechanics based on matrices. In 1927 he proposed the "uncertainty relation", setting limits for how precisely the position and velocity of a particle can be simultaneously determined. [Werner Heisenberg – Facts. NobelPrize.org. https://www.nobelprize.org/prizes/physics/1932/heisenberg/ facts/.]

> In his Nobel Prize lecture, Heisenberg noted that "the impossibility of harmonizing the Maxwellian theory with the pronouncedly visual concepts expressed in the hypothesis of light quanta subsequently compelled research workers to the conclusion that *radiation phenomena can only be understood by largely renouncing their immediate visualization.* … Classical physics seemed the limiting case of visualization of a fundamentally unvisualizable microphysics, the more accurately realizable the more Planck's constant vanishes relative to the parameters of the system". After describing the development of the current theory of quantum mechanics, he concluded that "a visual description for the atomic events is possible only within certain limits of accuracy - but within these limits the laws of classical physics also still apply. Owing to these limits of accuracy as defined by the uncertainty relations, moreover, a visual picture of the atom free from ambiguity has not been determined. On the contrary the corpuscular and the wave concepts are equally serviceable as a basis for visual interpretation. The laws of quantum mechanics are basically statistical". He noted that "the attention of the research workers was now primarily directed to *the problem of reconciling the claims of the special relativity theory with those of the quantum theory.* … The attempts made hitherto to achieve a *relativistic* formulation of the quantum theory are all based on visual concepts so close to those of classical physics that it seems impossible to determine the fine-structure constant within this system of concepts".

---

Quantum mechanics, on which I am to speak here, arose, in its formal content, from the endeavor to expand Bohr's principle of correspondence to a complete mathematical scheme by refining his assertions. The physically new viewpoints that distinguish quantum mechanics from classical physics were prepared by the researches of various investigators

303

engaged in analyzing the difficulties posed in Bohr's theory of atomic structure and in the radiation theory of light.

In 1900, through studying the law of black-body radiation which he had discovered, Planck had detected in optical phenomena a discontinuous phenomenon totally unknown to classical physics which, a few years later, was most precisely expressed in Einstein's hypothesis of light quanta. The impossibility of harmonizing the Maxwellian theory with the pronouncedly visual concepts expressed in the hypothesis of light quanta subsequently compelled research workers to the conclusion that *radiation phenomena can only be understood by largely renouncing their immediate visualization*. The fact, already found by Planck and used by Einstein, Debye, and others, that the element of discontinuity detected in radiation phenomena also plays an important part in material processes, was expressed systematically in Bohr's basic postulates of the quantum theory which, together with the Bohr-Sommerfeld quantum conditions of atomic structure, led to a qualitative interpretation of the chemical and optical properties of atoms. The acceptance of these basic postulates of the quantum theory contrasted uncompromisingly with the application of classical mechanics to atomic systems, which, however, at least in its qualitative affirmations, appeared indispensable for understanding the properties of atoms. This circumstance was a fresh argument in support of the assumption that the natural phenomena in which Planck's constant plays an important part can be understood only by largely foregoing a visual description of them. Classical physics seemed the limiting case of visualization of a fundamentally unvisualizable microphysics, the more accurately realizable the more Planck's constant vanishes relative to the parameters of the system. This view of classical mechanics as a limiting case of quantum mechanics also gave rise to Bohr's principle of correspondence which, at least in qualitative terms, transferred a number of conclusions formulated in classical mechanics to quantum mechanics. In connection with the principle of correspondence there was also discussion whether the quantum-mechanical laws could in principle be of a statistical nature; the possibility became particularly apparent in Einstein's derivation of Planck's law of radiation. Finally, the analysis of the relation between radiation theory and atomic theory by Bohr, Kramers, and Slater resulted in the following scientific situation:

According to the basic postulates of the quantum theory, an atomic system is capable of assuming discrete, stationary states, and therefore discrete energy values; in terms of the energy of the atom the emission and absorption of light by such a system occurs abruptly, in the form of impulses. On the other hand, the visualizable properties of the emitted radiation are described by a wave field, the frequency of which is associated with the difference in energy between the initial and final states of the atom by the relation

$$E_1 - E_2 = h\nu$$

To each stationary state of an atom corresponds a whole complex of parameters which specify the probability of transition from this state to another. There is no direct relation between the radiation classically emitted by an orbiting electron and those parameters defining the probability of emission; nevertheless, Bohr's principle of correspondence enables a specific term of the Fourier expansion of the classical path to be assigned to each transition of the atom, and the probability for the particular transition follows qualitatively similar laws as the intensity of those Fourier components. Although therefore in the researches carried out by Rutherford, Bohr, Sommerfeld and others, the comparison of the atom with a planetary system of electrons leads to a qualitative interpretation of the optical and chemical properties of atoms, nevertheless the fundamental dissimilarity between the atomic spectrum and the classical spectrum of an electron system imposes the need to relinquish the concept of an electron path and to forego a visual description of the atom.

The experiments necessary to define the electron-path concept also furnish an important aid in revising it. The most obvious answer to the question how the orbit of an electron in its path within the atom could be observed, namely …?, will perhaps be to use a microscope of extreme resolving power. But since the specimen in this microscope would have to be illuminated with light having an extremely short wavelength, the first light quantum from the light source to reach the electron and pass into the observer's eye would eject the electron completely from its path in accordance with the laws of the Compton effect. Consequently, only one point of the path would be observable experimentally at any one time.

In this situation, therefore, the obvious policy was to relinquish at first the concept of electron paths altogether, despite its substantiation by Wilson's experiments, and, as it were, to attempt subsequently how much of the electron-path concept can be carried over into quantum mechanics.

In the classical theory the specification of frequency, amplitude, and phase of all the light waves emitted by the atom would be fully equivalent to specifying its electron path. Since from the amplitude and phase of an emitted wave the coefficients of the appropriate term in the Fourier expansion of the electron path can be derived without ambiguity, the complete electron path therefore can be derived from a knowledge of all amplitudes and phases. Similarly, in quantum mechanics, too, the whole complex of amplitudes and phases of the radiation emitted by the atom can be regarded as a complete description of the atomic system, although its interpretation in the sense of an electron path inducing the radiation is impossible. In quantum mechanics, therefore, the place of the electron coordinates is taken by a complex of parameters corresponding to the Fourier coefficients of classical motion along a path. These, however, are no longer classified by the energy of state and the number of the corresponding harmonic vibration, but are in each case associated with two stationary states of the atom, and are a measure for the transition probability of the atom from one stationary state to another. A complex of coefficients of this type is comparable with a matrix such as occurs in linear algebra. In exactly the same way each parameter of classical mechanics, e.g. the momentum or the energy of the electrons, can then be assigned a corresponding matrix in quantum mechanics. To proceed from here beyond a mere description of the empirical state of affairs it was necessary to associate systematically the matrices assigned to the various parameters in the same way as the corresponding parameters in classical mechanics are associated by equations of motions. When, in the interest of achieving the closest possible correspondence between classical and quantum mechanics, the addition and multiplication of Fourier series were tentatively taken as the example for the addition and multiplication of the quantum-theory complexes, the product of two parameters represented by matrices appeared to be most naturally represented by the product matrix in the sense of linear algebra - an assumption already suggested by the formalism of the Kramers- Ladenburg dispersion theory.

It thus seemed consistent simply to adopt in quantum mechanics the equations of motion of classical physics, regarding them as a relation between the matrices representing the classical variables. The Bohr-Sommerfeld quantum conditions could also be reinterpreted in a relation between the matrices, and together with the equations of motion they were sufficient to define all matrices and hence the experimentally observable properties of the atom.

Born, Jordan, and Dirac deserve the credit for expanding the mathematical scheme outlined above into a consistent and practically usable theory. These investigators observed in the first place that the quantum conditions can be written as commutation relations between the matrices representing the momenta and the coordinates of the electrons, to yield the equations ($p_r$, momentum matrices; $q_r$, coordinate matrices):

$$p_r q_s - q_s p_r = h/2\pi i \; \delta_{rs} \qquad q_r q_s - q_s q_r = 0 \qquad p_r p_s - p_s p_r = 0$$

$$\delta_{rs} = I \text{ for } r = s; = 0 \text{ for } r \neq s.$$

By means of these commutation relations they were able to detect in quantum mechanics as well the laws which were fundamental to classical mechanics: the invariability in time of energy, momentum, and angular momentum.

The mathematical scheme so derived thus ultimately bears an extensive formal similarity to that of the classical theory, from which it differs outwardly by the commutation relations which, moreover, enabled the equations of motion to be derived from the Hamiltonian function.

In the physical consequences, however, there are very profound differences between quantum mechanics and classical mechanics which impose the need for a thorough discussion of the physical interpretation of quantum mechanics. As hitherto defined, quantum mechanics enables the radiation emitted by the atom, the energy values of the stationary states, and other parameters characteristic for the stationary states to be treated.

The theory hence complies with the experimental data contained in atomic spectra. In all those cases, however, where a visual description is required of a transient event, e.g. when interpreting Wilson photographs, the formalism of the theory does not seem to allow an adequate representation of the experimental state of affairs. At this point Schrödinger's wave mechanics, meanwhile developed on the basis of de Broglie's theses, came to the assistance of quantum mechanics.

In the course of the studies which Mr. Schrödinger will report here himself he converted the determination of the energy values of an atom into an eigenvalue problem defined by a boundary-value problem in the coordinate space of the particular atomic system. After Schrödinger had shown the mathematical equivalence of wave mechanics, which he had discovered, with quantum mechanics, the fruitful combination of these two different areas of physical ideas resulted in an extraordinary broadening and enrichment of the formalism of the quantum theory. Firstly, it was only wave mechanics which made possible the mathematical treatment of complex atomic systems, secondly analysis of the connection between the two theories led to what is known as the *transformation theory* developed by Dirac and Jordan. As it is impossible within the limits of the present lecture to give a detailed discussion of the mathematical structure of this theory, I should just like to point out its fundamental physical significance. Through the adoption of the physical principles of quantum mechanics into its expanded formalism, the *transformation theory* made it possible in completely general terms to calculate for atomic systems the probability for the occurrence of a particular, experimentally ascertainable, phenomenon under given experimental conditions. The hypothesis conjectured in the studies on the radiation theory and enunciated in precise terms in Born's collision theory, namely that the wave function governs the probability for the presence of a corpuscle, appeared to be a special case of a more general pattern of laws and to be a natural consequence of the fundamental assumptions of quantum mechanics. Schrödinger, and in later studies Jordan, Klein, and Wigner as well, had succeeded in developing as far as permitted by the principles of the quantum theory de Broglie's original concept of visualizable matter waves occurring in space and time, a concept formulated even before the development of quantum mechanics. But for that the connection between Schrödinger's concepts and de Broglie's original thesis would certainly have seemed a looser one by this statistical interpretation of wave mechanics and by the greater emphasis on the fact that Schrödinger's theory is concerned with waves in multidimensional space. Before proceeding to discuss the explicit significance of quantum mechanics it is perhaps right for me to deal briefly with this question as to the existence of matter waves in three-dimensional space, since the solution to this problem was only achieved by combining wave and quantum mechanics.

A long time before quantum mechanics was developed Pauli had inferred from the laws in the Periodic System of the elements the well-known principle that a particular quantum state can at all times be occupied by only a single electron. It proved possible to transfer this principle to quantum mechanics on the basis of what at first sight seemed a surprising result: the entire complex of stationary states which an atomic system is capable of adopting breaks down into definite classes such that an atom in a state belonging to one class can never change into a state belonging to another class under the action of whatever perturbations. As finally clarified beyond question by the studies of Wigner and Hund, such a class of states is characterized by a definite symmetry characteristic of the Schrödinger eigenfunction with respect to the transposition of the coordinates of two electrons. Owing to the fundamental identity of electrons, any external perturbation of the atom remains unchanged when two electrons are exchanged and hence causes no transitions between states of various classes. The Pauli principle and the Fermi-Dirac statistics derived from it are equivalent with the assumption that only that class of stationary states is achieved in nature in which the eigenfunction changes its sign when two electrons are exchanged. According to Dirac, selecting the symmetrical system of terms would lead not to the Pauli principle, but to Bose-Einstein electron statistics.

Between the classes of stationary states belonging to the Pauli principle or to Bose-Einstein statistics, and de Broglie's concept of matter waves there is a peculiar relation. A spatial wave phenomenon can be treated according to the principles of the quantum theory by analyzing it using the Fourier theorem and then applying to the individual Fourier component of the wave motion, as a system having one degree of freedom, the normal laws of quantum mechanics. Applying this procedure for treating wave phenomena by the quantum theory, a procedure that has also proved fruitful in Dirac's studies of the theory of radiation, to de Broglie's matter waves, exactly the same results are obtained as in treating a whole complex of material particles according to quantum mechanics and selecting the symmetrical system of terms. Jordan and Klein hold that the two methods are mathematically equivalent even if allowance is also made for the interaction of the electrons, i.e. if the field energy originating from the continuous space charge is included in the calculation in de Broglie's wave theory. Schrödinger's considerations of the energy-momentum tensor assigned to the matter waves can then also be adopted in this theory as consistent components of the formalism. The studies of Jordan and Wigner show that modifying the commutation relations underlying this quantum theory of waves results in a formalism equivalent to that of quantum mechanics based on the assumption of Pauli's exclusion principle.

These studies have established that the comparison of an atom with a planetary system composed of nucleus and electrons is not the only visual picture of how we can imagine the atom. On the contrary, it is apparently no less correct to compare the atom with a charge cloud and use the correspondence to the formalism of the quantum theory borne by this concept to derive qualitative conclusions about the behavior of the atom. However, it is the concern of wave mechanics to follow these consequences.

308

Reverting therefore to the formalism of quantum mechanics; its application to physical problems is justified partly by the original basic assumptions of the theory, partly by its expansion in the transformation theory on the basis of wave mechanics, and the question is now to expose the explicit significance of the theory by comparing it with classical physics.

In classical physics the aim of research was to investigate objective processes occurring in space and time, and to discover the laws governing their progress from the initial conditions. In classical physics a problem was considered solved when a particular phenomenon had been proved to occur objectively in space and time, and it had been shown to obey the general rules of classical physics as formulated by differential equations. The manner in which the knowledge of each process had been acquired, what observations may possibly have led to its experimental determination, was completely immaterial, and it was also immaterial for the consequences of the classical theory, which possible observations were to verify the predictions of the theory. In the quantum theory, however, the situation is completely different. The very fact that the formalism of quantum mechanics cannot be interpreted as visual description of a phenomenon occurring in space and time shows that quantum mechanics is in no way concerned with the objective determination of space-time phenomena. On the contrary, the formalism of quantum mechanics should be used in such a way that the probability for the outcome of a further experiment may be concluded from the determination of an experimental situation in an atomic system, providing that the system is subject to no perturbations other than those necessitated by performing the two experiments. The fact that the only definite known result to be ascertained after the fullest possible experimental investigation of the system is the probability for a certain outcome of a second experiment shows, however, that each observation must entail a discontinuous change in the formalism describing the atomic process and therefore also a discontinuous change in the physical phenomenon itself. Whereas in the classical theory the kind of observation has no bearing on the event, in the quantum theory the disturbance associated with each observation of the atomic phenomenon has a decisive role. Since, furthermore, the result of an observation as a rule leads only to assertions about the probability of certain results of subsequent observations, the fundamentally unverifiable part of each perturbation must, as shown by Bohr, be decisive for the non-contradictory operation of quantum mechanics. This difference between classical and atomic physics is understandable, of course, since for heavy bodies such as the planets moving around the sun the pressure of the sunlight which is reflected at their surface and which is necessary for them to be observed is negligible; for the smallest building units of matter, however, owing to their low mass, every observation has a decisive effect on their physical behavior.

The perturbation of the system to be observed caused by the observation is also an important factor in determining the limits within which a visual description of atomic phenomena is possible. If there were experiments which permitted accurate measurement of all the characteristics of an atomic system necessary to calculate classical motion, and which, for example, supplied accurate values for the location and velocity of each electron in the system at a particular time, the result of these experiments could not be utilized at all in the formalism, but rather it would directly contradict the formalism. Again, therefore, it is clearly that fundamentally unverifiable part of the perturbation of the system caused by the measurement itself which hampers accurate ascertainment of the classical characteristics and thus permits quantum mechanics to be applied. Closer examination of the formalism shows that between the accuracy with which the location of a particle can be ascertained and the accuracy with which its momentum can simultaneously be known, there is a relation according to which the product of the probable errors in the measurement of the location and momentum is invariably at least as large as Planck's constant divided by $4\pi$. In a very general form, therefore, we should have

$$\Delta p \, \Delta q \geq h/4\pi$$

where p and q are canonically conjugated variables. These uncertainty relations for the results of the measurement of classical variables form the necessary conditions for enabling the result of a measurement to be expressed in the formalism of the quantum theory. Bohr has shown in a series of examples how the perturbation necessarily associated with each observation indeed ensures that one cannot go below the limit set by the uncertainty relations. He contends that in the final analysis an uncertainty introduced by the concept of measurement itself is responsible for part of that perturbation remaining fundamentally unknown. The experimental determination of whatever space-time events invariably necessitates a fixed frame - say the system of coordinates in which the observer is at rest - to which all measurements are referred. The assumption that this frame is "fixed" implies neglecting its momentum from the outset, since "fixed" implies nothing other, of course, than that any transfer of momentum to it will evoke no perceptible effect. The fundamentally necessary uncertainty at this point is then transmitted via the measuring apparatus into the atomic event.

Since in connection with this situation it is tempting to consider the possibility of eliminating all uncertainties by amalgamating the object, the measuring apparatuses, and the observer into one quantum-mechanical system, it is important to emphasize that the act of measurement is necessarily visualizable, since, of course, physics is ultimately only concerned with the systematic description of space-time processes. The behavior of the observer as well as his measuring apparatus must therefore be discussed according to the laws of classical physics, as otherwise there is no further physical problem whatsoever. Within the measuring apparatus, as emphasized by Bohr, all events in the sense of the classical theory will therefore be regarded as determined, this also being a necessary condition before one can, from a result of measurements, unequivocally conclude what has happened. In quantum theory, too, the scheme of classical physics which objectifies the results of observation by assuming in space and time processes obeying laws is thus carried through up to the point where the fundamental limits are imposed by the unvisualizable character of the atomic events symbolized by Planck's constant. A visual description for the atomic events is possible only within certain limits of accuracy - but within these limits the laws of classical physics also still apply. Owing to these limits of accuracy as defined by the uncertainty relations, moreover, a visual picture of the atom free from ambiguity has not been determined. On the contrary the corpuscular and the wave concepts are equally serviceable as a basis for visual interpretation.

The laws of quantum mechanics are basically statistical. Although the parameters of an atomic system are determined in their entirety by an experiment, the result of a future observation of the system is not generally accurately predictable. But at any later point of time there are observations which yield accurately predictable results. For the other observations only the probability for a particular outcome of the experiment can be given. The degree of certainty which still attaches to the laws of quantum mechanics is, for example, responsible for the fact that the principles of conservation for energy and momentum still hold as strictly as ever. They can be checked with any desired accuracy and will then be valid according to the accuracy with which they are checked. The statistical character of the laws of quantum mechanics, however, becomes apparent in that an accurate study of the energetic conditions renders it impossible to pursue at the same time a particular event in space and time.

For the clearest analysis of the conceptual principles of quantum mechanics we are indebted to Bohr who, in particular, applied the concept of complementarity to interpret the validity of the quantum-mechanical laws. The uncertainty relations alone afford an instance of how in quantum mechanics the exact knowledge of one variable can exclude the exact knowledge of another. This complementary relationship between different aspects of one and the same physical process is indeed characteristic for the whole structure of quantum mechanics. I had just mentioned that, for example, the determination of energetic relations excludes the detailed description of space- time processes. Similarly, the study of the chemical properties of a molecule is complementary to the study of the motions of the individual electrons in the molecule, or the observation of interference phenomena complementary to the observation of individual light quanta. Finally, the areas of validity of classical and quantum mechanics can be marked off one from the other as follows: Classical physics represents that striving to learn about Nature in which essentially we seek to draw conclusions about objective processes from observations and so ignore the consideration of the influences which every observation has on the object to be observed; classical physics, therefore, has its limits at the point from which the influence of the observation on the event can no longer be ignored. Conversely, quantum mechanics makes possible the treatment of atomic processes by partially foregoing their space- time description and objectification.

So as not to dwell on assertions in excessively abstract terms about the interpretation of quantum mechanics, I would like briefly to explain with a well-known example how far it is possible through the atomic theory to achieve an understanding of the visual processes with which we are concerned in daily life. The interest of research workers has frequently been focused on the phenomenon of regularly shaped crystals suddenly forming from a liquid, e.g. a supersaturated salt solution. According to the atomic theory the forming force in this process is to a certain extent the symmetry characteristic of the solution to Schrödinger's wave equation, and to that extent crystallization is explained by the atomic theory. Nevertheless, this process retains a statistical and - one might almost say - historical element which cannot be further reduced: even when the state of the liquid is completely known before crystallization, the shape of the crystal is not determined by the laws of quantum mechanics. The formation of regular shapes is just far more probable than that of a shapeless lump. But the ultimate shape owes its genesis partly to an element of chance which in principle cannot be analyzed further.

Before closing this report on quantum mechanics, I may perhaps be allowed to discuss very briefly the hopes that may be attached to the further development of this branch of research. It would be superfluous to mention that the development must be continued, based equally on the studies of de Broglie, Schrödinger, Born, Jordan, and Dirac. Here *the attention of the research workers is primarily directed to the problem of reconciling the claims of the special relativity theory with those of the quantum theory*. The extraordinary advances made in this field by Dirac about which Mr. Dirac will speak here, meanwhile leave open the question *whether it will be possible to satisfy the claims of the two theories without at the same time determining the Sommerfeld fine-structure constant*. The attempts made hitherto to achieve a *relativistic* formulation of the quantum theory are all based on visual concepts so close to those of classical physics that it seems impossible to determine the fine-structure constant within this system of concepts. The expansion of the conceptual system under discussion here should, furthermore, be closely associated with the further development of the *quantum theory of wave fields*, and it appears to me as if this formalism, notwithstanding its thorough study by a number of workers (Dirac, Pauli, Jordan, Klein, Wigner, Fermi) has still not been completely exhausted. Important pointers for the further development of quantum mechanics also emerge from the experiments involving the structure of the atomic nuclei. From their analysis by means of the Gamow theory, it would appear that between the elementary particles of the atomic nucleus forces are at work which differ somewhat in type from the forces determining the structure of the atomic shell; Stem's experiments seem, furthermore, to indicate that the behavior of the heavy elementary particles cannot be represented by the formalism of Dirac's theory of the electron. Future research will thus have to be prepared for surprises which may otherwise come both from the field of experience of nuclear physics as well as from that of cosmic radiation. But however, the development proceeds in detail, the path so far traced by the quantum theory indicates that an understanding of those still unclarified features of atomic physics can only be acquired by foregoing visualization and objectification to an extent greater than that customary hitherto. We have probably no reason to regret this, because the thought of the great epistemological difficulties with which the visual atom concept of earlier physics had to contend gives us the hope that the abstracter atomic physics developing at present will one day fit more harmoniously into the great edifice of Science.

## Chadwick, J. & Goldhaber, M. (August, 1934). A Nuclear Photo-effect: Disintegration of the Diplon* by γ-Rays.

*Nature* **134**, 237–8 (1934). https://doi.org/10.1038/134237a0.

*a diplon is a nucleus consisting of a proton and a neutron.

## Abstract

*By analogy with the excitation and ionization of atoms by light, one might expect that any complex nucleus should be excited or 'ionized', that is, disintegrated, by γ-rays of suitable energy.* Disintegration would be much easier to detect than excitation. *The necessary condition to make disintegration possible is that the energy of the γ-ray must be greater than the binding energy of the emitted particle.* The γ-rays of thorium C″ of hv $= 2.62 \times 10^6$ electron volts are the most energetic which are available in sufficient intensity, and therefore one might expect to produce disintegration with emission of a heavy particle, *such as a neutron, proton*, etc., only of those nuclei which have a small or negative mass defect; for example, $D_2$, $Be_9$, and the radioactive nuclei which emit α-particles. The emission of a positive or negative electron from a nucleus under the influence of γ-rays would be difficult to detect unless the resulting nucleus were radioactive.

146        H. Bethe and R. Peierls

glide plane in preference to the latter, on which glide has never been detected.

Double and triple glide can take place, according to the ordinary laws governing the glide plane and glide direction. Hardening on one set of glide planes hardens the whole crystal.

The critical shear stress at − 43° C is 9·3 gm wt. per sq. mm.

The very purest mercury gives, in the early stages of stretch, an average spacing of the glide planes which is 0·0054 cm, agreeing closely with that for ordinary pure mercury, and for mercury deliberately contaminated. The appearance of preferred glide planes separated by about 15,000 lattice spacings is thus not due to impurities or dissolved gas.

Under constant stress the single crystal flows at a rate which diminishes to a constant value.

---

Quantum Theory of the Diplon

By H. BETHE and R. PEIERLS, University of Manchester

*(Communicated by D. R. Hartree, F.R.S. Received July 26, 1934)*

1.—INTRODUCTION

The work of Heisenberg,[†] Majorana,[‡] and Wigner[§] seems to show that the behaviour of protons and neutrons and their interaction in the nucleus may be described by the ordinary methods of quantum mechanics. It is of particular interest to study the simplest nuclear system, *i.e.*, the diplon, which almost certainly consists of a proton and a neutron. In dealing with such a two-body problem, the wave equation can be rigorously solved if the forces are known, and this problem therefore has the same importance for nuclear mechanics as the hydrogen atom has for atomic theory.

The force acting between a proton and a neutron has been investigated by Wigner (*loc. cit.*) who showed that in order to understand the high mass defect of $He^4$ compared with $H^2$ one must assume interaction forces with a range much smaller than the radius of $H^2$. Without knowing about these forces more than the binding energy of $H^2$, one can, then, investigate quantitatively the behaviour of $H^2$ against various perturbations.

[†] 'Z. Physik,' vol. 77, p. 1 (1932) ; vol. 78, p. 156 (1932) ; vol. 80, p. 587 (1933).
[‡] 'Z. Physik,' vol. 82, p. 137 (1932).
[§] 'Phys. Rev.,' vol. 43, p. 252 (1933) ; 'Z. Physik,' vol. 83, p. 253 (1933).

## Chadwick, J. & Goldhaber, M. (September, 1935). The Nuclear Photoelectric Effect.

*Proc. R. Soc. Lond. A*; 151, 873, 479-93; doi: 10.1098/rspa.1935.0162.

Some time ago we reported in '*Nature*'* the observation of a ***nuclear photo-effect***, the disintegration of the ***deuteron*** [ a +1 charged particle of mass 2, and a neutral particle of mass 1] by γ-rays.

*Chadwick, J. & Goldhaber, M. (1934). *Nature*, 134, 237.

### 1. Introduction.

An effect of γ-rays upon complex nuclei might be expected to occur from analogy with the phenomena of excitation and ionization of atoms by light, and such an effect has been looked for from time to time by various investigators. A necessary condition to make disintegration possible is that the energy of the γ-ray quantum must be greater than the binding energy of the particle which is to be removed from the nucleus. The most energetic γ-rays which are readily available in sufficient intensity are those of thorium C", which have an energy $h\nu = 2 \cdot 62 \times 10^6$ electron volts. One can hope, therefore, using these γ-rays, to produce disintegration with the emission of a heavy particle, such as a ***neutron, proton, etc.***, ***only in those nuclei which have a small or negative mass defect***, such as the nuclei of deuterium, beryllium, and those radioactive elements which emit α-particles. In fact, only the nuclei of deuterium and beryllium have so far been disintegrated in this way. The disintegration of beryllium by the γ-rays of radium was first reported by Szilard and Chalmers**.

** Szilard, L. & Chalmers, T. (September, 1934). *Detection of Neutrons Liberated from Beryllium by Gamma Rays: a New Technique for Inducing Radioactivity.* *Nature,* 134, 494–5; https://doi.org/10.1038/134494b0.

No evidence of a photo-electric disintegration amongst the radioactive elements has yet been found.

### 2—Experiments with Deuterium

The first element to be examined for this nuclear photo-effect was the heavy isotope of hydrogen, ***deuterium***. This was chosen because it was well established that the ***deuteron*** has a small mass defect. The disintegration to be expected is

$$_1D^2 + h\nu \rightarrow {}_1H^1 + {}_en^1. \tag{1}$$

If W is the binding energy of the *deuteron*, the energy available for the *proton* and *neutron* will be hv – W. Unless this energy is very small, the kinetic energies of the *proton* and *neutron* will be nearly equal since their *masses* are nearly the same and the momentum of the quantum is small. To establish the reaction (1) completely it is desirable to detect both the *proton* and the *neutron* released in the process.

**(a) Detection of the "Photo" protons**—The experimental arrangement for the detection of the protons released from deuterium, which we may for convenience call "photo''-protons, was as follows. An ionization chamber, 4 x 6 x 8 cm in dimension, was filled with heavy hydrogen of about 95% purity to atmospheric pressure. The chamber was connected to a linear valve amplifier and oscillograph in the usual way. A source of radiothorium, equivalent in γ-ray activity as measured through 6 mm of lead to 9 mg of radium, was placed at distances varying from 12 cm to 30 cm from the chamber. Blocks of lead were inserted between the source and the chamber to absorb the soft γ-rays emitted by the source; these would be ineffective in promoting the desired photo-effect while giving rise to a disturbing background due to the ionization produced in the chamber. The oscillograph records showed above the background due to the γ-radiation a number of deflections which could only be due to a heavy ionizing particle. When the heavy hydrogen was replaced by ordinary hydrogen the background remained the same, but very few deflections were observed and their number was not appreciably greater than the natural effect of the chamber, 15 "kicks" per hour on the average. Similarly, when the chamber was filled with nitrogen the numbers of kicks observed with and without the radiothorium source were about the same. *These observations show that the kicks observed with heavy hydrogen were not produced by the collisions of any neutrons which might perhaps be emitted by the source.* *

> * Radioactive sources often emit *neutrons* produced by the disintegration under $\alpha$-particle bombardment of light elements, e.g., boron, present in the material or container of the source.

They must be attributed to protons resulting from the splitting of the *deuteron* by the absorption of a γ-ray quantum.

The kicks were, allowing for the natural effect, all of about the same size; this is to be expected from the reaction (1) since the γ-radiation entering the chamber is mainly of the one energy hv = 2.62 x $10^6$ e.v.**

> ** A small variation in size will arise from the fact that the energy of the *proton* will vary with its direction of emission relative to the direction of the γ-ray by a noticeable amount, as a simple calculation shows.

316

An estimate of the energy of the photo-protons can be deduced from the measurement of the size of the oscillograph kicks. From experiments in which α-particles of known residual range were admitted into the ionization chamber it was found that an oscillograph deflection of 1 mm on the record corresponded to the production of 1,200 ion pairs. The average size of the kicks due to the photo-protons was about 6 mm, corresponding to the production of 7,200 ion pairs. Assuming that the amount of energy lost by a proton in producing 1 ion pair is the same as that lost by an α particle, viz., 33 e.v., we find that the energy of the photo-protons in our experiments was about 240,000 e.v. This estimate of the proton energy, although rough, is not likely to be in error by more than 80,000 e.v.

The sum of the kinetic energies of the proton and neutron resulting from the division of the deuteron is thus nearly 500,000 e.v. The **binding energy of the deuteron** is therefore about $2.1 \times 10^6$ e.v. or 0.0023 in mass units.

If the masses of the atoms of hydrogen and deuterium were accurately known we could now fix the mass of the neutron with almost equal precision. Unfortunately, there is still some uncertainty about these masses. During the past few years, it has become apparent that there was some discrepancy between the masses of the light atoms as compared in the mass-spectrograph measurements of Aston and Bainbridge and in the data obtained from nuclear transformations. It has been shown by Oliphant, Kempton, and Rutherford,* and by Bethe** that the discrepancies probably arise by the cumulative effect of a small error in the ratio of the masses of $He^4 : O^{16}$.

* (1935), *Proc. Roy. Soc. A*, 150, p. 241.
** (1935), *Phys. Rev.*, 47, p. 634.

This suggestion seems to be confirmed by direct measurements of Aston. ***

*** (1935), *Nature*, 135, p. 541.

The values derived from disintegration data for the masses of the atoms of hydrogen and deuterium are 1.0081 and 2.0142 respectively. Adopting these values, we obtain a value for the **mass of the neutron** of 1.0084. If, however, we take Aston's recent provisional values, 1.0081 and 2.0148, we obtain a value for the **mass of the neutron** of 1.0090.

**(b) Probability of Disintegration**—The next point of interest is the **probability of the "photo"-disintegration**. This was measured approximately both in the experiments already described and also in others in which the source of radiation was the active deposit, thorium (B + C). In the latter experiments, a source equivalent in γ-ray activity to 3-4 mg radium was placed 30 cm away from the center of the chamber. A block of lead 26 mm thick was placed in front of the chamber to absorb the soft γ-rays. Thus, the number of γ-ray quanta of hv = 2-6 x $10^6$ e.v. passing through the chamber was about 8 x $10^4$ per minute, per sq cm of surface. The number of kicks due to the photo-protons liberated under these conditions was about 30 per hour. The volume of the chamber was about 190 cc, and thus the number of deuterons was $10^{22}$. We find, therefore, that the cross-section for disintegration of the deuteron by a γ-ray of 2.62 x $10^6$ e.v. is about 6 6 x $10^{-28}$ sq cm. The average value for this cross-section obtained from this and the previous experiments was about 5 x $10^{-28}$ sq cm with a possible error of a factor of 2*

* Owing to an arithmetical error, the value given for this cross-section in the note to *Nature* was 1 x $10^{-28}$ sq cm.

*The effects of the γ-rays of radium in producing the photo-disintegration was also examined*. A radium source of 7 mg was used and the numbers of kicks observed when the ionization chamber filled with deuterium was exposed alternately to this **radium** source and the **radiothorium** source of 9 mg were counted. With the radiothorium source $10^7$ kicks per hour were obtained, compared with 20 kicks per hour with the radium source, while the natural effect to be deducted was 15 per hour. *It is clear that the radium γ-rays are much less effective than those of radio thorium*. This is indeed only to be expected if the binding energy of the **deuterium** is as high as 2.1 x $10^6$ e.v. for in the γ-ray spectrum of radium (B + C) there are only a few weak lines with energies greater than this. The excess of 5 kicks per hour over the natural effect shown in the radium experiments is barely outside the statistical error of the observations, but, as we shall show later, a detectable effect is in fact produced by the γ-rays of radium.

**(c) The "Photo"- neutrons from Deuterium**—To show the liberation of **neutrons** from **deuterium** by the effect of γ-rays we made use in the first experiments of the phenomenon of induced activity. The source of radiothorium was placed in the inner tube of a double-walled glass vessel, the outer tube of which contained heavy water. In the earlier experiments only about 10 cc of water containing about 20% $D_2O$ were available, but later we were able to use about 20 cc of 98% $D_2O$. A cylinder of silver was placed round the glass vessel and thus exposed to the **neutrons** liberated from the deuterium. After 2 to 10 minutes exposure the silver cylinder was removed and placed so as to enclose a Geiger-Miiller tube counter which had a wall of copper sufficiently thin to admit β-rays. In some experiments we found it convenient to use a silver-walled counter as detector. Only a small and somewhat uncertain effect was obtained in this way. The source + heavy water + silver tube or silver counter were therefore surrounded during the exposure by a large cylinder of paraffin wax. As Fermi and his collaborators have shown* the induced effect produced in silver by **neutrons** may be largely increased in this way.

* Amaldi, D'Agostino, Fermi, Pontecorvo, Rasetti,& Segre, (1935), *Proc. Roy. Soc., A*, 149, p. 522.

Under these conditions the silver showed a strong activity which decayed with the known periods of 22 secs and 2.3 min, and the liberation of **neutrons** from **deuterium** by the γ-rays of radiothorium was clearly established. We were also able to show, using a source of radon of about 150 millicuries, that the γ-rays of radium (B + C) are able to disintegrate **deuterium**.

To compare the relative effects of the γ-rays from radiothorium and radium we found it more convenient to use the disintegration effects provoked by slow **neutrons** rather than the induced radioactivity. When lithium and boron are exposed to slow **neutrons** the following reactions take place*

* Chadwick & Goldhaber, (1935). *Nature*, 135, p. 65; Amaldi and others, *loc. cit.*; Taylor & Goldhaber, (1935). *Nature*, 135, p. 341.

$$Li^6 + n^1 \rightarrow He^4 + H^3,$$
$$B^{10} + n^1 \rightarrow Li^7 + He^4.$$

In these reactions the particles are emitted with considerable energy and are therefore readily detectable by their ionization effects. The cross-section for these reactions is very large and thus they afford a very sensitive means of detecting slow **neutrons**.

The inner surface of an ionization chamber was coated with a thin layer of lithium metal or boron powder and the electrode was connected to a valve amplifier and oscillograph. The capture of a **neutron** by the lithium or boron was recorded by a kick of the oscillograph due to one of the particles emitted in the disintegration process. In order to reduce the unsteady background due to the presence of γ-rays a block of lead of suitable thickness was interposed between the source and ionization chamber. (The absorption of **neutrons** in the lead block was small and could be neglected in these experiments.) The whole arrangement, consisting of source + heavy water, lead and ionization chamber, was surrounded by paraffin wax. For convenience we shall use the term "**slow neutron intensity**" to denote the effect of a **neutron** source measured in such an arrangement. Using a lithium-coated chamber having a total lithium surface of about 40 sq cm the natural effect—the number of kicks in the absence of the source—was about 40 per hour. With a source of radiothorium of 8 mg γ-ray activity surrounded by 15 cc of heavy water (98%), the number of deflections observed was more than 2,000 per hour. To observe the effect of the γ-rays of radium (B + C) a radon source of about 150 millicuries was used, contained in a copper tube of 1 mm inner diameter and 12 mm length. *

*Such a source was found to emit very few *neutrons* as compared with radon sources contained in glass tubes.

It was found that the slow *neutron* intensity of this source was about 1/27 of that for radiothorium source of equal γ-ray activity. One must not conclude from this observation that the number of photo-neutrons from radiothorium + D is 27 times larger than the number of photo-neutrons from radon + D, (for equal γ-ray intensities), since the initial velocities of the *neutrons*, and hence their mean free paths in paraffin, are different in the two cases. The velocities of the *neutrons* from radon + D are smaller than those of the *neutrons* from radiothorium + D and their mean free paths in paraffin will therefore be smaller, so that the number of photo-neutrons from radon + D will be somewhat greater than 1 /27 of the number from radiothorium + D.

When no paraffin wax was used around the source and ionization chamber some effect was still observed, about 1/50 of the effect with paraffin. This effect was shown to be due partly to disintegrations produced by photo-neutrons of the primary energy and partly to disintegrations produced by some slow *neutrons*. Owing to the smallness of the effect it was not possible to settle definitely how these slow *neutrons* were produced, but it seems probable that they were due to scattering of photo-neutrons in the heavy water and were not produced by γ-rays of energy very close to the threshold value.

**(d) The Angular Distribution of the "Photo" Neutrons**—The probability of the direction of emission of the photo-proton or neutron with respect to the direction of the γ-ray is a point of some importance in the quantum theory of the *deuteron*. A detailed study of the angular distribution of the protons is possible with the aid of an expansion chamber filled with gas containing **deuterium**, and this is now in progress. In the meantime, however, we have obtained some rough information about the angular distribution of the neutrons in a plane containing the γ-ray beam by using the arrangement shown in fig. 1.

A cylindrical ionization chamber 1 of 2 cm diameter, with its inner wall and collecting rod coated with boron powder, was used as the *neutron detector*. It was enclosed in a block of paraffin wax. The radiothorium source was placed in positions A or B and in both cases, observations were made with and without the vessel of heavy water, 3 cm in diameter) containing 23 cc of $D_2O$ (98%), at C. The thickness of lead, *a*, was 3.5 cm, and the distances, b and c, were 7.5 cm and 3 cm respectively. The *neutron* intensity found with the source at B was about twice as great as with the source at A, showing that more photo-neutrons are emitted in a direction at right angles to the incident γ-ray beam than in the forward direction. The effects observed were too small to allow the geometry of the arrangement to be made more definite, but the result is sufficient to show that the angular distribution of the **neutrons** is not spherically symmetrical with probably a maximum at right angles to the γ-ray beam.

### 3—Experiments with Beryllium

(a) [As noted above,] the disintegration of beryllium by γ-rays was first shown by Szilard and Chalmers. *

> \* Szilard, L. & Chalmers, T. (September, 1934). *Detection of Neutrons Liberated from Beryllium by Gamma Rays: a New Technique for Inducing Radioactivity.* *Nature,* 134, 494–5; https://doi.org/10.1038/134494b0.

They used the γ-radiation of radium and observed the photo-neutrons by means of the induced activity produced in iodine. Two processes seem to be possible energetically with the γ-rays used: —

$$Be^9 + hv \rightarrow Be^8 + n^1, \qquad (2)$$
$$Be^9 + hv \rightarrow He^4 + He^4 + n^1. \qquad (3)$$

It may be that both reactions take place and that their relative probability depends on the energy of the γ-radiation. The experiments we shall now describe lead indirectly to the conclusion that reaction (2) probably predominates.

In our first experiments to detect the disintegration of beryllium by γ-rays we exposed an ionization chamber, the face of which was covered with a beryllium foil or beryllium powder, to the γ-rays of radium and radio-thorium. The chamber was connected to a linear amplifier in order to detect the charged particles, whether $Be^8$ or $He^4$, resulting from the disintegration. No definite effect was found. This might have been due either to the small probability of the disintegration process or to the small range of the charged particles liberated.

The detection of the **neutrons** is possible by a variety of means. In some experiments we have made use of the induced activity of silver, in others of the disintegration effects produced in a boron- or lithium-coated ionization chamber. *In general, the effects are small unless paraffin wax or water is used to slow down the neutrons*. Using a similar arrangement to that described in § 2 for the photo-neutrons from deuterium—lithium-coated chamber, source and chamber surrounded by paraffin—we have compared the intensities of the slow **neutrons** obtained when beryllium is irradiated by radium or radiothorium. The ratio found was about 2:1 for sources of equal γ-ray intensity, agreeing with a result of Gentner. *

* (1934). *C.R. Acad. Sci. Paris*, 199, p. 1211.

One must not conclude from this result that the effective cross-section for disintegration of beryllium by radon γ-rays is larger than that for disintegration by radiothorium γ-rays, for the **neutrons** will be liberated with different energies in the two cases and will therefore have different mean free paths in paraffin. Although the difference between the relative effects of radiothorium and radium γ-rays in deuterium and beryllium is very striking; it is not possible to make any certain deduction from the above observations about the dependence of the probability of disintegration with the energy of the γ-ray quantum.

**(b) The Threshold Value of Beryllium**—A matter of some interest is the minimum energy of the γ-ray quantum which will disintegrate beryllium. In some preliminary experiments we attempted to obtain some information on this point by interposing lead absorbers between the radon source and the beryllium, and measuring the decrease in the slow **neutron** intensity as the thickness of lead was increased. *The results showed that no rays of energy less than $10^6$ e.v. are able to disintegrate beryllium*. The threshold value for the energy could not be fixed with any accuracy in this way, for when the geometry of the arrangement was made reasonably definite the effects obtained were rather small.

We therefore proceeded to use the following method to determine the energy necessary to remove a **neutron** from beryllium. In an ionization chamber filled with either hydrogen or helium the effects due to the recoil atoms produced by impact of the photo-neutrons from radon-Be or radiothorium-Be can be readily observed. We measured the maximum energy of the recoil atoms produced in a helium chamber when bombarded by photo-neutrons from radiothorium-Be. With a source of radio-thorium the photo-neutrons are almost entirely due to the strong line of hv = 2.62 x $10^6$ e.v., and this source was therefore preferred to a radon source. Helium is in this case more suitable than hydrogen because the maximum range of the recoil atoms, which must be small compared with the dimensions of the chamber, is shorter; moreover, the oscillograph deflections due to the recoil atoms can be compared directly with those produced when α-particles of known range are admitted into the ionization chamber.

The radiothorium source was surrounded by pieces of beryllium metal of a total weight of 450 gm. To reduce the γ-ray background a suitable thickness of lead (from 3 to 6 cm) was interposed between the photo-neutron source and the helium chamber. By plotting a diagram of the number of deflections against their size and allowing for the small natural effect it was found that the maximum deflection was about 10 mm. This corresponded to a maximum energy of the recoil atoms of 580,000 e.v. Since in a collision with a *neutron* the helium atom receives a maximum of 16/25 of the neutron energy this leads to a maximum energy of the photo-neutrons of about 900,000 e.v. As at the most 8/9 of the kinetic energy of the particles resulting from the disintegration can be acquired by the *neutron* we obtain for the total surplus energy about $1 \times 10^6$ e.v. Since the $\gamma$-radiation used has an energy of $2.6 \times 10^6$ e.v. this means that an energy of about $1.6 \times 10^6$ e.v. is required to remove a *neutron* from $Be^9$. This is in general agreement with the experiments of Brasch, Lange, and others, * who found that beryllium can be disintegrated by x-rays of energy less than $2 \times 10^6$ e.v., and those of Arzimovitch and Palibin, ** who found that x-rays of $1.35 \times 10^6$ e.v. were ineffective.

* Brasch, Lange, Waly, Banks, Chalmers, Szilard, & Hopwood, (1934). *Nature*, 134, p. 880.
** (1935). *Z. Phys. Sowjet*, 7, p. 245.

**(c) The Disintegration Process**—The above result is consistent with the reaction

$$Be^9 + h\nu \rightarrow Be^8 + n^1. \tag{2}$$

The mass difference between $Be^9$ and $Be^8$ obtained from this reaction using our result agrees well with that determined by Oliphant, Kempton, and Rutherford * from the reaction

$$Be^9 + H^1 \rightarrow Be^8 + D^2. \tag{3}$$

* (1935), *Proc. Roy. Soc. A*, 150, p. 241.

Other evidence points indirectly to the same process (2). The failure to observe charged recoil atoms in the photo-disintegration of beryllium would be immediately explained, for recoil atoms of $Be^8$ would be short in range and of small energy. Moreover, if reaction (3) were the disintegration process one would expect to obtain in a three-body reaction some *neutrons* of small energy. **

** The term "small energy" here means energies up to a few thousand volts. Some unpublished experiments indicate that *neutrons* of these energies have still a high probability of disintegrating lithium.

This does not appear to be the case. Using lithium-and boron-coated ionization chambers without paraffin, we found a small but measurable effect which was not appreciably reduced when a plate of silver 3-5 mm thick was placed between the photo-neutron source, either radon-Be or radiothorium-Be, and the chamber. Such a silver plate is sufficient to absorb to a large extent any slow *neutrons* present. By placing a paraffin block behind the chamber, the relative effect of *neutrons* of small energy would be increased owing to the larger scattering in the paraffin. No appreciable difference, however, could be observed with and without the silver absorber in the primary *neutron* beam. Absorbers of lithium fluoride and borax were also used with the same negative result. These experiments indicate that the proportion of photo-neutrons of low energy is small.

Assuming that the reaction (2) is correct, which implies that the photo-neutrons from beryllium with a radiothorium source are homogeneous in velocity, a lower limit for the probability of the photo-disintegration by these γ-rays can be obtained in the following way. We found that the slow *neutron* intensities per atom Be or D when a radiothorium source was surrounded by beryllium powder or heavy water—keeping the geometry as far as possible the same—were in the ratio of 1 to 6. ***

    *** This ratio depends, of course, to some degree on the geometry of the general arrangement (paraffin, etc.) and on the indicator used for the slow *neutrons*.

As the mean free path in paraffin of the Be photo-neutrons will be greater than that of the slower D photo-neutrons we conclude that the cross-section for the disintegration of beryllium is greater than one-sixth of the cross-section for *deuterium*, i.e.,
$\sigma_{Be} > 1/6 \, \sigma_D \sim 10^{-28}$ sq cm using γ-rays of $h\nu = 2.62 \times 10^6$ e.v.

**(d) Angular Distribution**—The *angular distribution of the photo-neutrons* from beryllium was investigated, using an arrangement similar to that used for the D photo-neutrons and described in §2 and fig. 1. The vessel with the heavy water was replaced by a beryllium block of 98 gm weight. The block was of irregular shape with a maximum linear dimension of 6 cm. A radon source of about 150 millicuries was used, and the effects were large enough to allow a better-defined geometry than with heavy water. The distances a, b, c, fig. 1, were now a = 9 cm, b = 12 cm, c = 7.5 cm. The effects found with the source in positions A and B were equal within the probable error of 15%. Thus, the angular distribution of the photo-neutrons from beryllium differs markedly from that found for the photo-neutrons from *deuterium*.

**4**—We have given in the course of this paper measurements of the relative effects of the photo-neutrons from *deuterium* and beryllium under the action of radium and radiothorium γ-rays, as determined by the relative numbers of slow *neutrons* obtained in a paraffin scatterer. For convenience the results are collected in Table I in which the slow *neutron* intensities per mg γ-ray activity per nucleus D or Be are given relative to the intensity obtained from radiothorium-D.

Table I

| Source of γ-rays\ | \Nucleus | D | Be |
|---|---|---|---|
| RdTh | | 1 | 1/6 |
| Rn | | 1/27 | 1/3 |

These four sources of **neutrons** are useful in many experiments in which **neutrons** of fairly homogeneous or not too high velocity are required. They give **neutrons** which are not slow in the sense of the "slow" **neutrons** produced by paraffin scattering but give **neutrons** of energies from a few $10^4$ e.v. for radon-D up to $9 \times 10^6$ e.v. for radio-thorium-Be. By using **neutrons** of different energies, it may be possible to throw light on the processes which take place in the passage of **neutrons** through matter. It is interesting in this connection to note that when one of these photo-neutron sources was used no increase in effect could be obtained in a boron-coated chamber by surrounding it with lead; that is, no "slow" **neutrons** were produced by scattering in lead. On the other hand, we confirmed the result of Amaldi and others (*loc. cit..*) that lead slows down the **neutrons** from the usual beryllium-radon source. This suggests that these **neutrons** have not sufficient energy to excite a lead nucleus, for which process an energy of about $1.5 \times 10^6$ e.v. seems from Lea's experiments* to be required.

* (1935). *Proc. Roy. Soc. A*, 150, p. 637.

## 5—Search for Photo-effect in other Elements

*We have found no evidence of a nuclear photo-disintegration in any other elements*.

We have examined the following elements for the emission of charged particles, using a radiothorium source and sometimes also a radon source: lithium, boron, carbon, nitrogen, oxygen, fluorine, neon, aluminum, nickel, copper, zinc, and uranium. With uranium a foil of aluminum of 3 cm air equivalent was placed over the target of uranium oxide in order to prevent the α-particles of uranium from entering the chamber. It was expected that any α-particles emitted under the action of the γ-rays would have greater energy than the natural α-particles and would therefore pass through the aluminum screen. No evidence of such particles could be detected.

The following elements have been examined for the emission of **neutrons**: lithium, boron, carbon, nitrogen, oxygen, fluorine, magnesium, aluminum, phosphorus, chlorine, potassium, calcium, manganese, copper, zinc, bromine, tin, rhodium, silver, cadmium, tungsten, thallium, lead, bismuth, thorium, and uranium.

## 6—Discussion

The main importance of the *nuclear photo-electric effect* is that it gives a means of studying the interaction of electromagnetic radiation and heavy particles. The *deuteron* is of special interest because this is the simplest of all nuclear systems. As our experiments have shown directly, it consists of *a proton and a neutron bound together* with an energy of about $2.1 \times 10^6$ e.v. *The role of the deuteron in nuclear theory thus corresponds to the role of the hydrogen atom in atomic theory.* If the forces between the proton and the *neutron* were known exactly the properties of the deuteron-binding energy, probability of disintegration by γ-rays, etc.—could be calculated. On the assumption that the interaction between a proton and a *neutron* is very strong at short distances and is confined to a range of the order of $10^{-13}$ cm, Bethe and Peierls, * and also Massey and Mohr, ** have calculated the probability of disintegration of the *deuteron* by γ-rays.

* (1935). *Proc. Roy. Soc. A*, 148, p. 146.
** (1935). *Ibid.*, 148, p. 206.

*They assume further that the laws of non-relativistic quantum mechanics are valid for the interaction of the proton and neutron with each other and with radiation*; the proton is treated as a *point charge* and it is assumed that there is no interaction between the *neutron* and radiation. On this basis Bethe and Peierls obtained an excitation curve for the photo-disintegration of the *deuteron* in which the cross-section for disintegration is a function of the ratio of the energy of the quantum to the binding energy of the *deuteron*.

The maximum cross-section occurs when the quantum energy hv is equal to twice the binding energy W, and is given by

$$\sigma_{max} = 1/6 \ he^2/Mc \ . \ 1/W,$$

where ½ M is the reduced mass of the proton and *neutron*. The cross-section for a quantum of energy hv = γ. W is

$$\sigma = 8 \ . \ (\gamma - 1)^{3/2}/\gamma^3 . \ \sigma_{max}.$$

The cross-section for disintegration by a given γ-ray is thus completely determined when the binding energy of the *deuteron* is known. With the value given by the experiments of § 2 (a), $W = 2.1 \times 10^6$ e.v., the above formula gives $\sigma = 8 \times 10^{-28}$ sq cm for a γ -ray of hv $= 2.62 \times 10^6$ e.v. The direct measurement of the cross-section, § 2 (b), gave a value for σ of about $5 \times 10^{-28}$ sq cm with an uncertainty of a factor 2. This is a satisfactory agreement with the theoretical value.

An asymmetry of the angular distribution as found for the photo-neutrons from **deuterium**, § 2 (d), is to be expected on simple theoretical reasoning. If one assumes that the range of the forces between **neutron** and proton is small compared with the wave-length of the $\gamma$-ray used, namely 4.7 x $10^{-11}$ cm, and that the ground state of the **deuteron** has an orbital momentum $l = 0$, *

* Wigner, (1933). *Phys. Rev.*, 43, p. 252; Bethe & Peierls, *loc. cit.*

then the $\cos^2 \phi$ law should hold for the probability of emission of the photo-protons or **neutrons** at an angle $\phi$ with the direction of the electric vector, ** neglecting the momentum of the $\gamma$-ray.

** Cf. Wentzel, (1926). *Z. Physik*, 40, p. 574; Bethe, (1933). *Handb. der Physik*, 24, pt. 1, p. 482.

The apparently spherically symmetrical distribution of the photo-neutrons from beryllium, § 3 (d), suggests that the **neutron** in beryllium is probably bound in a state with orbital momentum $l = 1$.

A few remarks may be made about the **mass of the neutron**. As we have shown, our experiments fix the binding energy of the **deuteron** as about 2.1 x $10^6$ e.v. This leads to a value for the **mass of the neutron** of 1.0084 if we take the value for the **deutron** mass given by Oliphant, Kempton, and Rutherford, or Bethe, and 1.0090 if we take the new value of Aston. *It seems that the neutron is definitely heavier than the hydrogen atom (1.0081).*

*It is generally assumed that the neutron and proton behave in many ways like elementary particles* but yet can under certain conditions change into each other with the emission of an electron, negative or positive, as the case may be. *In order to conserve energy and spin it appears necessary to assume the existence of a neutrino*, a particle of small mass, no charge, and spin ½. *

* The non-conservation of energy is well known from the $\beta$-ray spectra. The non-conservation of spin appears from reactions such as

$$Mg^{24} + n^1 \rightarrow Na^{24} + H^1 \rightarrow Mg^{24} + e + H^1.$$

*If the neutron is definitely heavier than the hydrogen atom, then one must conclude that a free neutron is unstable*, i.e., it can change spontaneously into a proton + electron + neutrino, unless the neutrino has a mass of the order of the mass of the electron. These speculations must, however, depend on more exact measurements of the masses of hydrogen and deuterium.

In conclusion we wish to thank here Dr. D. E. Lea for the loan of the amplifier used in the Geiger counter experiments and Professor A. I. Leipunski for his help in some of the observations and for the loan of beryllium. One of us (M. G.) desires to acknowledge the financial assistance received from the International Student Service, London, and Magdalene College, Cambridge.

## Summary

An account is given of experiments on the nuclear photo-electric effect. With heavy hydrogen, the reaction

$$D^2 + h\nu \rightarrow H^1 + n^1$$

has been established, both protons and *neutrons* having been detected. From the energy of the protons released by a $\gamma$-ray of $h\nu = 2.62 \times 10^6$ e.v. the binding energy of the *deuteron* has been determined; it is about $2.1 \times 10^6$ e.v. This fixes the *mass of the neutron* with respect to the masses of H and D. Using the masses of H and D from disintegration data the *mass of the neutron* is 1.0084; using Aston's new values, the mass is 1.0090. The *neutron* appears to be definitely heavier than the hydrogen atom.

The probability of disintegration of the **deuteron** by $\gamma$-rays agrees satisfactorily with theoretical calculations.

Experiments with beryllium suggest that the main reaction is

$$Be^9 + h\nu \rightarrow Be^8 + n^1.$$

The energy necessary to remove a neutron from beryllium is about $1.6 \times 10^6$ e.v.

Some information about the angular distribution of the photo-neutrons is given for both cases. No evidence of a nuclear photo-effect has been observed for several other elements which were examined.

**Chadwick, J. (1935). The neutron and its properties.**

Nobel Lecture, December 12, 1935; https://www.nobelprize.org/uploads/2018/06/chadwick-lecture.pdf

The Nobel Prize in Physics 1935 was awarded to James Chadwick "for the discovery of the *neutron*". The Nobel Prize in Physics 1935. NobelPrize.org; https://www.nobelprize.org/prizes/physics/1935/summary/>. When Herbert Becker and Walter Bothe directed alpha particles (helium nuclei) at beryllium in 1930, a strong, penetrating radiation was emitted. One hypothesis was that this could be high-energy electromagnetic radiation. In 1932, however, James Chadwick proved that it consisted of a neutral particle with about the same mass as a proton. Ernest Rutherford had earlier proposed that such a particle might exist in atomic nuclei. Its existence now proven, it was called a "*neutron*". James Chadwick – Facts. NobelPrize.org. https://www.nobelprize.org/prizes/physics/1935/chadwick/facts/.

The idea that there might exist small particles with no electrical charge has been put forward several times. Nernst, for example, suggested that a neutral particle might be formed by a negative electron and an equal positive charge, and that these "*neutrons*" might possess many of the properties of the ether; while Bragg at one time suggested that the γ-rays emitted by radioactive substances consisted of small neutral particles, which, on breaking up, released a negative electron.

***The first suggestion of a neutral particle with the properties of the neutron we now know, was made by Rutherford in 1920.*** He thought that a proton and an electron might unite in a much more intimate way than they do in the hydrogen atom, and so form a particle of no net charge and with a mass nearly the same as that of the hydrogen atom. His view was that with such a particle as the first step in the formation of atomic nuclei from the two elementary units in the structure of matter - the proton and the electron - it would be much easier to picture how heavy complex nuclei can be gradually built up from the simpler ones. He pointed out that this neutral particle would have peculiar and interesting properties. It may be of interest to quote his remarks:

"Under some conditions, however, it may be possible for an electron to combine much more closely with the H nucleus, forming a kind of neutral doublet. Such an atom would have very novel properties. Its external field would be practically zero, except very close to the nucleus, and in consequence it should be able to move freely through matter. Its presence would probably be difficult to detect by the spectroscope, and it may be impossible to contain it in a sealed vessel. On the other hand, it should enter readily the structure of atoms, and may either unite with the nucleus or be disintegrated by its intense field.

The existence of such atoms seems almost necessary to explain the building up of the nuclei of heavy elements; for unless we suppose the production of charged particles of very high velocities it is difficult to see how any positively charged particle can reach the nucleus of a heavy atom against its intense repulsive field."

Rutherford's conception of closely combined proton and electron was adopted in pictures of nuclear structure developed by Ono (1926), by Fournier and others, but nothing essentially new was added to it.

No experimental evidence for the existence of neutral particles could be obtained for years. Some experiments were made in the Cavendish Laboratory in 1921 by Glasson and by Roberts, hoping to detect the formation of such particles when an electric discharge was passed through hydrogen. Their results were negative.

The possibility that neutral particles might exist was, nevertheless, not lost sight of. I myself made several attempts to detect them - in discharge tubes actuated in different ways, in the disintegration of radioactive substances, and in artificial disintegrations produced by α-particles. *

* Cf. Rutherford & Chadwick, (1929). *Proc. Cambridge Phil. Soc.*, 25, 186.

No doubt similar experiments were made in other laboratories, with the same result.

Later, Bothe and Becker showed that γ-radiations were excited in some light elements when bombarded by α-particles. Mr. H. C. Webster, in the Cavendish Laboratory had also been making similar experiments, and he proceeded to examine closely the production of these radiations. The radiation emitted by beryllium showed some rather peculiar features, which were very difficult to explain. I suggested therefore that the radiation might consist of neutral particles and that a test of this hypothesis might be made by passing the radiation into an expansion chamber. Several photographs were taken: some β-particle tracks - presumably recoil electrons - were observed, but nothing unexpected. **

** The failure was partly due to the weakness of the polonium source.

The first real step towards the discovery of the neutron was given by a very beautiful experiment of Mme. and M. Joliot-Curie, who were also investigating the properties of this beryllium radiation. They passed the radiation through a very thin window into an ionization vessel containing air. When paraffin wax or any other matter containing hydrogen was placed in front of the window the ionization in the vessel increased. They showed that this increase was due to the ejection from the wax of **proton**s, moving with very high velocities.

This behavior of the beryllium radiation was very difficult to explain if it were a quantum radiation. I therefore began immediately the study of this new effect using different methods - the counter, the expansion chamber, and the high-pressure ionization chamber.

It appeared at once that the beryllium radiation could eject particles not only from paraffin wax but also from other light substances, such as lithium, beryllium, boron, etc., though in these cases the particles had a range of only a few millimeters in air. The experiments showed that the particles are recoil atoms of the element through which the radiation passes, set in motion by the impact of the radiation.

The occurrence of these recoil atoms can be shown most strikingly by means of the expansion chamber. These experiments were carried out by Dr. Feather and Mr. Dee. Fig. 1 is a photograph taken by Dee, which shows the tracks of protons ejected from gelatine on the roof of the expansion chamber. Fig. 2 shows two photographs taken by Feather, using an expansion chamber filled with nitrogen. Two short dense tracks are seen. Each is due to an atom of nitrogen which has been struck by the radiation. One track (Fig. 2b) shows a short spur, due to collision with a nitrogen atom; the angle between the spurs is 90°, as it should be if the initial track is due to a nitrogen atom.

The beryllium radiation thus behaved very differently from a quantum radiation. This property of setting in motion the atoms of matter in its path suggests that the radiation consists of particles. Let us suppose that the radiation consists of particles of mass M moving with velocities up to a maximum velocity V. Then the maximum velocity which can be imparted to a hydrogen atom, mass 1, by the impart of such a particle will be

$$U_p = 2M/(M + 1) V$$

and the maximum velocity imparted to a nitrogen atom will be

$$U_n = 2M/(M + 14) V.$$

Then

$$(M + 14)/(M + 1) = U_p/U_n.$$

The velocities $U_p$ and $U_n$ were found by experiment. The maximum range of the protons ejected from paraffin wax was measured and also the ranges of the recoil atoms produced in an expansion chamber filled with nitrogen. From these ranges the velocities $U_p$ and $U_n$ can be deduced approximately: $U_p = ca.\ 3.7 \times 10^9$ cm/sec, $U_n = ca.\ 4.7 \times 10^8$ cm/sec. Thus, we find M = 0.9.

We must conclude that the beryllium radiation does in fact consist of particles, and that these particles have a mass about the same as that of a proton. Now the experiments further showed that these particles can pass easily through thicknesses of matter, e.g. 10 or even 20 cm lead. But a proton of the same velocity as this particle is stopped by a thickness of ¼ mm of lead. Since the penetrating power of particles of the same mass and speed depends only on the charge carried by the particle, it was clear that the particle of the beryllium radiation must have a very small charge compared with that of the proton. It was simplest to assume that it has no charge at all. All the properties of the beryllium radiation could be readily explained on this assumption, *that the radiation consists of particles of mass 1 and charge 0, or neutrons*.

### The nature of the neutron

I have already mentioned Rutherford's suggestion that there might exist a neutral particle formed by the close combination of a proton and an electron, and it was at first natural to suppose that the neutron might be such a complex particle. On the other hand, a structure of this kind cannot be fitted into the scheme of the quantum mechanics, in which the hydrogen atom represents the only possible combination of a proton and an electron. Moreover, an argument derived from the spins of the particles is against this view. *The statistics and spins of the lighter elements can only be given a consistent description if we assume that the neutron is an elementary particle*.

Similar arguments make it difficult to suppose that the proton is a combination of neutron and positive electron. It seems at present useless to discuss whether the neutron and proton are elementary particles or not; it may be that they are two different states of the fundamental heavy particle.

In the present view of the β-transformations of radioactive bodies the hypothesis is made that a neutron in the nucleus may transform into a proton and a negative electron with the emission of the electron, or conversely a proton in the nucleus may transform into a neutron and a positive electron with the emission of the positron. Thus,

$$n \rightarrow p + e^-$$
$$p \rightarrow n + e^+.$$

If spin is to be conserved in this process we must invoke the aid of another particle - Pauli's neutrino; we then write

$$n \rightarrow p + e^- + \text{neutrino}$$
$$p \rightarrow n + e^+ + \text{antineutrino},$$

where the neutrino is a particle of very small mass, no charge, and spin ½.

If we knew the masses of the neutron and proton accurately, these considerations would give the mass of the hypothetical neutrino.

As I have shown, observations of the momenta transferred in collisions of a neutron with atomic nuclei lead to a value of the mass of the neutron but the measurements cannot be made with precision. To obtain an accurate estimate of the neutron mass we must use the energy relations in a disintegration process in which a neutron is liberated from an atomic nucleus. The best estimate at present is obtained from the disintegration of the deuteron by the photoelectric effect of a γ-ray

$$^2_1D + h\nu \rightarrow {}^1_1p + {}^1_0n.$$

The energy of the protons liberated by a γ-ray quantum of $h\nu = 2.62 \times 10^6$ eV has been measured recently by Feather, Bretscher, and myself. It is 180,000 eV. Thus, the total kinetic energy set free is 360,000 eV, giving a binding energy of the deuteron of $2.26 \times 10^6$ eV. Using the value of the deuteron mass given by Oliphant, Kempton, and Rutherford, we then obtain a value for the mass of the neutron of I.0085*.

> \* Recent measurements of the mass of deuterium lead to a value of 1.0090 for the mass of the neutron.

The mass of the hydrogen atom is 1.0081. It would seem therefore that a free neutron should be un stable, i.e. it can change spontaneously into a proton + electron + neutrino, unless the neutrino has a mass of the order of the mass of an electron. On the other hand, an argument from the shape of the ray spectra suggests that the mass of the neutrino is zero. One must await more exact measurements of the masses of hydrogen and deuterium before speculating further on this matter.

**Passage of neutrons through matter**

The neutron in its passage through matter loses its energy in collisions with the atomic nuclei and not with the electrons. The experiments of Dee showed that the primary ionization along the track of a neutron in air could not be as much as 1 ion pair in 3 meters' path, while Massey has calculated that it may be as low as 1 ion pair in 105 km. This behavior is very different from that of a charged particle, such as a proton, which dissipates its energy almost entirely in electron collisions. The collision of a neutron with an atomic nucleus, although much more frequent than with an electron, is also a rare event, for the forces between a neutron and a nucleus are very small except at distances of the order of $10^{-12}$ cm. In a close collision the neutron may be deflected from its path and the struck nucleus may acquire sufficient energy to produce ions. The recoiling nucleus can then be detected either in an ionization chamber or by its track in an expansion chamber. In some of these collisions, however, the neutron enters the nucleus and a disintegration is produced. Such disintegrations were first observed by Feather in his observations on the passage of neutrons through an expansion chamber filled with nitrogen. An example is shown in Fig. 3. The disintegration process is

$$^{14}_{7}N + {}^{1}_{0}n \rightarrow {}^{11}_{5}B + {}^{4}_{2}He.$$

Since these early experiments many examples of this type of disintegration have been observed by different workers.

Fermi and his collaborators have also shown that the phenomenon of artificial radioactivity can be provoked in the great majority of all elements, even in those of large atomic number, by the bombardment of neutrons. They have also shown that neutrons of very small kinetic energy are peculiarly effective in many cases.

In some cases, an $\alpha$-particle is emitted in the disintegration process; in others a proton is emitted; while in others an unstable species of nucleus is formed by the simple capture of the neutron.

Examples of these types are:

$$^{31}_{15}P + {}^{1}_{0}n \rightarrow {}^{28}_{13}Al + {}^{4}_{2}He$$
$$^{28}_{14}Si + {}^{1}_{0}n \rightarrow {}^{28}_{13}Al + {}^{1}_{1}H$$
$$^{27}_{13}Al + {}^{1}_{0}n \rightarrow {}^{28}_{13}Al$$
$$^{127}_{53}I + {}^{1}_{0}n \rightarrow {}^{128}_{53}I.$$

In the cases just cited the nuclei formed in the reaction are unstable, showing the phenomenon of induced activity discovered by Mme. and M. Joliot Curie, and return to a stable form with the emission of negative electrons.

In the transformations produced in heavy elements by neutrons, the process is, with very few exceptions, one of simple capture. The nucleus so formed, an isotope of the original nucleus, is often unstable but not invariably so. For example, the reaction

$$_{48}Cd + _{0}n \rightarrow _{48}Cd + h\nu.$$

The cadmium isotope formed is stable, but a $\gamma$-ray quantum is emitted of energy corresponding to the binding energy of the neutron.

Other cases of this type of transformation are known.

The great effectiveness of the neutron in producing nuclear transmutations is not difficult to explain. In the collisions of a charged particle with a nucleus, the chance of entry is limited by the Coulomb forces between the particle and the nucleus; these impose a minimum distance of approach which increases with the atomic number of the nucleus and soon becomes so large that the chance of the particle entering the nucleus is very small. In the case of collisions of a neutron with a nucleus there is no limitation of this kind. The force between a neutron and a nucleus is inappreciable except at very small distances, when it increases very rapidly and is attractive. Instead of the potential wall in the case of the charged particle, the neutron encounters a potential hole. Thus, even neutrons of very small energy can penetrate into a nucleus. Indeed, slow neutrons may be enormously more effective than fast neutrons, for they spend a longer time in the nucleus. The calculations of Bethe show that the chance of capture of a neutron may be inversely proportional to its velocity. The possibility of capture will depend on whether the nucleus possesses an unoccupied p-level or a level with azimuthal quantum number $l = 1$.

In cases where a particle ($\alpha$-particle or proton) is ejected from the nucleus, the possibility of disintegration will depend on whether the particle can escape through the potential barrier. This will be easier the greater the energy set free in the disintegration process. As a rule, disintegration by neutrons will take place with absorption of kinetic energy if a proton is released in the transformation, and may take place with release of kinetic energy if one at least of the products is an $\alpha$-particle. Thus, processes in which a proton is emitted can only occur with fast neutrons, even in collisions with elements of low atomic number; while processes in which $\alpha$-particles are emitted can occur with slow neutrons in elements of low atomic number, but again only with fast neutrons in elements of higher atomic number. If the atomic number is sufficiently high, the neutrons at present at our disposal have insufficient energy and the particles cannot escape through the potential barrier. Thus, with elements of high atomic number, only capture processes are observed, although there may be a few exceptions. There may be, however, special cases in which the particles escape through a resonance level. These would be characterized by the phenomenon that the energy of the escaping particle would be independent of the energy of the incident neutron. These special cases may explain the exceptional disintegrations in which a particle is emitted from a heavy nucleus. They may be of particular interest in giving information about the resonance levels of atomic nuclei.

There is also the possibility of resonance capture of the neutrons, more particularly with very slow neutrons. The capture of neutrons of a certain energy may take place with very great frequency in one species of nucleus while for another neighboring nucleus the same neutrons may have a long free path. These resonance regions may perhaps be rather broad and therefore comparatively easy to observe experimentally.

**The structure of the nucleus**

***Before the discovery of the neutron, we had to assume that the fundamental particles from which an atomic nucleus was built up were the proton and the electron, with the α-particle as a secondary unit.*** The behavior of an electron in a space of nuclear dimensions cannot be described on present theory; and other difficulties, e.g. the statistics of the nitrogen nucleus, the peculiarities in the mass defect curve in the region of the heavy elements, also arose. ***These difficulties are removed if we suppose that the nuclei are built up from protons and neutrons.*** The forces which determine the stability of a nucleus will then be of three types, the interactions between proton and proton, between proton and neutron, and between neutron and neutron. ***It is assumed, with Heisenberg and Majorana, that the interaction between neutron and proton is of the exchange type*** - similar to that between the hydrogen atom and the hydrogen ion - and that the interaction between neutron and neutron is small.

For a nucleus of mass number A and charge Ze we shall have

$$N_n + N_p = A \qquad N_p = Z$$
$$N_n/N_p = (A - Z)/Z.$$

The value of $N_n/N_p$ for the most stable nucleus of a given mass number will be determined by the condition that the binding energy is a maximum. The repulsive Coulomb force between the protons tends to diminish the number of protons in a nucleus, while the neutron-proton interaction tends to make $N_n = N_p$ $Z = A/2$; the neutron-neutron interaction is probably very small. Now in existing nuclei $N_p \sim N_n$, and therefore the neutron-proton interaction must be the predominating force in the nucleus. In heavy elements $N_n > N_p$. This relative increase in the number of neutrons may be due either to an attractive force between neutron-neutron, or more probably to the Coulomb forces between proton-proton.

Thus, it appears that *the interaction between proton and neutron is of the highest significance in nuclear structure and governs the stability of a nucleus*. It is most important to obtain all experimental evidence about the nature of this interaction. The information we have at present is very meagre, but I think that it does to some degree support the view that the interaction is of the *exchange type*. Dr. Feather and I hope to obtain more definite information on this subject by an extensive study of the collisions of neutrons and protons.

Heisenberg's considerations of nuclear structure point very strongly to this *exchange interaction*. Such an interaction provides an attractive force at large distances between the particles and a repulsive force at very small distances, thus giving the effect of a more or less definite radius of the particles. A system of particles interacting with exchange forces will keep together due to the attraction, but there will be a minimum distance of approach of the particles; thus, the system will not collapse together but will have a more or less definite "radius".

The exchange forces between a hydrogen atom and a hydrogen ion are large compared with the forces between neutral atoms; by analogy we explain why the neutron-proton interaction is so much stronger than the proton-proton or neutron-neutron interactions.

By a suitable choice of the exchange forces, it is possible to obtain a saturation effect, analogous to the saturation of valency bindings between two atoms, when each neutron is bound to two protons and each proton to two neutrons. Thus, two neutrons and two protons form a closed system - the α-particle. These ideas thus explain the general features of the structure of atomic nuclei and it can be confidently expected that further work on these lines may reveal the elementary laws which govern the structure of matter.